Viral Membrane Proteins: Structure, Function, and Drug Design

PROTEIN REVIEWS

Recent Volumes in this Series

VIRAL MEMBRANE PROTEINS: STRUCTURE, FUNCTION, AND DRUG DESIGN
Edited by Wolfgang B. Fischer

A Continuation Order Plan is available for this series. A continuation order will bring delivery of each new volume immediately upon publication. Volumes are billed only upon actual shipment. For further information, please contact the publisher.

Viral Membrane Proteins: Structure, Function, and Drug Design

Edited by

WOLFGANG B. FISCHER
Department of Biochemistry, University of Oxford, United Kingdom

Kluwer Academic / Plenum Publishers
New York, Boston, Dordrecht, London, Moscow

Library of Congress Cataloging-in-Publication Data

Viral membrane proteins: structure, function, and drug design / edited by
Wolfgang Fischer.
p. cm. – (Protein reviews)
Includes bibliographical references and index.

1. Viral proteins. I. Fischer, Wolfgang (Wolfgang B.) II. Series.

QR460.V55 2005
616.9′101–dc22

2004042175

ISBN 978-1-4419-3453-6 e-ISBN 978-0-387-28146-9

233 Spring Street, New York, N.Y. 10013
Softcover reprint of the hardcover 1st edition 2005

http://www.kluweronline.com

10 9 8 7 6 5 4 3 2 1

A C.I.P. record for this book is available from the Library of Congress.

Contents

Part II. Fusion Proteins

3. HIV gp41: A Viral Membrane Fusion Machine

Sergio G. Peisajovich and Yechiel Shai

4. Diversity of Coronavirus Spikes: Relationship to Pathogen Entry and Dissemination

Edward B. Thorp and Thomas M. Gallagher

5. Aspects of the Fusogenic Activity of Influenza Hemagglutinin Peptides by Molecular Dynamics Simulations

L. Vaccaro, K.J. Cross, S.A. Wharton, J.J. Skehel, and F. Fraternali

Part III. Viral Ion Channels/viroporins

6. Viral Proteins that Enhance Membrane Permeability

María Eugenia González and Luis Carrasco

7. FTIR Studies of Viral Ion Channels

Itamar Kass and Isaiah T. Arkin

8. The M2 Proteins of Influenza A and B Viruses are Single-Pass Proton Channels

Yajun Tang, Padmavati Venkataraman, Jared Knopman, Robert A. Lamb, and Lawrence H. Pinto

9. Influenza A Virus M2 Protein: Proton Selectivity of the Ion Channel, Cytotoxicity, and a Hypothesis on Peripheral Raft Association and Virus Budding

Cornelia Schroeder and Tse-I Lin

10. Computer Simulations of Proton Transport Through the M2 Channel of the Influenza A Virus

Yujie Wu and Gregory A. Voth

Part IV. Membrane-Spanning/Membrane Associated

17. The Structure, Function, and Inhibition of Influenza Virus Neuraminidase

Elspeth Garman and Graeme Laver

18. Interaction of HIV-1 Nef with Human CD4 and Lck

Dieter Willbold

List of Contributors

Michael J. Adams, Plant Pathogen Interactions Division, Rothamsted Research, Harpenden, Herts AL5 2JQ, UK.

John F. Antoniw, Plant Pathogen Interactions Division, Rothamsted Research, Harpenden, Herts AL5 2JQ, UK.

Isaiah T. Arkin, The Alexander Silberman Institute of Life Science, Department of Biological Chemistry, The Hebrew University, Givat-Ram, Jerusalem, 91904, Israel.

Burkhard Bechinger, Université Louis Pasteur, Faculté de chimie, ILB, 4 rue Blaise Pascal, 67070 Strasbourg, France.

S. Bour, Bioinformatics and Cyber Technology Center, Office of Technology and Information Systems, National Institute of Allergy and Infectious Diseases, National Institutes of Health, 4 Center Drive, Bethesda, MD, USA.

Luis Carrasco, Centro de Biología Molecular Severo Ochoa, Facultad de Ciencias, Universidad Autónoma, Cantoblanco, 28049 Madrid, Spain.

K. Cross, National Research for Medical Research, Mill Hill, London NW7 1AA, UK.

Gary Ewart, Division of Molecular Bioscience, John Curtin School of Medical Research, Australian National University, Canberra, Australia.

D. Fischer, Biomembrane Structure Unit, Department of Biochemistry, Oxford University, South Parks Road, Oxford OX1 3QU, UK.

W.B. Fischer, Biomembrane Structure Unit, Department of Biochemistry, Oxford University, South Parks Road, Oxford OX1 3QU, UK, and Bionanotechnology Interdisciplinary Research Consortium, Clarendon Laboratory, Department of Physics, Oxford University, Parks Road, Oxford OX1 3SU, UK.

F. Fraternali, National Research for Medical Research, Mill Hill, London NW7 1AA, UK.

Peter W. Gage, Division of Molecular Bioscience, John Curtin School of Medical Research, Australian National University, Canberra, Australia.

Thomas M. Gallagher, Department of Microbiology and Immunology, Loyola University Medical Center, Maywood, IL, USA.

Elspeth Garman, Laboratory of Molecular Biophysics, Department of Biochemistry, University of Oxford, South Parks Road, Oxford OX1 3QU, UK.

María Eugenia González, Unidad de Expresión Viral, Centro Nacional de Microbiologia, Instituto de Salud Carlos III, Carretera de Majadahonda-Pozuelo, Km 2, Majadahonda 28220, Madrid, Spain.

Peter Henklein, Humboldt Universität, Institut für Biochemie, Hessische Str 3-4, 10115 Berlin, Germany.

D. Jones, Department of Chemistry and Biochemistry, University of California, San Diego, 9500 Gilman Drive, La Jolla, CA, USA.

Itamar Kass, The Alexander Silberman Institute of Life Science, Department of Biological Chemistry, The Hebrew University, Givat-Ram, Jerusalem 91904, Israel.

C.G. Kim, Biomembrane Structure Unit, Department of Biochemistry, Oxford University, South Parks Road, Oxford OX1 3QU, UK.

Jared Knopman, Department of Neurobiology and Physiology, Northwestern University, Evanston, IL, USA.

Y.H. Lam, Biomembrane Structure Unit, Department of Biochemistry, Oxford University, South Parks Road, Oxford OX1 3QU, UK.

Robert A. Lamb, Department of Biochemistry, Molecular Biology and Cell Biology, and Howard Hughes Medical Institute, Northwestern University, Evanston, IL, USA.
Graeme Laver, Barton Highway, Murrumbateman, NSW 2582, Australia.
S. Lee, Department of Chemistry and Biochemistry, University of California, San Diego, 9500 Gilman Drive, La Jolla, CA, USA.
V. Lemaitre, Biomembrane Structure Unit, Department of Biochemistry, Oxford University, South Parks Road, Oxford OX1 3QU, UK, and Nestec S.A., BioAnalytical Science Department, Vers-Chez-Les-Blanc, CH-1000 Lausanne 26, Switzerland.
Tse-I Lin, Tibotec BVDV, Gen. De. Wittelaan 11B-3, B-2800 Mechelen, Belgium.
V. Madan, Centro de Biología Molecular Severo Ochoa (CSIC-UAM), Facultad de Ciencias, Universidad Autónoma, Cantoblanco, 28049 Madrid, Spain.
F.M. Marassi, The Burnham Institute, 10901 North Torrey Pines Road, La Jolla, CA, USA.
Julian Melton, Division of Molecular Bioscience, John Curtin School of Medical Research, Australian National University, Canberra, Australia.
M. Mesleh, Department of Chemistry and Biochemistry, University of California, San Diego, 9500 Gilman Drive, La Jolla, CA, USA.
M. Montal, Section of Neurobiology Division of Biology, University of California, San Diego, 9500 Gilman Drive, La Jolla, CA, USA.
Anna Moroni, Dipartimento di Biologia and CNR—IBF, Unità di Milano; Istituto Nazionale di Fisica della Materia, Unità di Milano-Università, Milano, Italy.
A. Mrse, Department of Chemistry and Biochemistry, University of California, San Diego, 9500 Gilman Drive, La Jolla, CA, USA.
A. Nevzorov, Department of Chemistry and Biochemistry, University of California, San Diego, 9500 Gilman Drive, La Jolla, CA, USA.
J.L. Nieva, Unidad de Biofísica (CSIC-UPV/EHU), Departamento de Bioquímica, Universidad del País Vasco, Aptdo. 644, 48080 Bilbao, Spain.
M. Oblatt-Montal, Section of Neurobiology Division of Biology, University of California, San Diego, 9500 Gilman Drive, La Jolla, CA, USA.
S.J. Opella, Department of Chemistry and Biochemistry, University of California, San Diego, 9500 Gilman Drive, La Jolla, CA, USA.
S. Park, Department of Chemistry and Biochemistry, University of California, San Diego, 9500 Gilman Drive, La Jolla, CA, USA.
Sergio G. Peisajovich, Department of Biological Chemistry, Weizmann Institute of Science, Rehovot 76100, Israel.
Lawrence H. Pinto, Department of Neurobiology and Physiology, Northwestern University, Evanston, IL, USA.
Anita Premkumar, Division of Molecular Bioscience, John Curtin School of Medical Research, Australian National University, Canberra, Australia.
M.A. Sanz, Centro de Biología Molecular Severo Ochoa (CSIC-UAM), Facultad de Ciencias, Universidad Autónoma, Cantoblanco, 28049 Madrid, Spain.
Cornelia Schroeder, Abteilung Virologie, Institut für Mikrobiologie und Hygiene, Universitätskliniken Homburg/Saar, Germany.
Ulrich Schubert, Institute for Clinical and Molecular Virology, University of Erlangen-Nürnberg, Schlossgarten 4, D-91054 Erlangen, Germany.
Yechiel Shai, Department of Biological Chemistry, Weizmann Institute of Science, Rehovot 76100, Israel.
J.J. Skehel, National Research for Medical Research, Mill Hill, London NW7 1AA, UK.
K. Strebel, Bioinformatics Core Facility, Laboratory of Molecular Microbiology, National Institute of Allergy and Infectious Diseases, National Institutes of Health, 4 Center Drive, Bethesda, MD, USA.
Yajun Tang, Department of Neurobiology and Physiology, Northwestern University, Evanston, IL, USA.

Gerhard Thiel, Institute for Botany, University of Technology, Darmstadt, Germany.
Edward B. Thorp, Department of Microbiology and Immunology, Loyola University Medical Center, Maywood, IL, USA.
L. Vaccaro, National Research for Medical Research, Mill Hill, London NW7 1AA, UK.
James Van Etten, Department of Plant Pathology, Nebraska Center for Virology, University of Nebraska, Lincoln, NE, USA.
Padmavati Venkataraman, Department of Neurobiology and Physiology, Northwestern University, Evanston, IL, USA.
Gregory A. Voth, Department of Chemistry and Henry Eyring Center for Theoretical Chemistry, University of Utah, 315 S. 1400 E. Rm 2020, Salt Lake City, UT, USA.
A. Watts, Biomembrane Structure Unit, Department of Biochemistry, Oxford University, South Parks Road, Oxford OX1 3QU, UK.
S.A. Wharton, National Research for Medical Research, Mill Hill, London NW7 1AA, UK.
Dieter Willbold, Institut für Physikalische Biologie, Heinrich-Heine-Universität, 40225 Düsseldorf, Germany und Forschungszentrum Jülich, IBI-2, 52425 Jülich, Germany.
Victor Wray, Department of Structural Biology, German Research Centre for Biotechnology, Mascheroder Weg 1, D-38124 Braunschweig, Germany.
Yujie Wu, Department of Chemistry and Henry Eyring Center for Theoretical Chemistry, University of Utah, 315 S. 1400 E. Rm 2020, Salt Lake City, UT, USA.

Preface

Viruses enter cells and modulate the biosynthetic machinery of the host for the synthesis of their own building blocks. These building blocks assemble in an organized fashion and large numbers of viral copies finally leave the host, ready to enter the next host cell. This life style would be of little concern if it were not that many viruses have a fatal side effect to the host, leading to the death of the host cell and sometimes to the death of the whole organism. Under these circumstances, the first two sentences could be rewritten and the words "enter" and "leave" replaced by more drastic words such as "attack," "invade," or "kill." However, not all the attacks lead to an immediate cell death or to a phenotypic manifestation in the organism.

Viruses are under constant survival pressure and have evolved mechanisms to resist environmental pressure by having for example a high mutation rate. This may lead finally also to an increased spreading to novel hosts, which can have a devastating effect on the invaded organism, including humans, especially if the species barrier is crossed in an unpredicted way. In the modern world, the large density of populations and travel habits can lead to a rapid spread of the virus, with a possible major impact on our social behavior and the economy. The recent appearance of the SARS virus or avian influenza viruses in humans represents such an immediate threat. Once within a host, some viruses, such as HIV-1, replicate but rather than produce faithful copies of the parent virus, constantly mutate making it almost impossible to produce a vaccine and limiting the success of drug therapy. We are also directly or indirectly affected by animal or plant viruses. The last foot and mouth outbreak within the United Kingdom and other European countries resulted in the slaughter of large numbers of farm animals to prevent the rapid spread of the disease. As an indirect effect, the country side, dependent to a very large extent on tourism, had to be closed down to avoid any further spread of the disease. Plant viruses threatening our annual harvest and can through price rises add to inflation. However, not all viruses cause harm to us, some of the plant viruses may even cheer us up such as the tulip mosaic virus that causes the striping pattern of tulip petals.

In this book, we aim to summarize the current knowledge on a special class of viral molecules, the membrane proteins, from the full range of viruses, including plant viruses. Research on these membrane proteins has been limited by a number of technical difficulties, and rate of progress compared with globular proteins has been slow. Membrane proteins are involved at the stage of viral entry into the host cell, in modulating subcellular electrochemical gradients and/or shuffling proteins across cell membranes.

The first section is dedicated to viral membrane proteins from plant viruses with the most recent computational research on the viral genome revealing the first experimental evidence of a K^+ channel encoded by a plant virus. The second section in the book is dedicated to the proteins involved in the early event of the life cycle of the viruses in the host cell, the fusion proteins. The third section summarizes, in several chapters, the current state of the research on ion channels and viroporins, which are known to modulate the electrochemical balance in the virus itself and subcellular compartments in the host cell. The fourth section describes membrane-bound and membrane-associated viral proteins. All chapters include functional and structural data and address, where possible, the development of antiviral drugs. A large number of techniques are described by the authors, revealing the way in which a wide range of approaches are required to shed light on the molecular life of viruses.

I wish to thank the editorial team at Kluwer Academic Publishers for their enthusiasm and physical support during the generation of this synthesis of our recent advances in viral membrane protein research. Thanks go also to all the authors for their willingness and patience while working on the book. My acknowledgment includes also Judy Armitage (Oxford) and my colleagues in the lab for stimulating and helpful discussions.

Part I

Membrane Proteins from Plant Viruses

1

Membrane Proteins in Plant Viruses

Michael J. Adams and John F. Antoniw

1. Introduction

The plant cell wall is a substantial barrier preventing direct entry of viruses and therefore, unlike many animal viruses, plant viruses cannot initiate infection by any independent ability to cross membranes. A few plant viruses enter through microscopic wounds but most are introduced into their hosts by a vector, most frequently a leaf-feeding insect. Small pores between adjacent cells (plasmodesmata) provide cytoplasmic continuity and thus a channel for transport of nutrients and some larger molecules, and viruses exploit this route for cell-to-cell transport. While some viruses remain restricted to a small area around the initial site of infection and may be limited to certain cell types, many exhibit long distance movement via the plant vascular system.

Among the plant viruses, cell-to-cell movement depends on one or more virus-encoded movement proteins (MPs) and many of these are integral membrane proteins that interact with the endoplasmic reticulum (ER). Many RNA viruses multiply within the cell cytoplasm and there is recent evidence that replication proteins of such viruses are also targeted to membranes. In addition, membrane proteins may play a role in plant virus transmission for those viruses that enter the cells of their vectors. In this chapter, we survey the occurrence of membrane proteins among all plant viruses and review the literature on their biological role. We also present and discuss the limited structural information on plant virus integral membrane proteins.

2. Survey of Transmembrane Proteins in Plant Viruses

In preparation for this chapter, we have used a plant virus sequence database that we developed to make a comprehensive survey of all published complete gene sequences of all plant viruses.

2.1. The Database

The database used was developed from files originally prepared for the electronic version of the Association of Applied Biologists (AAB) Descriptions of Plant Viruses

Michael J. Adams and John F. Antoniw • Plant Pathogen Interactions Division, Rothamsted Research, Harpenden, United Kingdom.

Viral Membrane Proteins: Structure, Function, and Drug Design, edited by Wolfgang Fischer.
Kluwer Academic / Plenum Publishers, New York, 2005.

(Adams *et al.*, 1998). As part of that project, we provided software (DPVMap) to display selected virus sequences interactively. A separate enhanced feature table (EFT) file was written for each sequence containing the start and end nucleotide positions of the features (e.g., open-reading frames (ORFs), untranslated regions) within the sequence. In DPVMap any of the features of the sequence could be dragged into a sequence editor to display either its nucleotide sequence (as RNA or DNA), or the predicted amino acid sequence of an ORF. Annotations provided for the correct display of reverse complementary sequences and of those incorporating a frameshift or intron. Sequence features were checked for accuracy and, as far as possible, nomenclature for genes and proteins were standardized within genera and families to make it easier to compare features from different viruses. From a modest beginning, the number of sequences provided has been increased and now includes all complete sequences of plant viruses, viroids and satellites, and all sequences that contain at least one complete gene. The information contained in the individual EFT files is valuable because it has been checked for accuracy and is often more detailed than that provided in the original sequence file from EMBL or Genbank. However, the EFT files can only be used with DPVMap and to examine one sequence at a time. We therefore decided to transfer this information together with the sequences themselves into a database table, so that multiple data sets could be selected and extracted easily and then used for further analysis.

The database was prepared in MySQL on a Linux PC and includes up to date taxonomic information and a table of sequence data containing all the information from the individual EFT files. The version used here was based on sequences available from the public databases at the end of November 2002 and includes a total of 4,687 accessions. It therefore records the start and end positions of all important features and genes in every one of the significant plant virus sequences. The database has been placed on a public Internet site (DPVweb at http://www.dpvweb.net) where it may be accessed using client software.

2.2. Software

A web-enabled Windows client application was written in Delphi for IBM-compatible PCs to scan the database tables, translate each complete ORF into its amino acid sequence, and then to predict transmembrane (TM) regions using TMPRED (Hofmann and Stoffel, 1993). A summary of the results was exported to a Microsoft®Excel spreadsheet and examined for consistency within species and genera. The results have been used to inform the presentation and discussion of the different types and function of plant virus membrane proteins (below) and some ambiguous results were checked using the web-based software HMMTOP (Tusnády and Simon, 1998), TMHMM (Sonnhammer *et al.*, 1998), and TopPred 2 (von Heijne, 1992).

3. Cell-to-Cell Movement Proteins

Most plant viruses encode one or more specific MPs that are required for the virus to spread between adjacent host cells. Functions assigned to these proteins include nucleic-acid binding (some viruses move as nucleic acid–MP complexes), modification of the size exclusion limit of the plasmodesmata (the connections between adjacent cells), and targeting to the inter- and intracellular membrane system, the ER. A number of groups of MPs have been identified and at least some of these are integral membrane proteins.

3.1. The "30K" Superfamily

A very large number of plant viruses have MPs that share common structural features, which led Mushegian and Koonin (1993) to propose the name "30K" superfamily for them. This grouping has most recently been reviewed and defended by Melcher (2000). It includes a surprisingly diverse range of viruses including those with DNA genomes (the pararetroviruses and the ssDNA viruses in the genus *Begomovirus*) and many different groups of both positive sense (*Bromoviridae, Comoviridae, Capillovirus, Dianthovirus, Furovirus, Idaeovirus, Tobamovirus, Tobravirus, Tombusvirus, Trichovirus, Umbravirus*) and negative sense (*Nucleorhabdovirus, Tospovirus*) RNA viruses. These have been assigned by computer predictions showing the presence of a core domain consisting of two α-helices separated by a series of β-elements.

The best-studied virus from this group is *Tobacco mosaic virus* (TMV, genus *Tobamovirus*). Its MP has been shown to increase the size exclusion limit of plasmodesmata, and specifically at the leading edge of expanding lesions (Oparka *et al.*, 1997). It has non-specific RNA-binding activity, forming a viral RNA–MP complex that moves between cells (Citovsky and Zambryski, 1991). It can also bind to the cytoskeleton (Heinlein *et al.*, 1998; Reichel and Beachy, 1998; Reichel *et al.*, 1999; Boyko *et al.*, 2000) but it remains uncertain whether this property is essential for cell-to-cell movement as recent evidence suggests that TMV can replicate and move in the absence of microtubules (Gillespie *et al.*, 2002). There remains much to be discovered about the interaction of the MP with host cell components and how this facilitates cell-to-cell movement of the viral RNA, but a combination of CD spectroscopy, trypsin treatment, and mass spectroscopy has helped to develop a topological model (Brill *et al.*, 2000). This confirms the role of the two core α-helices as TM domains resistant to trypsin, and indicates that the N- and C-termini would be exposed in the cytoplasm and a short loop in the ER lumen (Figure 1.1).

There is less experimental information for the other MPs in this group but they are likely to have a similar association with membranes. For example, the movement protein (3a) of *Alfalfa mosaic virus* (genus *Alfamovirus*, family *Bromoviridae*), used as a MP–GFP (green fluorescent protein) construct, co-localized with ER in tobacco protoplasts and onion cells and moved between adjacent onion cells. Fractionation and biochemical studies in insect cells demonstrated that the MP–GFP was an integral membrane protein (Mei and Lee, 1999) although no ER targeting signal has been identified. Some other "30K" superfamily MPs that have been shown to interact with membranes are the ORF3 products of *Grapevine virus A* and *Grapevine virus B* (genus *Vitivirus*) (Saldarelli *et al.*, 2000), the P22 of *Tomato bushy stunt virus* (genus *Tombusvirus*, family *Tombusviridae*) (Desvoyes *et al.*, 2002), and the BC1 protein of *Abutilon mosaic virus* (genus *Begomovirus*, family *Geminiviridae*) (Zhang *et al.*, 2001, 2002; Aberle *et al.*, 2002).

Some of the superfamily member MPs act in a rather different fashion by producing tubules that extend through the plasmodesmata. This has been best studied in *Cowpea mosaic virus* (CPMV, genus *Comovirus*, family *Comoviridae*) (Van Lent *et al.*, 1991). In these examples, the virus has been shown to move as intact virions and therefore to require the coat protein (CP), but it appears that some "30K" superfamily MPs have both tubule-forming and RNA-binding activities (Perbal *et al.*, 1993; Jansen *et al.*, 1998; Canto and Palukaitis, 1999; Nurkiyanova *et al.*, 2001). Unlike TMV, the CPMV MP does not localize to either the microtubules or the ER and the mechanism of its delivery to the cell periphery is not known. The tubules themselves are thought to arise from the host protein plasma membrane (Pouwels *et al.*, 2002).

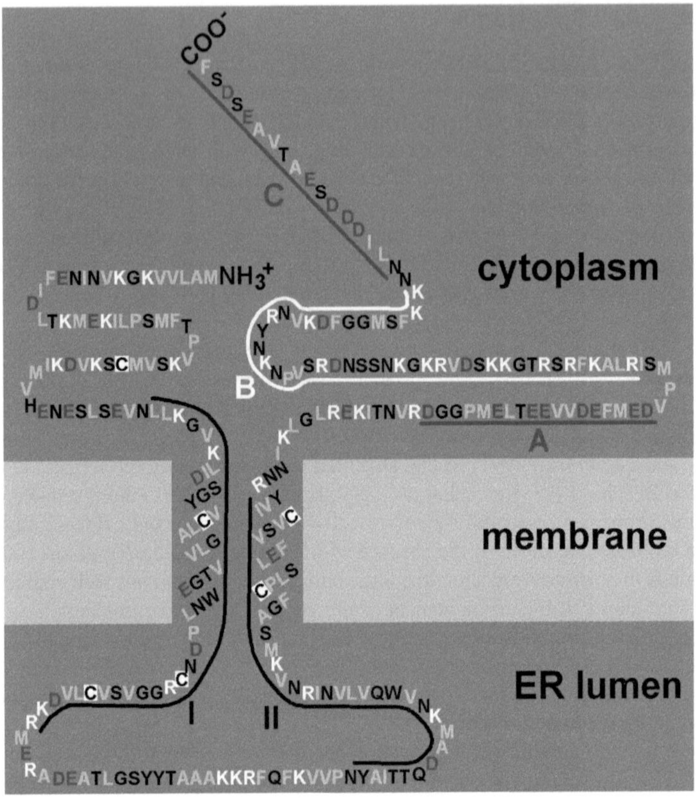

Figure 1.1. Topological model of the *Tobacco mosaic virus* movement protein, re-drawn from Brill *et al.* (2000), and used by kind permission of Prof. R.N. Beachy. Hydrophobic amino acid residues are shown in pale grey. Basic residues (in white) are concentrated in domain B and acidic residues, shown in dark grey are concentrated in domains A and C; Cys residues are shown with white background. Domains I and II are conserved amongst tobamoviruses.

TMPRED correctly identified the position and orientation of the two TM domains of TMV and 7 other tobamoviruses (out of 16 different species sequenced). Among the "30K" superfamily generally, a TM domain was identified in most viruses, but two domains were predicted in only 20 out of more than 60 species.

3.2. Triple Gene Block

Some positive-sense ssRNA filamentous and rod-shaped viruses do not have the single MP exemplified by TMV or CPMV but a group of three, partially overlapping, proteins known as the "triple gene block" (TGB). The structure and function of the TGB has been recently reviewed by Morozov and Solovyev (2003). All three TGB proteins are required for movement and the two smaller proteins, TGBp2 and TGBp3, are TM proteins. These were not only strongly predicted by computer analyses but also by *in vitro* studies (e.g., Morozov *et al.*, 1991), localization to membrane fractions of infected plant tissues (e.g., Gorshkova *et al.*, 2003), and by microscopical studies of proteins fused to GFP showing them to be localized to the ER or membrane bodies as well as to plasmodesmata (e.g., Solovyev *et al.*, 2000; Cowan *et al.*, 2002; Zamyatnin *et al.*, 2002; Gorshkova *et al.*, 2003). No detailed structural studies have been reported, but all TGBp2 molecules (11–14 kDa) contain two TM segments and it is predicted that the N- and C-termini are in the cytoplasm. TGBp3 molecules are of three different types. Those of the filamentous viruses (genera *Allexivirus, Carlavirus, Foveavirus, Potexvirus*) are 6–13 kDa in size and have a single TM segment, while those of the rod-shaped viruses are larger (15–24 kDa) and have two segments, and those of the genus *Benyvirus*, having a different arrangement to those in the genera *Hordeivirus, Pecluvirus*, and *Pomovirus*. The C-termini of TGBp3 molecules are predicted to be in the cytoplasm.

Transiently expressed TGBp2–GFP fusions localize to the ER, while TGBp3 fusions are found in membrane bodies near the plant cell periphery but in the presence of TGBp3, TGBp2 is re-targeted to the peripheral bodies (Solovyev *et al.*, 2000; Tamai and Meshi, 2001; Cowan *et al.*, 2002; Zamyatnin *et al.*, 2002; Gorshkova *et al.*, 2003). TGBp2 and TGBp3 together appear to be responsible for targeting rod-shaped virus TGBp1 to plasmodesmata (Erhardt *et al.*, 1999, 2000; Lawrence and Jackson, 2001), but the (smaller) TGBp1 of the filamentous viruses can move independently (e.g., Morozov *et al.*, 1999). TGBp2 is also involved in increasing the size exclusion limit of plasmodesmata and it has been suggested that this may occur via the regulation of callose deposition or degradation. Recent evidence that TGBp2 interacts with TIP, a host protein regulator of β-1,3-glucanase (a key enzyme of callose turnover), strengthens this hypothesis (Fridborg *et al.*, 2003).

In TMPRED, the expected TM domains were consistently and strongly detected in all TGBp2 and TGBp3 sequences. TGBp2 proteins were 104–154 amino acids (aa) long, with a loop between the two predicted TM domains of 39–61 aa. The two classes of TGBp3 proteins were correctly identified; in the rod-shaped viruses with two TM domains, the second domain is consistently at the C-terminus (within 2–5 aa).

3.3. *Carmovirus*-Like

Members of the genus *Carmovirus* are among the smallest RNA viruses (genome ~4 kb). They do not have MPs of the "30K" superfamily, nor a TGB, but two small, overlapping, internal ORFs are involved in cell-to-cell movement (Hacker *et al.*, 1992). The first, and slightly smaller, of these proteins is a soluble protein with RNA-binding capacity, while the second contains two potential TM domains. In experiments using the type member, *Carnation mottle virus* (CarMV), the two putative TM domains of the p9 protein were inserted into the *Escherichia coli* inner membrane protein Lep and then tested for insertion into dog pancreas microsomes. The experiments demonstrated TM activity and that the N- and C-termini of the protein were located in the cytoplasm. It was proposed that the charged C-terminus of p9 would interact with the C-terminal domain of the smaller p7 protein that had already bound to viral genomic RNA (Vilar *et al.*, 2002). Results of a spatio-temporal analysis are consistent with this hypothesis (Garcia-Castillo *et al.*, 2003). Our analysis confirmed the consistent presence of two TM domains in most members of the genus, but only indicated one such domain in *Melon necrotic spot virus* and in members of the related genus *Necrovirus*, where the protein seems to be smaller than in CarMV.

3.4. Other Movement Proteins

In *Maize streak virus* (genus *Mastrevirus*, family *Geminiviridae*), the MP is encoded by ORF V2, the smaller of the two ORFs translated in the positive sense and a central α-helical domain has been predicted to have TM properties (Boulton *et al.*, 1993). This is supported by studies showing its localization to plasmodesmata (Dickinson *et al.*, 1996) and by the occurrence of similar domains in other members of the genus (confirmed by our analyses), but it has not yet been proved experimentally that the hydrophobic domain is required for membrane association (Boulton, 2002).

In *Banana bunchy top virus* (genus *Babuvirus*, family *Nanoviridae*), GFP-tagging showed that the protein encoded by DNA-4, which possesses a hydrophobic N-terminus, was found to localize exclusively to the cell periphery. Deletion of the N-terminal region abolished its ability to localize to the cell periphery (Wanitchakorn *et al.*, 2000). Our analyses show similar domains in other viruses of this family.

Within the genus *Tymovirus*, the first ORF, which almost completely overlaps with the large replication protein, has been identified as a MP (Bozarth *et al.*, 1992). This protein is much larger than those discussed above (69–85 kDa) and is proline-rich. Its localization within cells has not been reported. There were no strongly hydrophobic regions in the sequences of this gene for any of the members of the genus and the few possible TM regions identified in our analysis were not strongly supported and did not appear at a consistent position within an alignment of the MPs of the genus members.

In *Beet yellows virus* (genus *Closterovirus*, family *Closteroviridae*), the 70K HSP70h (heat shock protein 70 homolog) has been shown to be absolutely required for cell-to-cell movement (Peremyslov *et al.*, 1999) and can be localized in plasmodesmatal channels (Medina *et al.*, 1999). The protein acts as a molecular chaperone and is incorporated into the tail of the functional virion (Alzhanova *et al.*, 2001). This activity appears to be related to its ATPase activity and it is not clear whether any membrane-targeting activity is involved, although our studies show several potential TM domains within the protein, one of which appears to be fairly consistent among all members of the family.

3.5. General Comments

At least for the better studied viral MPs (TMV, TGB proteins, *Carmovirus*), it seems probable that they enter the ER co-translationally and that the hydrophobic regions then migrate into the ER membrane. Movement to the cell periphery probably occurs as complexes with virions (or other nucleic acid–protein associations) in membrane-bound bodies and may use the cytoskeleton-based pathway. The complexes are thus delivered to the neck of the plasmodesmata. None of the plant host proteins that interact with viral MPs have yet been unequivocally identified but it is interesting that the NS_M movement protein of *Tomato spotted wilt virus* (genus *Tospovirus*, family *Bunyaviridae*), which has been classified in the "30K" superfamily, has been shown to interact with the viral CP, to bind viral RNA and, in a yeast two-hybrid screen, to bind to two plant proteins of the DnaJ family, that are in turn known to bind plant HSP70s (Soellick *et al.*, 2000). There are at least hints here of common links between what appear to be very dissimilar viral MPs. It is also interesting that there is increasing evidence that some plant proteins ("non-cell-autonomously replicating proteins," NCAPs) have properties similar to viral MPs in their effects upon plasmodesmatal size exclusion limits and in transporting RNA (see, for example, the detailed review by Roberts and

Oparka, 2003). It therefore appears likely that plant virus MPs mimic various aspects of the plant's own machinery for trafficking of large molecules.

4. Replication Proteins

Positive-strand RNA viruses assemble their RNA replication complexes on intracellular membranes and some progress has been made in identifying the proteins and sequences responsible.

In the genus *Tombusvirus* (family *Tombusviridae*), ORF1 encodes a polymerase with a readthrough (RT) domain and the smaller product contains an N-terminal hydrophilic portion followed by two predicted hydrophobic TM segments. In the type member, *Tomato bushy stunt virus*, the protein is localized to membrane fractions of cell extracts (Scholthof *et al.*, 1995). Infection of *Nicotiana benthamiana* cells with *Cymbidium ringspot virus* (CymRSV) or *Carnation Italian ringspot virus* (CIRV) results in the formation of conspicuous membranous bodies, which develop from modified peroxisomes or mitochondria, respectively. The ORF1 proteins can be localized in these membranous bodies (Bleve-Zacheo *et al.*, 1997) and have been shown to be integral membrane proteins with their N- and C-termini in the cytoplasm (Rubino and Russo, 1998; Rubino *et al.*, 2000, 2001; Weber-Lotfi *et al.*, 2002). These domains were consistently identified in all sequenced members of the genus by our TMPRED analysis; in the other genera of the *Tombusviridae*, although TM domains were identified they did not appear to be in corresponding positions, or at similar spacing, within the protein.

Members of the family *Bromoviridae* have three RNAs and the major products of both RNA1 and RNA2 (1a and 2a proteins) are required for replication. In both *Brome mosaic virus* (BMV, genus *Bromovirus*) and *Alfalfa mosaic virus* (genus *Alfamovirus*) proteins 1a and 2a co-localize to membranes, but respectively to the ER and tonoplast (Restrepo-Hartwig and Ahlquist, 1999; Heijden *et al.*, 2001). In BMV, the 1a protein is primarily responsible for this localization and a region, C-terminal to the core methyltransferase motif, has been identified by membrane floatation gradient analysis as sufficient for high-affinity ER membrane association although other regions are probably also involved (den Boon *et al.*, 2001). The 1a protein is fully susceptible to proteolytic digestion in the absence of detergent, suggesting that it does not span the membrane, but has an association with membranes that is stronger (resistant to high salt and high pH conditions) than is usual for a peripheral membrane protein. The 2a protein is then recruited to the membrane through its interaction with 1a and the N-terminal 120 amino-acid segment of 2a is sufficient for this (Chen and Ahlquist, 2000). Neither experimental evidence, nor computer predictions, suggest that a TM domain is involved with this interaction, although TMPRED does identify some (rather weak) regions in most *Bromoviridae* 1a proteins.

Members of the family *Comoviridae* have two RNAs, each of which encodes a polyprotein. Products of RNA1 are involved in replication, which has been associated with ER membranes in CPMV (Carette *et al.*, 2000, 2002) and in *Grapevine fanleaf virus* (genus *Nepovirus*) (Ritzenthaler *et al.*, 2002). In particular, the nucleoside triphosphate binding protein is believed to act as a membrane anchor for the replication complex and in *Tomato ringspot virus* (genus *Nepovirus*) a region at its C-terminus has been shown to have TM properties (Han and Sanfaçon, 2003). This is strongly confirmed by our TMPRED analyses for viruses in all genera of the family (*Comovirus, Fabavirus*, and *Nepovirus*).

Some progress in identifying the plant proteins with which the viral replication proteins interact has been made in the genus *Tobamovirus*. Western blot studies of membrane-bound *Tomato mosaic virus* (ToMV) replication complexes indicated the presence of a plant protein related to the 54.6-kDa GCD10 protein, the RNA-binding subunit of yeast eIF-3 (Osman and Buck, 1997). More recently, studies of *Arabidopsis* mutants have revealed several genes that are necessary for efficient multiplication of tobamoviruses. In particular TOM1 has been identified as a 7-pass TM protein of 291aa that interacts with the helicase domain of tobamovirus replication proteins and TOM2A, a 4-pass TM protein of 280 aa that interacts with TOM1. GFP-tagging had demonstrated that these proteins co-localize with the replication proteins to vacuolar (tonoplast) membranes in plant cells (Yamanaka *et al.*, 2000, 2002; Hagiwara *et al.*, 2003; Tsujimoto *et al.*, 2003).

There is less detailed evidence for the involvement of membrane targeting in the replication of other plant viruses but the replication proteins of *Peanut clump virus* (genus *Pecluvirus*) have been localized to membranes (Dunoyer *et al.*, 2002). In the genus *Potyvirus*, there have been suggestions that the 6K2 product of the polyprotein of *Tobacco etch virus* is involved with replication and that it binds to membranes (Restrepo-Hartwig and Carrington, 1994) and this is supported by recent results showing that the CI-6K2 protein of *Potato virus A* was associated with membrane fractions but that fully processed CI was not (Merits *et al.*, 2002). Our analyses show that there is a strongly predicted TM domain in all 6K2 proteins in the family *Potyviridae*.

5. Proteins Involved in Transmission by Vectors

To initiate infection of a host plant, viruses have to be introduced into a cell across the substantial barrier posed by the cell wall. Many plant viruses are dependent upon vectors for this step. Some virus–vector interactions involve adsorption onto, and release from, an external surface and this is typified by the nonpersistent, stylet-borne transmission by aphids of many viruses, for example in the genus *Potyvirus*. In other viruses, there is a more intimate and lasting ("persistent") relationship with the vector, in which the virus enters the host cells of its vector ("circulative") and, in some cases may replicate within it ("propagative") as well as within the plant host. Viral membrane proteins may therefore play an important role in the transmission of some viruses.

5.1. Insect Transmission

5.1.1. Persistent Transmission by Aphids

Persistent (circulative but not propagative) transmission has been best studied in members of the family *Luteoviridae*. Electron microscopy indicates that virus particles cross the gut into the aphid haemocoel in coated vesicles by receptor-mediated endocytosis (Gildow, 1993; Garret *et al.*, 1996). While the aphid gut acts as a barrier against the uptake of some morphologically similar viruses, uptake of different luteoviruses is not always related to the efficiency of virus transmission and it therefore appears that endocytosis is only partially selective. It is likely that the CP is primarily involved in interactions with the receptor but evidence for the role of the CP–RT is not entirely consistent. Mutants of *Barley yellow dwarf virus-PAV* lacking the RT were taken up through the aphid gut (although not subsequently transmitted) (Chay *et al.*, 1996) but some mutations in the *Beet western yellows virus* RT domain apparently affect acquisition across the gut membrane (Brault *et al.*, 2000). Changes in the CP and/or the RT of *Potato leafroll virus* (PLRV) have also been shown to hinder passage across the gut membrane (Rouze-Jouan *et al.*, 2001). Virions taken up into the haemocoel appear to be bound to a protein (symbionin) produced by endosymbiotic bacteria of the genus *Buchnera*. This appears to be important for virus survival within the vector (see review of Reavy and Mayo, 2002). If a virus is to be transmitted, it must then cross a membrane into the accessory salivary gland and this, also, is a specific, receptor-mediated process. The aphid and virus determinants of this process have not been characterized in detail but virus-like particles of PLRV consisting of CP (without the RT) and no genomic RNA could be exported into the salivary duct canal suggesting that the virus determinants are located within the CP alone (Gildow *et al.*, 2000). Our analyses do not suggest that there are TM domains in the CP or RT and it is likely, therefore, that their association with membranes is peripheral.

5.1.2. Transmission by Hoppers

Viruses transmitted by leafhoppers, planthoppers, and treehoppers include members of the genera *Mastrevirus, Curtovirus*, and *Topocuvirus* (family *Geminiviridae*) which have circulative, but not propagative, transmission. There is little experimental work to determine how these enter their vector, but chimerical clones based on the whitefly-transmitted *African cassava mosaic virus* (genus *Begomovirus*, family *Geminiviridae*) with the CP of the leafhopper transmitted *Beet curly top virus* (genus *Curtovirus*) could be transmitted by the leafhopper, demonstrating that the CP was the major determinant of vector specificity (Briddon *et al.*, 1990). A single TM domain is predicted in the CP of all these viruses by TMPRED (but not in the whitefly-transmitted geminiviruses) but it is not known whether this is related to any role in vector transmission.

Hopper-transmitted viruses that are propagative include members of the genera *Marafivirus* (family *Tymoviridae*) and *Tenuivirus*, some plant rhabdoviruses and all plant-infecting members of the family *Reoviridae*. In *Rice dwarf virus* (genus *Phytoreovirus*), a nontransmissible isolate that could not infect cells of the vector was shown to lack the P2 outer capsid protein, one of the six structural proteins of the virus (Tomaru *et al.*, 1997). It was subsequently shown that this protein was required for adsorption to cells of the insect vector (Omura *et al.*, 1998). In another reovirus, *Rice ragged stunt virus* (genus *Oryzavirus*), the spike protein encoded by S9 was expressed in bacteria, fed to the vector, and shown to inhibit transmission. Its ability to bind a 32-kDa insect membrane protein indicated that this might be a virus receptor that interacts with the spike protein (Zhou *et al.*, 1999). Within the genus *Tenuivirus*, the larger RNA2 product pC2, encoded in a negative sense, has several typical features of viral membrane glycoproteins (Takahashi *et al.*, 1993; Miranda *et al.*, 1996) and these are strongly detected by TMPRED, but its structure and function have not been studied in detail.

5.1.3. Transmission by Thrips

Viruses in the genus *Tospovirus* (family *Bunyaviridae*) are transmitted by thrips in a propagative manner, and the best studied is the type member, *Tomato spotted wilt virus* (TSWV). Virus enters its vector after ingestion of infected plant material and involves endocytosis by fusion at the apical plasmalemma of midgut epithelial cells. It is believed that one or both of the membrane glycoproteins (GP1 and GP2) serve as virus attachment proteins, binding to vector receptor proteins. The evidence for this, largely derived from electron microscopy has recently been summarized by Ullman *et al.* (2002). The use of anti-idiotypic antibodies has indicated that GP1 and GP2 bind thrips proteins of about 50 kDa (Bandla *et al.*, 1998; Meideros *et al.*, 2000) but the receptors have not been characterized in detail. Experiments in mammalian cells show that transporting and targeting of TSWV glycoproteins is probably very similar to that in animal-infecting bunyaviruses (e.g., *Uukeniemi virus* and *Bunyamwera virus*). The glycoprotein precursor was efficiently cleaved and the resulting GP1 and GP2 glycoproteins were transported from the ER to the Golgi complex, where they were retained. GP2 alone was retained in the Golgi complex, while GP1 alone was retained in the ER, irrespective of whether it contained the precursor's signal sequence or its own N-terminal hydrophobic sequence (Kikkert *et al.*, 2001). TMPRED predicts 5–10 TM segments in the precursor glycoprotein of different tospoviruses.

5.1.4. Persistent Transmission by Whiteflies

Viruses in the genus *Begomovirus* (family *Geminiviridae*) are transmitted by whiteflies in a circulative, but not propagative, manner. The route of transmission is similar to that described above for aphids (Section 5.1.1.) and it is therefore likely that receptor-mediated endocytosis is involved, both in crossing the gut into the haemocoel and then in viral transmission through the salivary glands. Several experiments indicate that the specificity for this resides in the CP. For example, *Abutilon mosaic virus* has lost its ability to be transmitted by whiteflies (probably because it has been maintained in plants by cuttings) and does not move into the haemocoel (Morin *et al.*, 2000). However, this ability can be restored by substitution of the CP by that of *Sida golden mosaic virus* (Hofer *et al.*, 1997) or by mutation at 2 or 3 positions (aa 124, 149, 174) in the CP (Hohnle *et al.*, 2001). Conversely, replacement of two amino acids (129 Q to P, 134 Q to H) in the CP of *Tomato yellow leaf curl Sardinia virus* was sufficient to abolish transmission (Norris *et al.*, 1998). There is not yet any detailed information on the interaction between the CP and putative whitefly receptors but our TMPRED results show that this is unlikely to involve a TM protein.

5.2. Fungus Transmission

A range of single-stranded RNA viruses are transmitted by plasmodiophorid "fungi," obligate intracellular parasites that are confined to plant roots. Although traditionally regarded as fungi by plant pathologists, these organisms are of uncertain taxonomic affinity but appear to be more closely related to protists than to the true fungi. In some of these, the viruses are carried within the vector and both acquisition and transmission involves transport across the membrane that separates the cytoplasm of the vector from that of its host (Adams, 2002; Kanyuka *et al.*, 2003). For rod-shaped viruses of the genera *Benyvirus, Furovirus*, and *Pomovirus*, deletions in the CP–RT domain abolish transmissibility (Tamada and Kusume, 1991; Schmitt *et al.*, 1992; Reavy *et al.*, 1998), while for filamentous viruses of the genus *Bymovirus* (family *Potyviridae*), deletions in the P2 domain have a similar effect (Adams *et al.*, 1988; Jacobi *et al.*, 1995; Peerenboom *et al.*, 1996). In *Beet necrotic yellow vein virus* (BNYVV, genus *Benyvirus*), substitution of two amino acids (KTER to ATAR at 553–556) in the CP–RT prevented transmission by the vector, *Polymyxa betae* (Tamada *et al.*, 1996). Computer predictions by TMPRED and other software suggest that all the CP–RTs and P2 proteins have two hydrophobic regions. Directional alignment of these two helices also shows

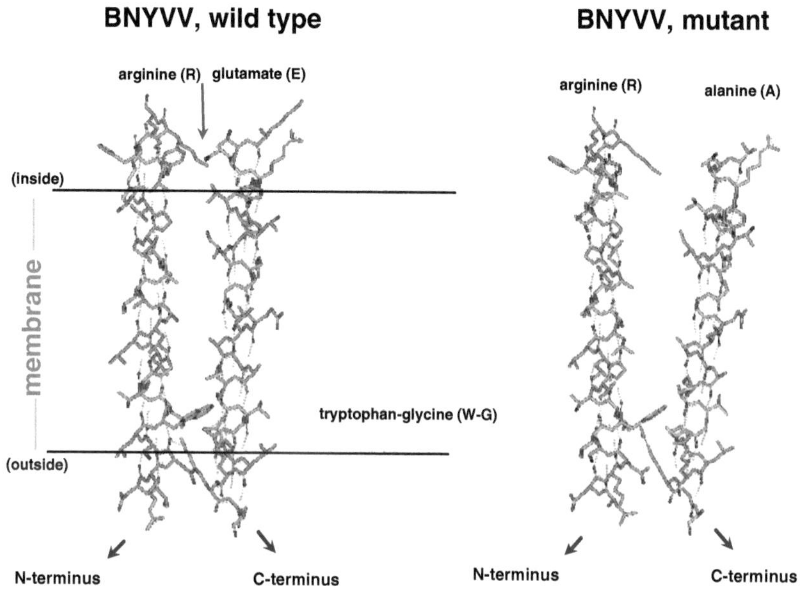

Figure 1.2. Models of the predicted helices and interfacial regions of the TM domains in the CP–RT of *Beet necrotic yellow vein virus* (BNYVV), showing the effects of the KTER>ATAR substitution that abolishes transmission by the plasmodiophorid vector *Polymyxa betae*, modeled using MOLMOL (Ver. 2.6) and displayed using the WebLab Viewer (from Adams *et al.*, 2001). Electrostatic interactions are shown dotted.

evidence of compatibility between their amino acids, with groupings of amino acids that are either identical or in the same hydrophobicity group and evidence of possible fits between the small residues on one helix and the larger aromatic ones on the other. From these patterns, and from calculation of relative helix tilts, structural arrangements consistent with tight packing of TM helices were detected. These included ridge/groove arrangements between the two helices and strong electrostatic associations at the interfacial regions of the membrane. This suggests that the two TM domains could be paired within a membrane and with their C- and N-termini on the outside of the membrane. Nontransmissible deletion mutants lack the second of these putative TM regions and modeling of the BNYVV substitution suggests that it would disrupt the alignment of the polypeptide at a critical position adjacent to the second TM domain (Adams *et al.*, 2001) (Figure 1.2). As there are few other similarities between the genomes of some of these viruses, it seems probable that the TM regions are instrumental in assisting virus particles to move across the vector membrane.

6. Other Membrane Proteins

Studies with *Southern cowpea mosaic virus* (genus *Sobemovirus*) have investigated the interaction of the CP with artificial membranes using a liposome dye-release assay and circular dichroism. The native CP and the R domain (which binds RNA and is usually on the inside of the spherical particle, but which is externalized under certain pH and salt conditions) were shown to interact with liposomes *in vitro*. Studies of mutants, mapped the region responsible to residues 1–30 and analysis of this region by circular dichroism indicated that it assumes an alpha-helical structure when exposed to liposomes composed of anionic lipids (Lee *et al.*, 2001). It has not yet been shown if this occurs *in vivo* but, if it does, the authors suggest that it could be related to cell-to-cell movement, to replication, to beetle transmission, or alteration of ion flux into or out of the cytoplasm. In our analyses, the region is not predicted to have TM properties.

Our TMPRED analyses indicate a few other plant virus proteins with strong TM properties, for which functions have not been assigned. These include the nonstructural protein P9-2 in *Rice black streaked dwarf virus* (RBSDV, genus *Fijivirus*, family *Reoviridae*) and its homologs in other members of the genus (P9-2 of *Nilaparvarta lugens reovirus*, P9-2 of *Fiji disease virus*, P10-2 of *Oat sterile dwarf virus*, and P8-2 of *Maize rough dwarf virus*). The TM domains occur in similar positions in the middle of the protein with the N- and C-termini exposed to the outside and a loop of 20–25 aa between them. No protein with similar properties can be identified in other plant-infecting reoviruses and it has not been detected within infected plants (Isogai *et al.*, 1998). A further example is the small P6 protein encoded by Barley yellow dwarf viruses of the genus *Luteovirus* (family *Luteoviridae*), which all contain a single, strongly predicted TM region. Viruses assigned to other genera in the family do not appear to have this ORF and its function is not known.

7. Conclusions

It is clear that MPs play an essential role in the pathogenesis and movement within the plant of many plant viruses. However, studies of the structure and function of such proteins are still in their infancy. Substantial progress may be expected in the next few years, particularly in the area of cell-to-cell movement where viruses are proving useful tools to study the basic processes of macromolecular trafficking between adjacent plant cells.

Acknowledgments

We thank Prof. R.N. Beachy for permission to reproduce Figure 1.1 and Drs A. Tymon and K. Hammond-Kosack for helpful comments on the manuscript. Rothamsted Research receives grant-aided support from the Biotechnology and Biological Sciences Research Council of the United Kingdom.

References

Aberle, H.J., Rutz, M.L., Karayavuz, M., Frischmuth, S., Wege, C., Hulser, D. *et al.* (2002). Localizing the movement proteins of Abutilon mosaic geminivirus in yeast by subcellular fractionation and freeze-fracture immuno-labelling. *Arch. Virol.* **147**, 103–117.

Adams, M.J. (2002). Fungi. In R.T. Plumb (ed.), *Plant Virus Vector Interactions* (*Adv. Bot. Res.* 36). Academic Press, San Diego, CA, pp. 47–64.

Adams, M.J., Swaby, A.G., and Jones, P. (1988). Confirmation of the transmission of barley yellow mosaic virus (BaYMV) by the fungus *Polymyxa graminis*. *Ann. Appl. Biol.* **112**, 133–141.

Adams, M.J., Antoniw, J.F., and Mullins, J.G.L. (2001). Plant virus transmission by plasmodiophorid fungi is associated with distinctive transmembrane regions of virus-encoded proteins. *Arch. Virol.* **146**, 1139–1153.

Adams, M.J., Antoniw, J.F., Barker, H., Jones, A.T., Murant, A.F., and Robinson, D. (1998). *Descriptions of Plant Viruses on CD-ROM*. Association of Applied Biologists, Wellesbourne.

Alzhanova, D.V., Napuli, A.J., Creamer, R., and Dolja, V.V. (2001). Cell-to-cell movement and assembly of a plant closterovirus: Roles for the capsid proteins and Hsp70 homolog. *EMBO J.* **20**, 6997–7007.

Bandla, M.D., Campbell, L.R., Ullman, D.E., and Sherwood, J.L. (1998). Interaction of tomato spotted wilt tospovirus (TSWV) glycoproteins with a thrips midgut protein, a potential cellular receptor for TSWV. *Phytopathology* **88**, 98–104.

Bleve-Zacheo, T., Rubino, L., Melillo, M.T., and Russo, M. (1997). The 33K protein encoded by cymbidium ringspot tombusvirus localizes to modified peroxisomes of infected cells and of uninfected transgenic plants. *J. Plant Pathol.* **79**, 197–202.

Boulton, M.I. (2002). Functions and interactions of mastrevirus gene products. *Physiol. Mol. Plant Path.* **60**, 243–255.

Boulton, M.I., Pallaghy, C.K., Chatani, M., MacFarlane, S.A., and Davies, J.W. (1993). Replication of maize streak virus mutants in protoplasts – evidence for movement protein. *Virology* **192**, 85–93.

Boyko, V., Ferralli, J., Ashby, J., Schellenbaum, P., and Heinlein, M. (2000). Function of microtubules in intercellular transport of plant virus RNA. *Nat. Cell Biol.* **2**, 826–832.

Bozarth, C.S., Weiland, J.J., and Dreher, T.W. (1992). Expression of Orf-69 of turnip yellow mosaic-virus is necessary for viral spread in plants. *Virology* **187**, 124–130.

Brault, V., Mutterer, J., Scheidecker, D., Simonis, M.T., Herrbach, E., Richards, K. *et al.* (2000). Effects of point mutations in the readthrough domain of the beet western yellows virus minor capsid protein on virus accumulation in planta and on transmission by aphids. *J. Virol.* **74**, 1140–1148.

Briddon, R.W., Pinner, M.S., Stanley, J., and Markham, P.G. (1990). Geminivirus coat protein replacement alters insect specificity. *Virology* **177**, 85–94.

Brill, L.M., Nunn, R.S., Kahn, T.W., Yeager, M., and Beachy, R.N. (2000). Recombinant tobacco mosaic virus movement protein is an RNA-binding, alpha-helical membrane protein. *Proc. Natl. Acad. Sci. USA* **97**, 7112–7117.

Canto, T. and Palukaitis, P. (1999). Are tubules generated by the 3a protein necessary for cucumber mosaic virus movement? *Mol. Plant–Microbe Interact.* **12**, 985–993.

Carette, J.E., Stuiver, M., Van Lent, J., Wellink, J., and Van Kammen, A.B. (2000). Cowpea mosaic virus infection induces a massive proliferation of endoplasmic reticulum but not Golgi membranes and is dependent on de novo membrane synthesis. *J. Virol.* **74**, 6556–6563.

Carette, J.E., Verver, J., Martens, J., van Kampen, T., Wellink, J., and van Kammen, A. (2002). Characterization of plant proteins that interact with cowpea mosaic virus "60K" protein in the yeast two-hybrid system. *J. Gen. Virol.* **83**, 885–893.

Chay, C.A., Gunasinge, U.B., Dinesh-Kumar, S.P., Miller, W.A., and Gray, S.M. (1996). Aphid transmission and systemic plant infection determinants of barley yellow dwarf luteovirus—PAV are contained in the coat protein readthrough domain and 17-kDa protein, respectively. *Virology* **219**, 57–65.

Chen, J.B. and Ahlquist, P. (2000). Brome mosaic virus polymerase-like protein 2a is directed to the endoplasmic reticulum by helicase-like viral protein 1a. *J. Virol.* **74**, 4310–4318.

Citovsky, V. and Zambryski, P. (1991). How do plant virus nucleic acids move through intercellular connections? *BioEssays* **13**, 373–379.

Cowan, G.H., Lioliopoulou, F., Ziegler, A., and Torrance, L. (2002). Subcellular localisation, protein interactions, and RNA binding of potato mop-top virus triple gene block proteins. *Virology* **298**, 106–115.

den Boon, J.A., Chen, J.B., and Ahlquist, P. (2001). Identification of sequences in brome mosaic virus replicase protein 1a that mediate association with endoplasmic reticulum membranes. *J. Virol.* **75**, 12370–12381.

Desvoyes, B., Faure-Rabasse, S., Chen, M.H., Park, J.W., and Scholthof, H.B. (2002). A novel plant homeodomain protein interacts in a functionally relevant manner with a virus movement protein. *Plant Physiol.* **129**, 1521–1532.

Dickinson, V.J., Halder, J., and Woolston, C.J. (1996). The product of maize streak virus ORF V1 is associated with secondary plasmodesmata and is first detected with the onset of viral lesions. *Virology* **220**, 51–59.

Dunoyer, P., Ritzenthaler, C., Hemmer, O., Michler, P., and Fritsch, C. (2002). Intracellular localization of the *Peanut clump virus* replication complex in tobacco BY-2 protoplasts containing green fluorescent protein-labelled endoplasmic reticulum or golgi apparatus. *J. Virol.* **76**, 865–874.

Erhardt, M., Morant, M., Ritzenthaler, C., Stussi-Garaud, C., Guilley, H., Richards, K. *et al.* (2000). P42 movement protein of Beet necrotic yellow vein virus is targeted by the movement proteins P13 and P15 to puncuate bodies associated with plasmodesmata. *Mol. Plant–Microbe Interact.* **13**, 520–528.

Erhardt, M., Stussi-Garaud, C., Guilley, H., Richards, K.E., Jonard, G., and Bouzoubaa, S. (1999). The first triple gene block protein of peanut clump virus localizes to the plasmodesmata during virus infection. *Virology* **264**, 220–229.

Fridborg, I., Grainger, J., Page, A., Coleman, M., Findlay, K., and Angell, S. (2003). TIP, a novel host factor linking callose degradation with the cell-to-cell movement of Potato virus X. *Mol. Plant–Microbe Interact.* **16**, 132–140.

Garcia-Castillo, S., Sanchez-Pina, M.A., and Pallas, V. (2003). Spatio-temporal analysis of the RNAs, coat and movement (p7) proteins of Carnation mottle virus in Chenopodium quinoa plant. *J. Gen. Virol.* **84**, 745–749.

Garret, A., Kerlan, C., and Thomas, D. (1996). Ultrastructural study of acquisition and retention of potato leafroll luteovirus in the alimentary canal of its aphid vector, *Myzus persicae* Sulz. *Arch. Virol.* **141**, 1279–1292.

Gildow, F.E. (1993). Evidence for receptor-mediated endocytosis regulating luteovirus acquistion by aphids. *Phytopathology* **83**, 270–277.

Gildow, F.E., Reavy, B., Mayo, M.A., Duncan, G.H., Woodford, T., Lamb, J.W. *et al.* (2000). Aphid acquisition and cellular transport of Potato leafroll virus-like particles lacking P5 readthrough. *Phytopathology* **90**, 1153–1161.

Gillespie, T., Boevink, P., Haupt, S., Roberts, A.G., Toth, R., Valentine, T. *et al.* (2002). Functional analysis of a DNA-shuffled movement protein reveals that microtubules are dispensable for the cell-to-cell movement of Tobacco mosaic virus. *Plant Cell* **14**, 1207–1222.

Gorshkova, E.N., Erokhina, T.N., Stroganova, T.A., Yelina, N.E., Zamayatin, A.A. J., Kalinina, N.O. *et al.* (2003). Immunodetection and fluorescent microscopy of transgenically expressed hordeivirus TGBp3 movement protein reveals its association with endoplasmic reticulum elements in close proximity to plasmodesmata. *J. Gen. Virol.* **84**, 985–994.

Hacker, D.L., Petty, I., Wei, N., and Morris, T.J. (1992). Turnip crinkle virus genes required for RNA replication and virus movement. *Virology* **186**, 1–8.

Hagiwara, Y., Komoda, K., Yamanaka, T., Tamai, A., Meshi, T., Funada, R. *et al.* (2003). Subcellular localization of host and viral proteins associated with tobamovirus RNA replication. *EMBO J.* **22**, 344–353.

Han, S. and Sanfaçon, H. (2003). *Tomato ringspot virus* proteins containing the nucleotide triphosphate binding domain are transmembrane proteins that associate with the endoplasmic reticulum and cofractionate with replication complexes. *J. Virol.* **77**, 523–534.

Heijden, M.W. v. d., Carette, J.E., Reinhoud, P.J., Haegi, A., and Bol, J.F. (2001). Alfalfa mosaic virus replicase proteins P1 and P2 interact and colocalize at the vacuolar membrane. *J. Virol.* **75**, 1879–1887.

Heinlein, M., Padgett, H.S., Gens, J.S., Pickard, B.G., Casper, S.J., Epel, B.L. *et al.* (1998). Changing patterns of localization of the tobacco mosaic virus movement protein and replicase to the endoplasmic reticulum and microtubules during infection. *Plant Cell* **10**, 1107–1120.

Hofer, P., Bedford, I.D., Markham, P.G., Jeske, H., and Frischmuth, T. (1997). Coat protein gene replacement results in whitefly transmission of an insect non-transmissible geminivirus isolate. *Virology* **236**, 288–295.

Hofmann, K. and Stoffel, W. (1993). TMbase—a database of membrane spanning proteins segments. *Biol. Chem. Hoppe-Seyler* **374**, 166.

Hohnle, M., Hofer, P., Bedford, I.D., Briddon, R.W., Markham, P.G., and Frischmuth, T. (2001). Exchange of three amino acids in the coat protein results in efficient whitefly transmission of a nontransmissible Abutilon mosaic virus isolate. *Virology* **290**, 164–171.

Isogai, M., Uyeda, I., and Lee, B.-C. (1998). Detection and assignment of proteins encoded by rice black streaked dwarf fijivius S7, S8, S9 and S10. *J. Gen. Virol.* **79**, 1487–1494.

Jacobi, V., Peerenboom, E., Schenk, P.M., Antoniw, J.F., Steinbiss, H.-H., and Adams, M.J. (1995). Cloning and sequence analysis of RNA-2 of a mechanically-transmitted UK isolate of barley mild mosaic bymovirus (BaMMV). *Virus Res.* **37**, 99–111.

Jansen, K.A.J., Wolfs, C.J.A.M., Lohuis, H., Goldbach, R., and Verduin, B.J.M. (1998). Characterization of the brome mosaic virus movement protein expressed in *E. coli. Virology* **242**, 387–394.

Kanyuka, K., Ward, E., and Adams, M.J. (2003). *Polymyxa graminis* and the cereal viruses it transmits: A research challenge. *Mol. Plant Pathol.*, **4**, 393–406.

Kikkert, M., Verschoor, A., Kormelink, R., Rottier, P., and Goldbach, R. (2001). Tomato spotted wilt virus glycoproteins exhibit trafficking and localization signals that are functional in mammalian cells. *J. Virol.* **75**, 1004–1012.

Lawrence, D.M. and Jackson, A.O. (2001). Interactions of the TGB1 protein during cell-to-cell movement of Barley stripe mosaic virus. *J. Virol.* **75**, 8712–8723.

Lee, S.K., Dabney-Smith, C., Hacker, D.L., and Bruce, B.D. (2001). Membrane activity of the southern cowpea mosaic virus coat protein: The role of basic amino acids, helix-forming potential, and lipid composition. *Virology* **291**, 299–310.

Medina, V., Peremyslov, V.V., Hagiwara, Y., and Dolja, V.V. (1999). Subcellular localization of the HSP70-homolog encoded by beet yellows closterovirus. *Virology* **260**, 173–181.

Mei, H. and Lee, Z. (1999). Association of the movement protein of alfalfa mosaic virus with the endoplasmic reticulum and its trafficking in epidermal cells of onion bulb scales. *Mol. Plant–Microbe Interact.* **12**, 680–690.

Meideros, R.B., Ullman, D.E., Sherwood, J.L., and German, T.L. (2000). Immunoprecipitation of a 50 kDa protein: A candidate receptor for tomato spotted wilt virus topsovirus (*Bunyaviridae*) in its main vector, *Frankliniella occidentalis*. *Virus Res.* **67**, 109–118.

Melcher, U. (2000). The "30K" superfamily of viral movement proteins. *J. Gen. Virol.* **81**, 257–266.

Merits, A., Rajamaki, M.L., Lindholm, P., Runeberg-Roos, P., Kekarainen, T., Puustinen, P. *et al.* (2002). Proteolytic processing of potyviral proteins and polyprotein processing intermediates in insect and plant cells. *J. Gen. Virol.* **83**, 1211–1221.

Miranda, J.R.d., Munoz, M., Wu, R., Hull, R., and Espinoza, A.M. (1996). Sequence of rice hoja blanca tenuivirus RNA-2. *Virus Genes* **12**, 231–237.

Morin, S., Ghanim, M., Sobol, I., and Czosnek, H. (2000). The GroEL protein of the whitefly Bemisia tabaci interacts with the coat protein of transmissible and non-transmissible begomoviruses in the yet two-hybrid system. *Virology* **276**, 404–416.

Morozov, S.Y. and Solovyev, A.G. (2003). Triple gene block: Modular design of a multifunctional machine for plant virus movement. *J. Gen. Virol.* **84**, 1351–1366.

Morozov, S.Y., Solovyev, A.G., Kalinina, N.O., Fedorkin, O.N., Samuilova, O.V., Schiemann, J. *et al.* (1999). Evidence for two nonoverlapping functional domains in the potato virus X 25K movement protein. *Virology* **260**, 55–63.

Morozov, S.Y., Miroshnichenko, N.A., Solovyev, A.G., Zelenina, D.A., Fedorkin, O.N., Lukasheva *et al.* (1991). In vitro Membrane-Binding of the Translation Products of the Carlavirus 7-Kda Protein Genes. *Virology* **183**, 782–785.

Mushegian, A.R. and Koonin, E.V. (1993). Cell-to-cell movement of plant viruses. Insights from amino acid sequence comparisons of movement proteins and from analogies with cellular transport systems. *Arch. Virol.* **133**, 239–257.

Norris, E., Vaira, A.M., Caciagli, P., Masenga, V., Gronenborn, B., and Accotto, G.P. (1998). Amino acids in the capsid protein of tomato yellow leaf curl virus that are crucial for systemic infection, particle formation, and insect transmission. *J. Virol.* **72**, 10050–10057.

Nurkiyanova, K.M., Ryabov, E.V., Kalinina, N.O., Fan, Y., Andreev, I., Fitzgerald, A.G. *et al.* (2001). Umbravirus-encoded movement protein induces tubule formation on the surface of protoplasts and binds RNA incompletely and non-cooperatively. *J. Gen. Virol.* **82**, 2579–2588.

Omura, T., Yan, J., Zhong, B., Wada, M., Zhu, Y., Tomaru, M. *et al.* (1998). The P2 protein of rice dwarf phytoreovirus is required for adsorption of the virus to cells of the insect vector. *J. Virol.* **72**, 9370–9373.

Oparka, K.J., Prior, D.A.M., Santa Cruz, S., Padgett, H.S., and Beachy, R.N. (1997). Gating of epidermal plasmodesmata is restricted to the leading edge of expanding infection sites of tobacco mosaic virus (TMV). *Plant J.* **12**, 781–789.

Osman, T.A.M. and Buck, K.W. (1997). The tobacco mosaic virus RNA polymerase complex contains a plant protein related to the RNA-Binding subunit of yeast eIF-3. *J. Virol.* **71**, 6075–6082.

Peerenboom, E., Jacobi, V., Antoniw, J.F., Schlichter, U.H.A., Cartwright, E.J., Steinbiss, H.-H. *et al.* (1996). The complete nucleotide sequence of RNA-2 of a fungally-transmitted UK isolate of barley mild mosaic bymovirus (BaMMV) and identification of amino acid combinations possibly involved in fungus transmission. *Virus Res.* **40**, 149–159.

Perbal, M.-C., Thomas, C.L., and Maule, A.J. (1993). Cauliflower mosaic virus gene I product (P1) forms tubular structures which extend from the surface of infected protoplasts. *Virology* **195**, 281–285.

Peremyslov, V.V., Hagiwara, Y., and Dolja, V.V. (1999). HSP70 homolog functions in cell-to-cell movement of a plant virus. *Proc. Natl. Acad. Sci. USA* **96**, 14771–14776.

Pouwels, J., Van der Krogt, G.N.M., Van Lent, J., Bisseling, T., and Wellink, J. (2002). The cytoskeleton and the secretory pathway are not involved in targeting the cowpea mosaic virus movement protein to the cell periphery. *Virology* **297**, 48–56.

Reavy, B. and Mayo, M.A. (2002). Persistent transmission of luteoviruses by aphids. In R.T. Plumb (ed.), *Plant Virus Vector Interactions* (*Adv. Bot. Res.* 36). Academic Press, San Diego, CA, pp. 21–46.

Reavy, B., Arif, M., Cowan, G.H., and Torrance, L. (1998). Association of sequences in the coat protein/readthrough domain of potato mop-top virus with transmission by Spongospora subterranea. *J. Gen. Virol.* **79**, 2343–2347.

Reichel, C. and Beachy, R.N. (1998). Tobacco mosaic virus infection induces severe morphological changes of the endoplasmic reticulum. *Proc. Natl. Acad. Sci. USA* **95**, 11169–11174.

Reichel, C., Mas, P., and Beachy, R.N. (1999). The role of the ER and cytoskeleton in plant viral trafficking. *Trends Plant Sci.* **4**, 458–462.

Restrepo-Hartwig, M. and Ahlquist, P. (1999). Brome mosaic virus RNA replication proteins 1a and 2a colocalize and 1a independently localizes on the yeast endoplasmic reticulum. *J. Virol.* **73**, 10303–10309.

Restrepo-Hartwig, M.A. and Carrington, J.C. (1994). The tobacco etch potyvirus 6-kilodalton protein is membrane associated and involved in viral replication. *J. Virol.* **68**, 2388–2397.

Ritzenthaler, C., Laporte, C., Gaire, F., Dunoyer, P., Schmitt, C., Duval, S. *et al.* (2002). Grapevine fanleaf virus replication occurs on endoplasmic reticulum-derived membranes. *J. Virol.* **76**, 8808–8819.

Roberts, A.G. and Oparka, K.J. (2003). Plasmodesmata and the control of symplastic transport. *Plant, Cell Environ.* **26**, 103–124.

Rouze-Jouan, J., Terradot, L., Pasquer, F., Tanguy, S., and Ducray-Bourdin, D.G. (2001). The passage of Potato leafroll virus through *Myzus persicae* gut membrane regulates transmission efficiency. *J. Gen. Virol.* **82**, 17–23.

Rubino, L. and Russo, M. (1998). Membrane targeting sequences in Tombusvirus infections. *Virology* **252**, 431–437.

Rubino, L., Di Franco, A., and Russo, M. (2000). Expression of a plant virus non-structural protein in *Saccharomyces cerevisiae* causes membrane proliferation and altered mitochondrial morphology. *J. Gen. Virol.* **81**, 279–286.

Rubino, L., Weber-Lotfi, F., Dietrich, A., Stussi-Garaud, C., and Russo, M. (2001). The open reading frame 1-encoded ("36K") protein of Carnation Italian ringspot virus localizes to mitochondria. *J. Gen. Virol.* **82**, 29–34.

Saldarelli, P., Minafra, A., Castellano, M.A., and Martelli, G.P. (2000). Immunodetection and subcellular localization of the proteins encoded by ORF 3 of grapevine viruses A and B. *Arch. Virol.* **145**, 1535–1542.

Schmitt, C., Balmori, E., Guilley, H., Richards, K., and Jonard, G. (1992). In vitro mutagenesis of biologically active transcripts of beet necrotic yellow vein virus RNA 2: Evidence that a domain of the 75 kDa readthrough protein is important for efficient virus assembly. *Proc. Natl. Acad. Sci. USA* **89**, 5715–5719.

Scholthof, K.B.G., Scholthof, H.B., and Jackson, A.O. (1995). The tomato bushy stunt virus replicase proteins are coordinately expressed and membrane associated. *Virology* **208**, 365–369.

Soellick, T.R., Uhrig, J.F., Bucher, G.L., Kellmann, J.W., and Schreier, P.H. (2000). The movement protein NSm of tomato spotted wilt tospovirus (TSWV): RNA binding, interaction with the TSWV N protein, and identification of interacting plant proteins. *Proc. Natl. Acad. Sci. USA* **97**, 2373–2378.

Solovyev, A.G., Stroganova, T.A., Zamyatnin, A.A., Jr., Fedorkin, O.N., Schiemann, J., and Morozov, S.Y. (2000). Subcellular sorting of small membrane-associated triple gene block proteins: TGBp3-assisted targeting of TGBp2. *Virology* **269**, 113–127.

Sonnhammer, E.L.L., von Heijne, G., and Krogh, A. (1998). A hidden Markov model for predicting transmembrane helices in protein sequences. *Proceedings of the Sixth International Conference on intelligent systems for molecular biology*. AAAI Press, Menlo Park, CA, pp. 175–182.

Takahashi, M., Toriyama, S., Hamamatsu, C., and Ishihama, A. (1993). Nucleotide sequence and possible ambisense coding strategy of rice stripe virus RNA segment 2. *J. Gen. Virol.* **74**, 769–773.

Tamada, T. and Kusume, T. (1991). Evidence that the 75K readthrough protein of beet necrotic yellow vein virus RNA-2 is essential for transmission by the fungus *Polymyxa betae*. *J. Gen. Virol.* **72**, 1497–1504.

Tamada, T., Schmitt, C., Saito, M., Guilley, H., Richards, K., and Jonard, G. (1996). High resolution analysis of the readthrough domain of beet necrotic yellow vein virus readthrough protein: A KTER motif is important for efficient transmission of the virus by *Polymyxa betae*. *J. Gen. Virol.* **77**, 1359–1367.

Tamai, A. and Meshi, T. (2001). Cell-to-cell movement of Potato virus X: The role of p12 and p8 encoded by the second and third open reading frames of the triple gene block. *Mol. Plant–Microbe Interact.* **14**, 1158–1167.

Tomaru, M., Maruyama, W., Kikuchi, A., Yan, J., Zhu, Y., Suzuki, N. *et al.* (1997). The loss of outer capsid protein P2 results in nontransmissibility by the insect vector of rice dwarf phytoreovirus. *J. Gen. Virol.* **71**, 8019–8023.

Tsujimoto, Y., Numaga, T., Ohshima, K., Yano, M., Ohsawa, R., Goto, D.B. *et al.* (2003). Arabidopsis Tobamovirus Multiplication (TOM) 2 locus encodes a transmembrane protein that interacts with TOM1. *EMBO J.* **22**, 335–343.

Tusnády, G.E. and Simon, I. (1998). Principles governing amino acid composition of integral membrane proteins: Applications to topology prediction. *J. Mol. Biol.* **283**, 489–506.

Ullman, D.E., Meideros, R., Campbell, L.R., Whitfield, A.E., Sherwood, J.L., and German, T.L. (2002). Thrips as vectors of Tospoviruses. In R.T. Plumb (ed.), *Plant Virus Vector Interactions* (*Adv. Bot. Res.* 36). Academic Press, San Diego, CA, pp. 113–140.

Van Lent, J., Storms, M., van der Meer, F., Wellink, J., and Goldbach, R. (1991). Tubular structures involved in movement of cowpea mosaic virus are also formed in infected cowpea protoplasts. *J. Gen. Virol.* **72**, 2615–2623.

Vilar, M., Sauri, A., Monne, M., Marcos, J.F., von Heijne, G., Perez-Paya, E. *et al.* (2002). Insertion and topology of a plant viral movement protein in the endoplasmic reticulum membrane. *J. Biol. Chem.* **277**, 23447–23452.

von Heijne, G. (1992). Membrane protein structure prediction, hydrophobicity analysis and the positive-inside rule. *J. Mol. Biol.* **225**, 487–494.

Wanitchakorn, R., Hafner, G.J., Harding, R.M., and Dale, J.L. (2000). Functional analysis of proteins encoded by banana bunchy top virus DNA-4 to -6. *J. Gen. Virol.* **81**, 299–306.

Weber-Lotfi, F., Dietrich, A., Russo, M., and Rubino, L. (2002). Mitochondrial targeting and membrane anchoring of a viral replicase in plant and yeast cells. *J. Virol.* **76**, 10485–10496.

Yamanaka, T., Ohta, T., Takahashi, M., Meshi, T., Schmidt, R., Dean, C. *et al.* (2000). TOM1, an Arabidopsis gene required for efficient multiplication of a tobamovirus, encodes a putative transmembrane protein. *Proc. Natl. Acad. Sci. USA* **97**, 10107–10112.

Yamanaka, T., Imai, T., Satoh, R., Kawashima, A., Takahashi, M., Tomita, K. *et al.* (2002). Complete inhibition of tobamovirus multiplication by simultaneous mutations in two homologous host genes. *J. Virol.* **76**, 2491–2497.

Zamyatnin, A.A., Solovyev, A.G., Sablina, A.A., Agranovsky, A.A., Katul, L., Vetten, H.J. *et al.* (2002). Dual-colour imaging of membrane protein targeting directed by poa semilatent virus movement protein TGBp3 in plant and mammalian cells. *J. Gen. Virol.* **83**, 651–662.

Zhang, S.C., Ghosh, R., and Jeske, H. (2002). Subcellular targeting domains of Abutilon mosaic geminivirus movement protein BC1. *Arch. Virol.* **147**, 2349–2363.

Zhang, S.C., Wege, C., and Jeske, H. (2001). Movement proteins (BC1 and BV1) of Abutilon mosaic geminivirus are cotransported in and between cells of sink but not of source leaves as detected by green fluorescent protein tagging. *Virology* **290**, 249–260.

Zhou, G.Y., Lu, X.B., Lu, H.J., Lei, J.L., Chen, S.X., and Gong, Z.X. (1999). Rice Ragged Stunt Oryzavirus: Role of the viral spike protein in transmission by the insect vector. *Ann. Appl. Biol.* **135**, 573–578.

2

Structure and Function of a Viral Encoded K^+ Channel

Anna Moroni, James Van Etten, and Gerhard Thiel

1. Introduction

Since the pioneering work on the M2 protein from influenza virus A (Pinto *et al.*, 1992; Wang *et al.*, 1993), several other viruses have been discovered to encode proteins that function as ion channels. Typically these proteins are small, consisting of about 100 amino acids, and they contain at least one hydrophobic transmembrane (TM) domain. Oligomerization of these subunits produces a hydrophilic pore, which facilitates ions transport across host cell membranes (Carrasco, 1995; Plugge *et al.*, 2000; Fischer and Sansom, 2002; Mould *et al.*, 2003).

This chapter will focus on the properties of the Kcv protein encoded by chlorella virus PBCV-1, a protein with the structural and functional hallmarks of a K^+ channel protein. Other virus-encoded channel proteins or suspected channel proteins are described elsewhere in this book.

PBCV-1 is the prototype of a genus (called *Chlorovirus*) of large, icosahedral, plaque-forming, dsDNA viruses that replicate in certain unicellular, eukaryotic chlorella-like green algae. The chlorella viruses are in the family Phycodnaviridae, which has ~75 members (Van Etten, 2000). The PBCV-1 330-kb genome contains ~375 protein-encoding genes, including the *kcv* gene, and 11 tRNA genes. Cryo-electron microscopy and three-dimensional image reconstructions of PBCV-1 indicate that its outer glycoprotein capsid shell surrounds a lipid bilayer membrane (Yan *et al.*, 2000; Van Etten, 2003). The membrane is required for infectivity because the virus loses infectivity rapidly in chloroform and more slowly in ethyl ether or toluene (Skrdla *et al.*, 1984).

Anna Moroni • Dipartimento di Biologia and CNR—IBF, Unità di Milano; Istituto Nazionale di Fisica della Materia, Unità di Milano-Università, Milano, Italy. **James Van Etten** • Department of Plant Pathology, Nebraska Center for Virology, University of Nebraska, Lincoln, Nebraska. **Gerhard Thiel** • Institute for Botany, University of Technology Darmstadt, Germany.

Viral Membrane Proteins: Structure, Function, and Drug Design, edited by Wolfgang Fischer.
Kluwer Academic / Plenum Publishers, New York, 2005.

2. K^+ Channels are Highly Conserved Proteins with Important Physiological Functions

Eukaryotic K^+ channels are involved in many important physiological functions. These range from controlling heart beat and brain functions in animals (Hille, 2001) to the control of growth and development in plants (Blatt and Thiel, 1993; Very and Sentenac, 2002). Potassium channels also exist in prokaryotes (Schrempf *et al.*, 1995; Derst and Karschin, 1998), where they may be involved in osmoregulation.

All K^+ channels consist of tetramers of identical or similar subunits, arranged around a central ion-conducting pore (Miller, 2000). Within this central pore is a motif of eight amino acids, TxxTxG(Y/F)G, which is common to all K^+ channels; this motif forms the selectivity filter of the permeation pathway. Based on their subunit TM topology, K^+ channels are classified into 6, 4, and 2 TM domain channels. The smallest K^+ channels belong to the 2-TM class and are formed by two TM domains connected by a stretch of amino acids, the so-called pore (P) domain. This basic structure is conserved among the pores of the more complex 4-TM and 6-TM channels (for overview, see Wei *et al.*, 1996).

3. Structural Aspects of Viral K^+ Channel Proteins as Compared to those from Other Sources

The Kcv protein encoded by virus PBCV-1 is a 94 amino acid peptide, which shares the overall molecular architecture of K^+ channel proteins from prokaryotes and eukaryotes (Plugge *et al.*, 2000; Gazzarrini *et al.*, 2003). Hydropathy analysis of the Kcv primary amino acid sequence suggests two TM domains (M1 and M2). These domains are separated by 44 amino acids, which contains the above-mentioned signature sequence ThsTvGFG (Plugge *et al.*, 2000) common to all K^+ channels (Heginbotham *et al.*, 1994). The amino acid sequence immediately flanking the Kcv signature sequence resembles the pore domains of prokaryotic and eukaryotic K^+ channel proteins. A comparative analysis reveals that it displays on average 61% similarity and 38% identity to the corresponding domain of many K^+ channel proteins (Plugge *et al.*, 2000).

A gene (*ORF 223*) which encodes a protein with all the elements to form a K^+ channel was recently identified in another member of the Phycodnaviridae family, EsV-1 (*Ectocarpus siliculosus* virus) (Delaroque *et al.*, 2001). The derived amino acid sequence of this putative channel from virus EsV-1 predicts a 124 amino acid protein that resembles Kcv (Delaroque *et al.*, 2001). The identity is ~41% over a stretch of 77 amino acids (Van Etten *et al.*, 2002). The EsV-1 encoded protein also has the K^+ channel signature sequence. In fact, it has a Tyr in the selectivity filter GYG that makes it even more similar to other K^+ channels than Kcv, which has a Phe in this position.

A unique feature of the PBCV-1 K^+ channel protein in comparison with other prokaryotic and eukaryotic K^+ channels is the small size of its cytoplasmic N- and C-termini. The derived amino acid structure of Kcv predicts that it lacks a cytoplasmic C-terminus and that the N-terminus (^{1}MLVFSKFLTRTE^{12}) only consists of 12 amino acids (Plugge *et al.*, 2000). With this minimal structure, Kcv is the smallest K^+ channel known. Essentially, the entire Kcv protein is the size of the "pore module" of all K^+ channels consisting of three functional elements M1–pore domain–M2. The 124 amino acid EsV-1 virus homolog also has very short cytoplasmic tails. Compared to Kcv it has four and seven extra amino acids at the N- and C-termini, respectively.

Solving the crystal structure of bacterial K^+ channel proteins, KcsA, KirBac1.1, and MthK has provided an enormous amount of information on the structure of ion channels and on how channels function (Doyle *et al.*, 1998; Jiang *et al.*, 2002, 2003; Kuo *et al.*, 2003). Because the Kcv amino acid sequence aligns well with the pore module of KcsA (Gazzarrini *et al.*, 2003) and KirBac1.1 (Figure 2.1), their high-resolution structures provide a guide for modeling Kcv structure. Because of a good alignment of the entire Kcv channel (including the N-terminus) with KirBac1.1, we have used the latter structure as a template for modeling the Kcv structure (Figure 2.1). The predicted structure contains all the structural elements of a K^+ channel pore present in KcsA and KirBac1.1. These elements include the outer helix TM1, the turret, the pore helix, the selectivity filter, and the inner helix TM2. Interestingly, the model also predicts an N-terminal helix in Kcv. This prediction agrees with the functional importance of the Kcv N-terminus (see below) and the suggested role for this "slide helix" domain in KirBac1.1 (Kuo *et al.*, 2003).

Not only is the overall Kcv structure similar to KirBac1.1, functionally important amino acids occur in the same positions. For example, there are two structurally important aromatic amino acids conserved among K^+ channels (e.g., aa $F^{101}F^{102}$ in KirBac1), which are also present in the Kcv pore helix (residues 55 and 56) (Figure 2.1A). Crystallographic studies on

```
A KirBac1.1  48 WRDLYYWALK VSWPVFFASL AALFVVNNTL FALLYQLGDA PIANQSPPGF  97
  Kcv         1 MLVFSKFLTR TEPFMIHLFI LAMFVMIYKF FPGGFENNFS VANPDKKASW  50

                    **
  KirBac1.1     VGAFFFSVET LATVGYGDMH PQTVYAHAIA TLEIFVGMSG IALSTGLVFA 147
  Kcv           IDCIYFGVTT HSTVGFGDIL PKTT-GAKLC TIAHIVTVFF IVLTL----- 100

  KirBac1.1     RFA
  Kcv           ---
```

B

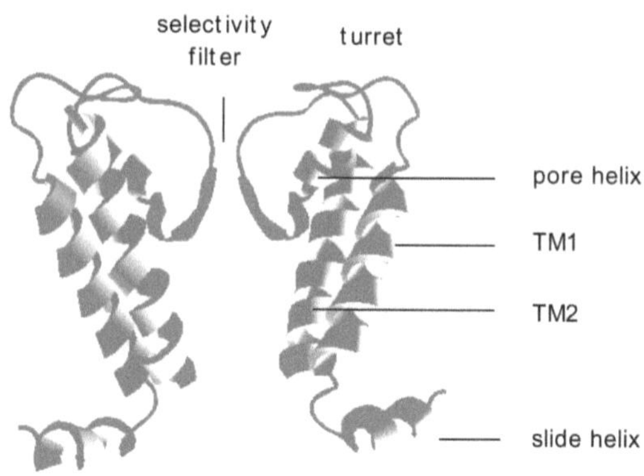

Figure 2.1. Putative structure of full-length Kcv. (A) Sequence alignment of the pore region of KirBac1.1 (aa 48–147) with full-length Kcv. Conserved substitutions are shown in gray. Pair of aromatic residues structurally important (see text) is indicated with asterisk. (B) Homology model of two of the four subunits of Kcv, generated by using the sequence alignment in A and the crystallographic coordinates of KirBac1.1 (Kuo *et al.*, 2003). In analogy to KirBac1.1 structure, the helical N-terminus is indicated as "slide helix." Model generated with automated homology modeling (Deep View Swiss-Pdb Viewer) within Swiss-Model.

KcsA indicate that this pair of aromatic residues is a part of a structure that acts as a cuff to keep the pore open at the appropriate diameter for K^+ passage (Doyle *et al.*, 1998).

However, differences also exist between the crystallized bacterial channels and the hypothetical Kcv structure. The main difference is that Kcv has a much shorter inner helix as compared to KcsA and KirBac1.1. This suggests that Kcv lacks the gating mechanism present in KcsA and KirBac1.1 that results from crossing the C-termini of the inner helixes.

4. The Short N-Terminus is Important for Kcv Function

Proteins encoded by virus PBCV-1 are often small and in many cases are the smallest of their kind when compared with their prokaryotic and eukaryotic homologs (Van Etten and Meins, 1999; Van Etten, 2003). Apparently, Kcv is another example of the ability of PBCV-1 to optimize a protein for structure and function. In this context, we tested the role of the small Kcv cytoplasmic N-terminus in channel functions. Deletion of 13 Kcv N-terminal amino acids (^{2}L–F^{13}) results in lose of conductance in *Xenopus* oocytes and mammalian HEK293 cells (Moroni *et al.*, 2002). The C-terminus of the wild-type and truncated proteins was also fused to the green fluorescent protein (GFP) in order to visualize its intracellular location in HEK293 cells. The truncated Kcv was synthesized to the same extent as the wild-type protein. Also, no obvious differences occurred in the intracellular distribution of the two Kcv : GFP chimeras in HEK293 cells. Thus, incorrect localization of Kcv does not explain the failure of the truncated protein to conduct a current (Moroni *et al.*, 2002). These results indicate that the N-terminus has an essential, but unknown, role in channel function.

A Prosite scan of the Kcv sequence identified two putative CK2 phosphorylation sites. One site (^{9}TRTE^{12}) is located in the cytoplasmic N-terminus. This same sequence is also a possible site for interaction with 14.3.3 proteins, a class of modulatory proteins that interact with K^+ channels in plant and animal cells (DeBoer, 2002; Kagan *et al.*, 2002). However, when the two threonins (^{9}T, ^{11}T) in Kcv were mutated to alanines, the channel properties in *Xenopus* oocytes were similar to the wild-type channel. This result rules out the N-terminus as an essential target for protein phosphorylation/dephosphorylation.

Other interesting features of the Kcv N-terminus include two positive-charged amino acids located at residues 6 and 10. To examine their role in channel function, the two amino acids were neutralized in a double mutant (Lys^{6}Ala Arg^{10}Ala). However, these amino acid substitutions had no obvious effect on Kcv activity when measured in *Xenopus* oocytes (Gazzarrini *et al.*, 2002). All together these results indicate that the N-terminus is required for CSV channel function but that there is some tolerance with respect to its amino acid composition.

Resolution of the crystal structure of bacterial K^+ inward rectifier KirBac1.1 identified an aforementioned structural element in the N-terminus preceding the first TM domain (Kuo *et al.*, 2003). This amphipathic helical element termed "slide helix" is positioned at the membrane/cytosol interface. The slide helix may be a more general structural element of K^+ channels because spin labeling electron paramagnetic resonance analysis identified a similar structure in KcsA (Cortes *et al.*, 2001; Li *et al.*, 2002).

The amphiphathic nature of the short Kcv N-terminus with its predicted helical structure (Figure 2.1) suggests that this domain might also function as a slide helix. This observation does not explain the functional role of this domain, but it is reasonable to speculate that it is involved in the proper anchoring and perhaps even moving of the channel protein in the plasma membrane.

5. Functional Properties of Kcv Conductance in Heterologous Expression Systems

Kcv is suitable for structure/function studies because it expresses in heterologous systems (Gazzarrini *et al.*, 2003). When Kcv cRNA is injected into *Xenopus* oocytes, the cells exhibit a specific K^+ conductance that is absent in water-injected cells (Plugge *et al.*, 2000; Gazzarrini *et al.*, 2002). Two control experiments established that this conductance was not due to upregulation of endogenous channels but resulted from expression of the viral encoded channel. First, a Kcv mutant was created in which the Phe^{66} in the selectivity filter was exchanged for ala. Other k^+ channels with this amino acid substitution are expressed in oocytes but do not conduct a current (Heginbotham *et al.*, 1994). The Kcv mutant also did not produce a current in *Xenopus* oocytes (Gazzarrini *et al.*, 2002) indicating that this critical amino acid is also required for the Kcv generated conductance. Second, Kcv forms a functional K^+ channel in two other heterologous systems. Using a chimera with GFP Kcv was successfully expressed in both mammalian HEK293 and CHO cells (Moroni *et al.*, 2002; Gazzarrini *et al.*, 2003) where it produced a conductance similar to that measured in Kcv expressing oocytes. The Kcv :: GFP chimera localized to membranes in both cell lines. The results of these two experiments indicate that the expression of the PBCV-1 *kcv* gene in heterologous cells results in a functional K^+ channel.

We have also tried to express the putative K^+ channel protein-encoding gene from virus EsV-1 in both *Xenopus* oocytes and HEK 293 cells. Although this protein resembles Kcv, it does not produce a K^+ conductance in either system. Instead, it produces nonspecific and variable currents in both systems (M. Mehmel, unpublished results). Currently it is not possible to distinguish whether these currents are due to an EsV-1 induced conductance or if they result from overexpression of a foreign protein in these cells with the consequent upregulation of endogenous channels (see Barhanin *et al.*, 1996; Coady *et al.*, 1998; Kelly *et al.*, 2003).

6. Kcv is a K^+ Selective Channel

The similarity in ion selectivity between different K^+ channels results from the highly conserved architecture of the selectivity filter (Heginbotham *et al.*, 1994; Doyle *et al.*, 1998; Miller, 2000). Usually K^+ channels conduct K^+ and Rb^+ equally well and Cs^+ with lesser efficiency. Permeability of Na^+ and Li^+ is generally low (Hille, 2001).

Kcv expressed in *Xenopus* oocytes or HEK293 cells also exhibits ion selectivity. The Kcv channel has about a 10-fold preference for K^+ over Na^+ in both systems. The complete selectivity sequence for monovalent cations is: $Rb^{+} \cdot K^{+} > Cs^{+} \gg Na^{+} > Li^{+}$ (Kang *et al.*, 2003).

7. Kcv has Some Voltage Dependency

Kcv currents produced in *Xenopus* oocytes consist of two kinetic components; during voltage-clamp steps, a small time-dependent component is superimposed on an instantaneous component (Gazzarrini *et al.*, 2002). Because channel blockers inhibit both of these components in a proportional manner, it is likely that they reflect distinct kinetic states of the same channel (Gazzarrini *et al.*, 2002). From these results, the activity of the channel can be described as follows: about 70% of the Kcv channels are always open at all voltages. At both extreme positive and negative voltages, the open Kcv channel conductance decreases in a voltage-dependent manner. The remaining channels (30%) open in a time- and voltage-dependent manner, reaching half-maximal activation at about -70 mV (Gazzarrini *et al.*, 2002).

The voltage dependency of the Kcv channel is not understood in the context of its simple channel structure. Mutation analyses eliminated the possibility that the two positively charged amino acids in the N-terminus are involved in voltage sensing (Gazzarrini *et al.*, 2002). Additional experiments indicate that the Kcv voltage dependency is not related to a voltage-dependent block by external divalent cations. Although Kcv conductance is sensitive to changes in extracellular Ca^{2+}(but not Mg^{2+}) this sensitivity is voltage independent (Gazzarrini *et al.*, 2002), and therefore, can not explain the voltage-dependent behavior of the channel.

8. Kcv has Distinct Sensitivity to K^+ Channel Blockers

Kcv currents are blocked by conventional K^+ channel inhibitors. For example, external Ba^{2+} and Sr^{2+} inhibit Kcv inward current in a strong voltage-dependent manner (Plugge *et al.*, 2000; Gazzarrini *et al.*, 2003). In contrast, the Kcv channel is only slightly inhibited by external Cs^+; 10 mM Cs^+ reduces the K^+ inward current by ~20% (Plugge *et al.*, 2000). Likewise, another K^+ channel blocker tetraethylammonium (TEA) (10 mM) only inhibits Kcv K^+ conductance 30–40% at all voltages (Gazzarrini *et al.*, 2003).

Scorpion and snake toxins block many prokaryotic and eukaryotic K^+ channels, including KcsA (MacKinnon *et al.*, 1998), by binding to conserved domains on the external side of the channel pore (i.e., turret). However, Kcv differs from KcsA in the turnet region that has most of the toxin-binding sites (Doyle *et al.*, 1998; MacKinnon *et al.*, 1998). We examined the sensitivity of Kcv to two scorpion toxins, Charybdotoxin and Agytoxin-2. Both toxins specifically block wild-type Shaker channels and Shaker channels in which the pore was replaced with the KcsA pore module (MacKinnon and Miller, 1989; Garcia *et al.*, 1994; MacKinnon *et al.*, 1998). However, neither toxin has any effect on Kcv conductance, consistent with Kcv's apparent lack of binding sites for these toxins.

The well-known antiviral drug amantadine was found to block the M2 channel from influenza virus A (Wang *et al.*, 1993) and the putative p7 channel from hepatitis C virus (Griffin *et al.*, 2003) at micromolar concentrations. This cyclic amine is also a good inhibitor of Kcv conductance (Figure 2.2B). But the inhibition of Kcv by amantadine can not be interpreted in analogy to the action of the cyclic amine on the other viral channels. While the affinity of the M2 channel to amantadine is in the nanomolar range (Wang *et al.*, 1993), Kcv has only a low affinity to this drug. In the case of Kcv, a K_i of 0.8 mM was obtained for the amantadine block in both oocytes and HEK293 cells (Gazzarrini *et al.*, 2003). Furthermore it appears that other K^+ channels are also inhibited by amantadine with a low affinity in the millimolar range (Blanpied *et al.*, 1997). These results suggest that amantadine is not a specific antiviral drug but also acts as a genuine K^+ channel blocker at high concentrations.

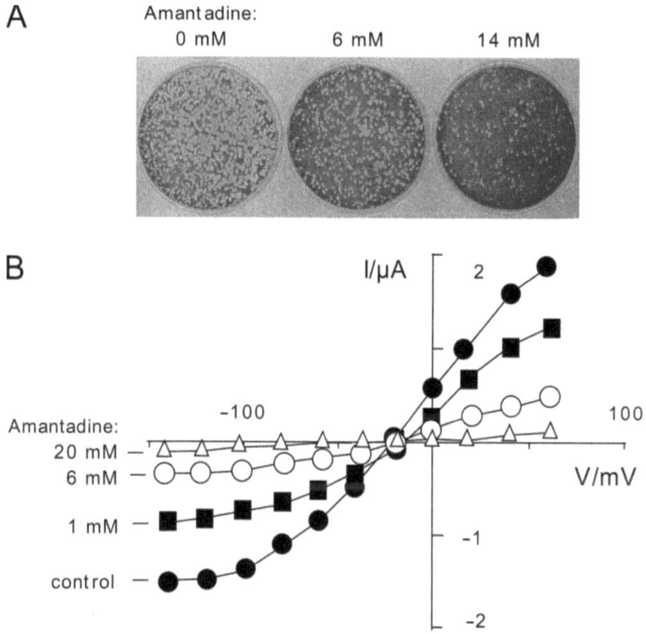

Figure 2.2. Sensitivity of PBCV-1 replication and Kcv conductance to amantadine. (A) Increasing concentrations of amantadine cause a decrease in the number of PBCV-1 generated plaques (here light spots) in lawns of *Chlorella NC64A* cells. (B) Steady-state current/voltage relation of Kcv expressed in *Xenopus* oocytes with 50 mM K^+ in the bath medium and increasing concentrations of amantadine. (For experimental details, see A: Van Etten *et al.*, 1983; B: Gazzarrini *et al.*, 2002).

9. Ion Channel Function in Viral Replication

With the exceptions of the M2 and BM2 proteins from influenza A and influenza B viruses, respectively, the physiological roles of viral channel proteins are unknown. It is well established that the M2 channel allows H^+ to enter the virion. This results in a pH-dependent fusion of endosomal and viral membranes (Martin and Helenius, 1991; Ciampor *et al.*, 1995). While membrane fusion is essential for influenza virus replication, the M2 protein is not absolutely required for this process. Influenza viruses also replicate, albeit at a lower rate, in the absence of a functional M2 channel (Watanabe *et al.*, 2001). The BM2 protein presumably serves a similar function in influenza B virus (Mould *et al.*, 2003). For other viral channels, or suspected viral channels, little is known about their physiological roles. Current hypotheses suggest a participation in infection, physiological modification of the host, or release of progeny viruses (Carrasco, 1995; Fischer and Sansom, 2002). However, in all these cases, the viral proteins accelerate replication but they are not obligatory (Gonzalez and Carrasco, 2003).

10. Kcv is Important for Viral Replication

The function of Kcv in the PBCV-1 life cycle is unknown. However, because the K^+ channel inhibitors amantadine and Ba^{2+}, but not Cs^+, inhibit K^+ conductance in oocytes, as well as PBCV-1 plaque formation, we assume that Kcv activity is important in one or more stages of virus replication. Furthermore, we predict that channel activity is involved in a very early phase of infection because the channel and viral infection/replication have the same sensitivity to Ba^{2+}. Since Ba^{2+} is not membrane permeable, presumably it can only inhibit viral replication in an early phase of infection, that is, when virus and host are exposed to the external medium.

One possible scenario assumes that Kcv is located in the internal membrane of the virion (Skrdla *et al.*, 1984; Yan *et al.*, 2000). For the early phase of infection, we hypothesize that after digestion of the cell wall, the viral membrane fuses with the host plasma membrane. This fusion of the Kcv containing viral membrane with the high resistance host membrane increases the K^+ conductance of the host membrane. The resulting membrane depolarization might prevent membrane fusion by a second virus, thus inhibiting infection by a second virus.

Several observations are consistent with this hypothesis: (a) Like other viruses which depolarize the host plasma membrane during infection (Boulanger and Letellier, 1988; Daugelavicius *et al.*, 1997; Piller *et al.*, 1998, 1999), PBCV-1 infection results in a rapid (within a few minutes) depolarization of the host plasma membrane as monitored by bisoxinol fluorescence (Mehmel *et al.*, 2003). Depolarization is host specific and its time course is correlated with the number of infecting viruses. Destruction of the viral membrane with organic solvents prior to adding the virus to the host prevents host membrane depolarization (Mehmel *et al.*, 2003). (b) The virus-induced membrane depolarization is correlated with an increase in host cell K^+ conductance (Figure 2.3). That is, increasing the extracellular concentration of K-glutamate prior to virus infection has no obvious effect on bisoxinol fluorescence. This result suggests that the resting conductance of the host cell to K^+ is low. However, when the same amount of K-glutamate is added after infection, an increase in fluorescence occurs, that is, membrane depolarization increases. This experiment indicates a postinfection increase in membrane conductance to K^+. (c) Membrane depolarization exhibits the same sensitivity to inhibitors as Kcv conductance in oocytes. When expressed in oocytes, the Kcv channel is inhibited by amantadine and Ba^{2+} but is relatively insensitive to Cs^+ and TEA (Plugge *et al.*, 2000; Gazzarrini *et al.*, 2003). This same pharmacological sensitivity occurs in PBCV-1 plaque assays; amantadine and Ba^{2+} strongly inhibit and TEA and Cs^+ weakly inhibit PBCV-1 plaque formation (Plugge *et al.*, 2000; Kang *et al.*, 2003; Van Etten, unpublished results). This similarity is not only qualitative but also quantitative; amantadine inhibits both processes with roughly the same half-maximal concentration (Figure 2.2 and Plugge *et al.*, 2000). (d) The correlation of Kcv activity and virus replication to inhibitors is supported by recent findings with a Kcv-like protein from another chlorella virus, MA-1D. The MA-1D Kcv-like protein and PBCV-1 Kcv differ by five amino acids (Kang *et al.*, 2003). However, unlike PBCV-1 Kcv, the MA-1D Kcv-like channel is sensitive to $Cs^+$when it is expressed in *Xenopus* oocyctes. Interestingly, MA-1D plaque formation is also inhibited by Cs^+ (Kang *et al.*, 2003). (e) Typically, proteins that are packaged in nascent virions are expressed late in infection (Van Etten *et al.*, 1984). The *kcv* gene is expressed as both an early and late gene and thus Kcv could be packaged in nascent virions (Mehmel and Kang, unpublished results). Unfortunately, attempts to produce antibodies to Kcv peptide fragments have been unsuccessful. Consequently, it is not known when the Kcv protein is synthesized and also its intracellular location, either in the cell or in the virion. However, it is plausible to

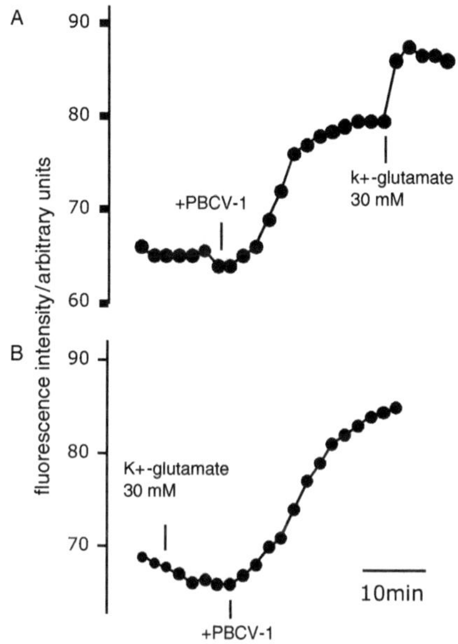

Figure 2.3. Effect of PBCV-1 and K^+ on the fluorescence of the voltage-sensitive dye bisoxinol in *Chlorella NC64A* cells before and after PBCV-1 infection. (A) Infection of *Chlorella NC64A* cells with PBCV-1 (m.o.i. of 20) results in an increase in bisoxinol fluorescence. The increase is indicative of membrane depolarization. Elevation of the external K^+ concentration from 1 to 30 mM causes an additional rise in bisoxinol fluorescence. (B) Elevation of the K^+ concentration prior to infection has no perceivable effect on bisoxinol fluorescence. The fluorescence increases only after mixing the chlorella cells with PBCV-1. (For experimental details, see Bronner and Landry [1991] and Mehmel *et al.* [2003].)

speculate that the channel is inserted in the internal membrane of the virion. Both PBCV-1 and EsV-1 have such an internal membrane located underneath their capsid shell (Skrdla *et al.*, 1984; Wolf *et al.*, 1998; Yan *et al.*, 2000). The origin of this membrane is unknown in PBCV-1. For EsV-1 there is good evidence that the membrane originates from the endoplasmic reticulum of the host cell (Wolf *et al.*, 1998; Van Etten *et al.*, 2002). (f) The chlorella host exhibits an exclusion phenomenon. That is, cells dually inoculated with different viruses typically only replicate one of the infecting viruses (Chase *et al.*, 1989).

Collectively, these results support the hypothesis that PBCV-1 induces depolarization of the host cell membrane in an early phase of infection. This depolarization arises from an increase in K^+ conductance of the host cell, possibly as a consequence of fusion of the virus membrane with the host plasma membrane. The physiological importance of such a host cell depolarization is still unknown, but presumably, the membrane depolarization alters the physiological state of the host cell so that virus specific functions are favored.

11. Evolutionary Aspects of the *kcv* Gene

The origin of the *kcv* gene is unknown. Phylogenetic comparison with certain K^+ channels (e.g., Kir, Kv, tandem K^+ eukaryotic and prokaryotic K^+ channels) suggests that

Kcv belongs to an independent cluster (Plugge *et al.*, 2000). Many viruses "acquire" genes from their host cells. The possibility that *kcv* has a close homolog in its host cell *Chlorella N64A* was eliminated by hybridizing a *kcv* gene probe with DNA isolated from the uninfected host, as well as several viruses that infect *Chlorella NC64A*. Hybridization occurred with DNA from all the viruses but not with DNA from the host cell (Kang *et al.*, 2003). Hence, while a *kcv*-like gene is present in all the viruses that infect *Chlorella NC64A*, the gene appears to be absent in the host cells.

As noted above, another Phycodnavirus, EsV-1, encodes an ORF with the hallmarks of a K^+ channel. The derived primary amino acid sequence of this protein resembles Kcv (see above). It is interesting to note that even though PBCV-1 and EsV-1 viruses have very large genomes and are obviously related, they only have 33 genes in common. Among the common genes are the ones encoding K^+ channel-like proteins (Van Etten *et al.*, 2002). This observation suggests that the K^+ channel may serve an essential function in the life cycle of viruses classified in the family Phycodnaviridae.

Acknowledgments

We thank Jack Dainty (Norwich) for help with the manuscript. This investigation was supported, in part, by Public Health Service Grant GM32441 (JVE) from the National Institute of General Medical Sciences and by the Deutsche Forschungsgemeinschaft.

References

Barhanin, J., Lesage, F., Guillemare, E., Fink, M., Lazdunski, M., and Romey, G. (1996). K(V)LQT1 and IsK (minK) proteins associate to form the I(Ks) cardiac potassium current. *Nature* **384**, 78–80.

Blanpied, T.A., Boeckman, F.A., Aizenman, E., and Johnson, J.W. (1997). Trapping channel block of NMDA-activated responses by amantadine and memantine. *J. Neurophysiol.* **77**, 309–323.

Blatt, M.R. and Thiel, G. (1993). Hormonal control of ion channel gating. *Annu. Rev. Plant Mol. Physiol.* **44**, 543–567.

Boulanger, P. and Letellier, L. (1988). Characterization of ion channels involved in the penetration of phage T4 DNA into *Escherichia coli cells*. *J. Biol. Chem.* **15**, 9767–9775.

Bronner, C. and Landry, Y. (1991). The use of the potential-sensitive fluorescent probe bisoxonol in mast cells. *Biochim. Biophys. Acta.* **1070**, 321–331.

Carrasco, L. (1995). Modification of membrane permeability by animal viruses. *Adv. Virus Res.* **45**, 61–112.

Chase, T.E., Nelson, J.A., Burbank, D.E., and Van Etten, J.L. (1989). Mutual exclusion occurs in a chlorella-like green alga inoculated with two viruses. *J. Gen. Virol.* **70**, 1829–1836.

Ciampor, F., Cmark, D., Cmarkova, J., and Zavodska, E. (1995). Influenza virus M2 protein and haemagglutinin conformation changes during intracellular transport. *Acta Virol.* **39**, 171–181.

Coady, M.J., Daniel, N.G., Tiganos, E., Allain, B., Friborg, J., Lapointe, J.Y. *et al.* (1998). Effects of Vpu expression on *Xenopus* oocyte membrane conductance. *Virology* **244**, 39–49.

Cortes, D.M., Cuello, L.G., and Perozo, E. (2001). Molecular architecture of a full-length Kcs: Role of cytoplasmic domains in ion permeation and activation gating. *J. Gen. Physiol.* **117**, 165–180.

Daugelavicius, R., Bamford, J.K.H., and Bamford, D.H. (1997). Changes in host cell energetics in response to bacteriophage PRD1 DNA entry. *J. Bacteriol.* **179**, 5203–5210.

DeBoer, A.H. (2002). Plant 14-3-3 proteins assist ion channels and pumps. *Biochem. Soc. Trans.* **30**, 416–421.

Delaroque, N., Muller, D.G., Bothe, G., Pohl, T., Knippers, R., and Boland, W. (2001). The complete DNA sequence of the *Ectocarpus siliculosus* virus Es V-1 genome. *Virology* **15**, 112–132.

Derst, C. and Karschin, A. (1998). Evolutionary link between prokaryotic and eukaryotic K^+ channels. *J. Exp. Biol.* **201**, 2791–2799.

Doyle, D.A., Cabral, J.M., Pfuetzner, R.A., Kuo, A., Gulbis, J.M., Cohen, S.L., *et al.* (1998). The structure of the potassium channel: Molecular basis of K^+ conduction and selectivity. *Science* **280**, 69–76.

Fischer, W.B. and Sansom, M.S. (2002). Viral ion channels: Structure and function. *Biochim. Biophys. Acta* **1561**, 27–45.

Garcia, M.L., Garcia-Calvo, M., Hidalgo, P., Lee, A., and MacKinnon, R. (1994). Purification and characterization of three inhibitors of voltage-dependent K^+ channels from *Leiurus quinquestriatus* var. *hebraeus* venom. *Biochemistry* **33**, 6834–6839.

Gazzarrini, S., Van Etten, J.L., DiFrancesco, D., Thiel, G., and Moroni, A. (2002). Voltage-dependence of virus-encoded miniature K^+ channel Kcv. *J. Memb. Biol.* **187**, 15–25.

Gazzarrini, S., Severino, M., Lombardi, M., Morandi, M., DiFrancesco, D., Van Etten, J.L. *et al.* (2003). The viral potassium channel Kcv: Structural and functional features. *FEBS Lett.* **552**, 12–16.

Gonzalez, M.E. and Carrasco, L. (2003). Viroporins. *FEBS Lett.* **552**, 28–34.

Griffin, S.D., L.P. Beales, D.S. Clark, O. Worsfold, S.D. Evans, J. Jaeger, *et al.* (2003). *FEBS Lett.* **535**, 34–38.

Heginbotham, L., Lu, Z., Abramson, T., and MacKinnon, R. (1994). Mutations in the K^+ channel signature sequence. *Biophys. J.* **66**, 1061–1067.

Hille, B. (2001). *Ion Channels of Excitable Membranes*, 3nd edn. Sinauer Associates Inc., Sunderland, MA.

Jiang, Y., Lee, A., Chen, J., Cadene, M., Chait, B.T., and MacKinnon, R. (2002). Crystal structure and mechanism of a calcium-gated potassium channel. *Nature* **417**, 515–522.

Jiang, Y., Lee, A., Chen, J., Ruta, V., Cadene, M., Chait, B.T. *et al.* (2003). X-ray structure of a voltage-dependent K^+ channel. *Nature* **423**, 33–41.

Kagan, A., Melman, Y.F., Krumerman, A., and McDonald, T.V. (2002). 14-3-3 amplifies and prolongs adrenergic stimulation of HERG K^+ channel activity. *EMBO J.* **21**, 1889–1898.

Kang, M., Moroni, A., Gazzarrini, S., DiFrancesco, D., Thiel, G., Severino, M., Van Etten, J.L. (2004). Small potassium ion channel proteins encoded by chlorella viruses. *PNAS* **101**, 5318–5324.

Kelly, M.L., Cook, J.A., Brown-Augsburger, P., Heinz, B.A., Smith, M.C., and Pinto, L.H. (2003). Demonstrating the intrinsic ion channel activity of virally encoded proteins. FEBS *Lett.* (in press).

Kuo, A., Gulbis, J.M., Antcliff, J.F., Rahman, T., Lowe, E., Zimmer, J. *et al.* (2003). Crystal structure of the potassium channel KirBac1.1 in the closed state. *Science* **300**, 1922–1926.

Li, J., Xu, Q., Cortes, D.M., Perozo, E., Laskey, A., and Karlin, A. (2002). Reactions of cysteins substituted in the amphipathic N-terminus tail of a bacterial potassium channel with hydrophilic and hydrophobic maleimides. *Proc. Natl. Sci. USA* **99**, 11605–11610.

MacKinnon, R., Cohen, S.L., Kuo, A., Lee, A., and Chait, B.T. (1998). Structural conservation in prokaryotic and eukaryotic potassium channels. *Science* **280**, 106–109.

MacKinnon, R. and Miller, C. (1989). Mutant potassium channels with altered binding of charybdotoxin, a pore-blocking peptide inhibitor. *Science* **245**, 1382–1385.

Martin, K. and Helenius, A. (1991). Nuclear transport of influenza virus ribonucleoproteins: The viral matrix protein (M1) promotes export and inhibits import. *Cell* **67**, 117–130.

Mehmel, M., Rothermel, M., Meckel, T., Van J.L., Etten, Moroni, A., and Thiel, G. (2003). Possible function for virus encoded K^+ channel Kcv in the replication of chlorella virus PBCV-1. *FEBS Lett.* **552**, 7–11.

Miller, C. (2000). Ion channels: Doing hard chemistry with hard ions. *Curr. Opin. Chem. Biol.* **4**, 148–151.

Moroni, A., Viscomi, C., Sangiorgio, V., Pagliuca, C., Meckel, T., Horvath, F. *et al.* (2002). The short N-terminus is required for functional expression of the virus-encoded miniature K^+ channel Kcv. *FEBS Lett.* **530**, 65–69.

Mould, J.A., Paterson, R.G., Takeda, M., Ohigashi, Y., Venkataraman, P., Lamb, R.A. *et al.* (2003). Influenza B virus BM2 protein has ion channel activity that conducts protons across membranes. *Dev. Cell* **5**, 175–184.

Pinto, L.H., Holsinger, L.J., and Lamb, R.A. (1992). Influenza virus M2protein has ion channel activity. *Cell* **69**, 517–528.

Piller, S.C., Jans, P., Gage, P.W., and Jans, D.A. (1998). Extracellular HIV-1 virus protein R causes a large inward current and cell death in cultured hippocampal neurons: Implications for AIDS pathology. *Proc. Natl. Acad. Sci. USA* **95**, 4595–4600.

Piller, S.C., Ewart, G.D., Jans, D.A., Gage, P.W., and Cox, G.B. (1999). The amino-terminal region of Vpr from human immunodeficient virus Type 1 forms ion channels and kills neurons. *J. Virol.* **73**, 4230–4238.

Plugge, B., Gazzarrini, S., Nelson, M., Cerana, R., Van Etten, J.L., Derst, C. *et al.* (2000). A potassium channel protein encoded by chlorella virus PBCV-1. *Science* **287**, 1641–1644.

Schrempf, H., Schimdt, O., Kümmerlen, R., Hinnah, S., Müller, D., Betzler, M. *et al.* (1995). *EMBO J.* **114**, 5170–5178.

Skrdla, M.P., D.E. Burbank, Y. Xia, R.H. Meints, and J.L. Van Etten, (1984). Structural proteins and lipids in a virus, PBCV-1, which replicates in a chlorella-like alga. *Virology* **135**, 308–315.

Van Etten, J.L. (2000). Phycodnaviridae. In M.H.V. Van Regenmortel, C.M. Fauquet, D.H.L. Bishop, E.B. Carstens, M.K. Estes, S.M. Lemon *et al.* (eds), *Virus Taxonomy, Classification and Nomenclature of Viruses*, 7th rep., Academic Press, San Diego: CA, pp. 183–193.

Van Etten, J.L. (2003). Unusual life style of giant chlorella viruses. *Annu. Rev. Genet.* **37**, 163–195.

Van Etten, J.L. and Meints, R.H. (1999). Giant viruses infecting algae. *Annu. Rev. Microbiol.* **53**, 447–494.

Van Etten J.L., Burbank, D.E., Joshi, J., and Meints, R.H. (1984). DNA synthesis in a chlorella-like alga following infection with the virus PBCV-1. *Virology* **134**, 443–449.

Van Etten, J.L., Burbank, D.E., Kuczmarski, D., and Meints, R.H. (1983). Virus infection of culturable Chlorella-like algae and development of a plaque assay. *Science* **219**, 994–997.

Van Etten, J.L., Graves, M.V., Müller, D.G., Boland, W., and Delaroque, N. (2002). Phycodnaviridae—large DNA viruses. *Arch. Virol.* **147**, 1479–1516.

Very, A.A. and Sentenac, H. (2002). Cation channels in the *Arabidopsis* plasma membrane. *Trends Plant Sci.* 7, 168–175.

Wang, C., Takeuchi, K., Pinto, L.H., and Lamb, R.A. (1993). Ion channel activity of influenza A virus M2 protein: Characterization of the amantadine block. *J. Virol.* **67**, 5585–5594.

Watanabe, T., Watanabe, S., Ito, H., Kida, H., and Kawaoka, Y. (2001). Influenza A virus can undergo multiple cycles of replication without M2 ion channel activity. *J. Virol.* **75**, 5656–5662.

Wei, A., Jegla, T., and Salkoff, L. (1996). Eight potassium channel families revealed by the *C. elegans* genome project. *Neuropharmacology* **37**, 805–829.

Wolf, S., Maier, I., Katsaros, C., and Müller, D.G. (1998). Virus assembly in *Hincksia hincksiae* (*Ectocarpales, Phaeophyceae*). An electron and fluorescence microscopic study. *Protoplasma* **203**, 153–167.

Yan, X., Olson, N.H., Van Etten, J.L., Bergoin, M., Rossmann, M.G., and Baker, T.S. (2000). Structure and assembly of large lipid-containing dsDNA viruses. *Nat. Struct. Biol.* **7**, 101–103.

Part II

Fusion Proteins

3

HIV gp41: A Viral Membrane Fusion Machine

Sergio G. Peisajovich and Yechiel Shai

Entry of HIV into its target cell requires the fusion of the virion envelope and the cell plasma membrane. This process is catalyzed by the viral envelope glycoproteins gp120, which mediates the binding to specific receptors on the target cell surface, and gp41, responsible for the actual membrane merging. This chapter will focus on the conformational changes that gp41 undergoes during the fusion process, on the mechanism by which these conformational changes are linked to membrane fusion, and on the design of inhibitors able to block different stages of this process.

1. Introduction

In order to deposit their genetic material in the interior of their target cells, primate lentiviruses, such as HIV and SIV, must fuse their lipidic envelopes with the target cell plasma membrane. The energetic barriers associated with this process are overcome by the action of the specific viral envelope glycoproteins gp120 and gp41. They are initially synthesized as an inactive precursor, gp160, that is cleaved by cellular proteases during its transport to the plasma membrane. Once cleaved, gp120 remains attached by noncovalent forces to the transmembrane (TM) subunit gp41 and, during budding, are incorporated to the envelope of the virions.

The activity of the gp41/gp120 complex is precisely regulated. This prevents its fusogenic properties from destabilizing the intracellular membranes of the infected cell as well as the viral envelope. Only the interaction of gp120 with specific receptors on the target cell (CD4 and a co-receptor, most commonly the chemokine receptors CXCR4 in T-cell tropic and CCR5 in M-tropic viruses) triggers a series of conformational changes in the gp120/gp41 complex that ultimately leads to the merging of the virion and cell membranes.

At the level of the amino acid sequence, gp41 shares many features with other viral envelope glycoproteins involved in membrane fusion (see Figure 3.1): (a) its N-terminus

Sergio G. Peisajovich and Yechiel Shai • Department of Biological Chemistry, Weizmann Institute of Science, Rehovot 76100, Israel.

Viral Membrane Proteins: Structure, Function, and Drug Design, edited by Wolfgang Fischer.
Kluwer Academic / Plenum Publishers, New York, 2005.

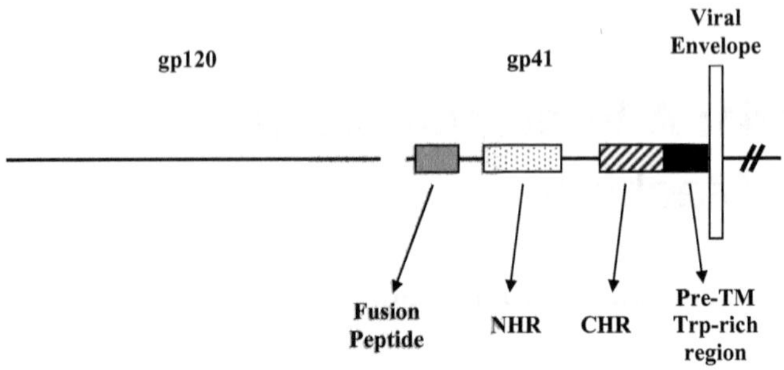

Figure 3.1. Schematic representation of gp120/gp41. gp41 is attached to the viral membrane, whereas gp120 and gp41 are held together by non-covalent interactions. gp41 ectodomain includes an N-terminal fusion peptide (in grey), an N-terminal heptad repeat region (in dots), a C-terminal heptad repeat region (in diagonal), and a pre-transmembrane Trp-rich region (in black).

contains a stretch of about 15 hydrophobic residues (Gallaher, 1987), named "the fusion peptide," which is believed to insert into and destabilize the target cell membrane during the fusion process, (b) following the fusion peptide and preceding the TM domain, there are two heptad repeat regions (Gallaher *et al.*, 1989), called NHR and CHR, respectively, and (c) both NMR and crystallographic studies have shown that three NHR fold into a trimeric coiled coil surrounded by three CHR helices, each NHR–CHR pair being connected by an extended loop (Chan *et al.*, 1997; Caffrey *et al.*, 1998; Skehel and Wiley, 1998). This six-helix bundle likely represents the post-fusion conformation (Melikyan *et al.*, 2000). This chapter will focus on the conformational changes that gp41 undergoes during the fusion process that lead to the post-fusion conformation, on the mechanism by which these conformational changes are linked to membrane fusion, and on the design of inhibitors able to block different stages of this process.

2. HIV Envelope Native Conformation

Despite initial discrepancies (Rey *et al.*, 1990; Rhodes *et al.*, 1994), it is now accepted that, in the surface of mature virions, gp41/gp120 forms a trimeric complex (Center *et al.*, 2001). The regions within the complex responsible for trimerization are likely to be located in the ectodomain, as soluble "gp140" constructs comprising gp120 and the extraviral region of gp41 (lacking gp41 TM and intraviral segments) were shown to be trimeric by chemical cross-linking, gel filtration chromatography, and analytical ultracentrifugation (Chen *et al.*, 2000; Zhang *et al.*, 2001). The number of trimeric gp41/gp120 complexes per virion has also been controversial, likely changing depending on the particular strain considered. Indeed, some HIV-1 isolates have been shown to contain an average of between 7 and 14 trimers, whereas some SIV isolates contained levels of gp41/gp120 trimers at least 10-fold greater (Chertova *et al.*, 2002). Furthermore, the large variations observed within strains as well as the low percentage of actually infective viral particles still cast doubts on the minimum number of trimers per virion necessary to sustain infection.

Drawing from extensive structural and functional studies done on the envelope protein from Influenza virus (Wilson *et al.*, 1981; Carr and Kim, 1993; Bullough *et al.*, 1994), it is currently believed that the trimeric gp41/gp120 "native or pre-fusion" conformation is metastable. Although the three-dimensional structure of fragments from gp120 has been determined (Ghiara *et al.*, 1994; Kwong *et al.*, 1998; Wyatt *et al.*, 1998), the pre-fusion conformation of gp41/gp120 remains unknown. Recently, low resolution STEM images showed that soluble gp140 complexes exhibit a triangular or "tri-lobed" morphology (Center *et al.*, 2001) not surprisingly considering their trimeric nature.

3. Receptor-Induced Conformational Changes

3.1. CD4 Interaction

For most HIV-1 strains, interaction of gp120 with the CD4 receptor present in the target cell is the first step in the viral entry reaction (see Figure 3.2). The interaction of gp120 with CD4 induces conformational changes in gp120 and gp41, both of which are reflected by changes in epitopes exposure. In gp41, there is an immunogenic region that corresponds to the loop connecting the NHR and CHR helices in the post-fusion conformation (residues 598–604), another one that includes part of the CHR helix (residues 644–663), and a third one

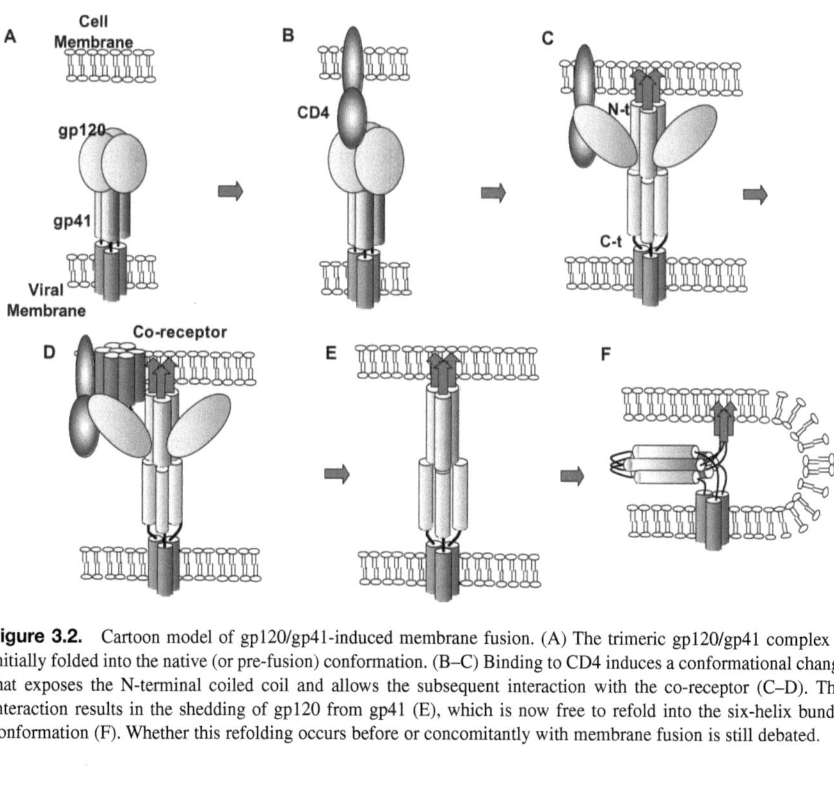

Figure 3.2. Cartoon model of gp120/gp41-induced membrane fusion. (A) The trimeric gp120/gp41 complex is initially folded into the native (or pre-fusion) conformation. (B–C) Binding to CD4 induces a conformational change that exposes the N-terminal coiled coil and allows the subsequent interaction with the co-receptor (C–D). This interaction results in the shedding of gp120 from gp41 (E), which is now free to refold into the six-helix bundle conformation (F). Whether this refolding occurs before or concomitantly with membrane fusion is still debated.

that corresponds to a region proximal to the TM domain (residues 656–671, not present in the solved structures). Interestingly, the first two epitopes become exposed only after gp120–CD4 interaction (Sattentau *et al.*, 1995), whereas the third epitope becomes less immunoreactive following CD4 interaction (Sattentau *et al.*, 1995).

Further structural information regarding the conformational changes induced by CD4 interaction was provided by the observation that synthetic peptides that partially overlap the CHR segment and a consecutive Trp-rich region potently block gp41-mediated membrane fusion. These peptides can be grouped in two distinctive classes: (a) C34 and its different analogs, which contain in their N-termini the sequence WMEW, believed to fit into a deep cavity located at the opposite end of the N-terminal coiled coil (Chan *et al.*, 1998), and (b) T20 (formerly DP178) and its analogs, which are shifted toward the C-terminus and, therefore, do not include these cavity-filling residues (Wild *et al.*, 1994). The similarity between the structural organizations of gp41 and Influenza virus hemagglutinin (Carr and Kim, 1993; Chan *et al.*, 1997), and the inhibitory activity of the C-terminal peptides led to postulate a "pre-hairpin intermediate" in gp41-induced membrane fusion, in which the N-terminal coiled coil is formed, but the C-terminal helices are not packed (Chan *et al.*, 1997; Weissenhorn *et al.*, 1997). At this stage, the C-peptide inhibitors can bind to the exposed coiled coil (see Figure 3.2B–C), thus preventing the subsequent refolding and blocking fusion (Furuta *et al.*, 1998).

Mirroring the inhibitory ability of CHR-derived peptides, it was early discovered that NHR-derived synthetic peptides were also capable of blocking gp41-mediated membrane fusion (Wild *et al.*, 1992). Indeed, the higher concentrations of NHR-derived peptides required to block fusion led to postulate that, whereas monomeric CHR-derived peptides interact with the exposed trimeric coiled coil, trimers of NHR-derived peptides were needed to interact with the monomeric CHR regions, presumably exposed in the pre-hairpin intermediate. This hypothesis was further supported by the observation that homodimerization of the NHR-derived peptides (by addition of an N-terminal cysteine) decreased the concentration needed for antiviral activity (Wild *et al.*, 1992).

3.2. Co-Receptor Interaction

Interaction of gp120 with CD4 results also in the exposure of several new epitopes in gp120 (Sattentau and Moore, 1991; Wyatt *et al.* 1995; Sullivan *et al.*, 1998), some of which are required for co-receptor engagement. Most commonly, T-cell tropic viruses utilize the chemokine receptor CXCR4, whereas M-tropic viruses utilize the chemokine receptor CCR5. Although there are differences that depend on the co-receptor and the particular virus involved, it has been reported that a lag time (that might be as long as tens of minutes in cell cultures for CXCR4-utilizing viruses) exists between CD4 binding and co-receptor engagement (Gallo *et al.*, 2001).

Co-receptor interactions result in further conformational changes, the most dramatic of which are gp120–gp41 dissociation (see Figure 3.2D–E, although this has not been confirmed for all HIV/SIV types) and formation of gp41 six-helix bundle. In a series of elegant experiments, Melikyan and coworkers (2000) have shown that gp41-induced cell–cell fusion is temperature sensitive and can be reversibly arrested by incubation at 25°C (Temperature Arrested State, TAS). Interestingly, addition of soluble CD4 during TAS does not inhibit fusion upon re-incubation at 37°C, whereas addition of T22 (a peptide that binds to CXCR4) does block fusion, indicating that the process has been arrested after CD4 binding but before co-receptor engagement. Furthermore, addition of CHR-derived peptides during the TAS inhibits fusion, confirming that the pre-hairpin intermediate had formed upon CD4 interaction. Release from the TAS leads to fast six-helix bundle formation (see Figure 3.2E–F) (Melikyan *et al.*, 2000) (as evidenced by the inability of CHR-derived peptides to block fusion at this stage, which indeed had been shown to lose the ability to interact with the coiled coil once the six-helix bundle has formed (Kliger and Shai, 2000). Whether six-helix bundle formation occurs concomitantly with membrane merger (and indeed this refolding is its driving force) or it actually precedes membrane fusion is the focus of an ongoing debate. On one hand, it has been shown that addition of lyso-phophatidyl choline (LPC), which arrests fusion reversibly by preventing deformation of the membrane into an intermediate state structure with negative curvature (as detailed in Section 4), prevents also the formation of the six-helix bundle (Melikyan *et al.*, 2000), suggesting that bundle formation and membrane fusion are each contingent on the other. On the other hand, it has been reported that anti-six-helix bundle antibodies bind to gp120/gp41 expressing cells attached to CD4/co-receptor bearing cells arrested at 32°C, prior to membrane fusion (Golding *et al.*, 2002), suggesting that membrane merger occurs only after six-helix bundle formation.

4. The Actual Membrane Fusion Step

Six-helix bundle formation might exert the force required for bringing the apposing membranes together. However, membrane merging needs to overcome the strong repulsive forces that exist between lipid bilayers at very short distances, thought to arise from the hydration of the polar headgroups, steric interactions, and elastic pressures (Blumenthal *et al.*, 2003). Viral fusion proteins contain specific regions that are believed to destabilize the apposing bilayers, helping to overcome these repulsion forces. In the case of gp41, two different complex regions, one located at the N-terminus and the other at the C-terminus of the ectodomain, have been postulated to play roles in membrane destabilization.

4.1. The Role of the N-Terminal Fusion Domain

At the N-terminus of gp41, there is a stretch of about 15 hydrophobic residues (initially identified by its similarity to the N-termini of other viral fusion proteins (Gallaher, 1987)) named the "fusion peptide," which is believed to insert into the target cell membrane upon interaction of gp120 with CD4 and the co-receptor. The role of fusion peptides in membrane fusion has been extensively analyzed using a protein dissection strategy that involves the study of short synthetic peptides and their interaction with model membranes, such as large unilamellar liposomes (Durell *et al.*, 1997). This strategy has revealed that fusion peptides insertion into lipidic membranes can result in their destabilization and fusion (Pereira *et al.*, 1997a, b). Specifically, it has been postulated that most fusion peptides have a hydrophobicity gradient along the axis of an ideal helical peptide that results in the insertion of the peptide into the membrane in an oblique orientation (Brasseur *et al.*, 1990). This insertion preferentially expands the acyl chain region, thus promoting the negative curvature of the membrane (Epand, 1998). Both theoretical and experimental studies have indicated that the membrane fusion proceeds through a series of intermediate steps (Markin *et al.*, 1984; Siegel, 1999; Kozlovsky and Kozlov, 2002; Yang and Huang, 2002). The first one is a highly negatively curved structure named "stalk," in which only the outer contacting monolayers have fused. Indeed, addition to the outer monolayer of compounds that disfavor negative curvature blocks fusion (Chernomordik *et al.*, 1995). In the second stage, rupture of the inner monolayers leads to a fusion pore that then expands, allowing the transfer of the viral contents into the cytoplasm of the infected cell. Fusion peptides are believed to stabilize these intermediates, thus facilitating fusion. In the case of HIV-1, a 16-residue peptide corresponding to the gp41 N-terminal fusion peptide was shown to induce lipid mixing of PC/PE/Cho (1:1:1) large unilamellar vesicles (Peisajovich *et al.*, 2000). Furthermore, a 33-residue peptide that included the fusion peptide as well as the following 17, mostly polar, residues had a significantly enhanced fusogenic activity, suggesting that the consecutive polar region might assist the destabilizing activity of the fusion peptide. As many other fusion peptides, the N-terminal 16-residue peptide decreased the bilayers to hexagonal phase transition temperature of dipalitoleoylphosphatidylethanolamine (T_H), suggesting its ability to obliquely insert into the membrane and promote negative curvature. The 17 consecutive residues presumably lie near the surface of the membrane, contributing to the correct oligomerization of the peptide, further enhancing its fusogenic activity (Peisajovich *et al.*, 2000). Supporting the role of gp41 N-terminal region in oligomerization, it has been reported that the wild-type 33-residue peptide mentioned above inhibits gp41-induced cell–cell fusion (Kliger *et al.*, 1997). Furthermore, a V2E mutant was shown to inhibit both gp41-mediated cell–cell fusion and the liposome fusion induced by the wild-type 33-residue peptide, again suggesting that oligomerization of gp41 N-terminus is needed for proper functioning (Pereira *et al.*, 1995; Kliger *et al.*, 1997). Interestingly, it has been shown that a synthetic D-enantiomer of the HIV-1 gp41-derived wild-type 33-residue peptide can potently inhibit HIV-1 gp41-mediated cell–cell fusion (but not HIV-2) (Pritsker *et al.*, 1998). The ability of the D-peptide to interact with its natural L-enantiomer in the membrane suggests that chirality plays a complex, yet unexplained role in protein–protein interactions within the hydrophobic membrane milieu (Pritsker *et al.*, 1998).

The NHR region, which follows the fusion peptide and the consecutive polar segment, was also shown to be involved in the process of membrane fusion. Indeed, elongation of the mentioned 33-residue peptide (that comprised the fusion peptide and the consecutive polar region) to include the full NHR region resulted in a 70-residue segment with dramatically increased fusogenic activity (Sackett and Shai, 2002). Interestingly, this 70-residue construct mimics the N-terminal half of gp41 during the pre-hairpin intermediate stage, in which the fusion peptide is inserted into the membrane and the consecutive coiled coil is exposed. Furthermore, a I62D mutation in the C-terminal region of the heptad repeat, far away from the N-terminal fusion peptide and known to render the virus noninfectious (Dubay *et al.*, 1992), resulted in a significant reduction of the fusogenicity of the 70-residue peptide. This mutation is believed to block the ability of the heptad repeat to form a stable trimeric coiled coil, suggesting that the correct structure and/or oligomeric state of the full-length peptide is necessary for its proper activity.

4.2. The Role of the C-Terminal Fusion Domain

HIV gp41, as many other viral fusion proteins, is attached to the viral membrane via a single TM domain. The TM is preceded by a membrane-proximal Trp-rich region. The importance of gp41 TM and pre-TM regions on the mechanism of viral entry has been initially postulated by Salzwedel *et al.* (1993) who showed that replacement of the TM segment of gp41 by a glycosyl-phosphatidylinositol anchor (GPI) abolished gp41-induced syncytium formation. Furthermore, mutational studies showed that the Trp-rich pre-TM region of gp41 ectodomain is crucial for fusion (Munoz-Barroso *et al.*, 1999; Salzwedel *et al.*, 1999). These studies highlighted the importance of the C-terminal regions of fusion proteins ectodomains in the process of membrane merging. More recently, a more direct role for the C-terminus of gp41 ectodomain in membrane destabilization has emerged. Nieva and coworkers (Suarez *et al.*, 2000) observed that a Trp-rich segment located at the C-terminus of HIV-1 gp41 ectodomain displayed a high tendency to partition into the membrane interface as revealed by the interfacial hydrophobicity scale developed by Wimley and White (1996). Furthermore, using a protein dissection approach, they showed that a 20-residue segment that immediately precedes the gp41 TM domain, and included the Trp-rich region, destabilized PC/PE/Cho (1 : 1 : 1) liposomes (i.e., caused permeabilization and lipid mixing). Moreover, mutations within this region, known to render gp41 nonfunctional resulted in inactive peptides. Based on these observations, they postulated that the pre-TM region plays a role equivalent to that of the polar segment that follows the N-terminal fusion peptide (Suarez *et al.*, 2000). In other words, the structural organization of the N- and C-termini of gp41 ectodomain has two elements in common: A hydrophobic segment that inserts into the membrane (the fusion peptide at the N-terminus and the TM domain at the C-terminus), followed by (in the case of the N-terminus) or preceded by (in the case of the C-terminus) more polar helical regions that participate in the fusion process. Their inherent location at opposite ends of the envelope proteins ectodomains suggests that N-terminal fusion peptides exert their effect on the target cell membrane, whereas pre-TM regions are more likely to act on the viral envelope. Indeed, it is tempting to speculate that, given the differences between the lipid composition of cell and viral membranes, these two regions might have been optimized by evolution, thus better performing their task on the particular lipid composition of their respective membrane environments.

5. HIV Entry Inhibitors

The entry of HIV into susceptible cells is a multistep process that offers many targets for the development of putative inhibitors. Indeed, some inhibitors played crucial roles in our understanding of the HIV entry mechanism. They can be grouped in three main categories, those that prevent CD4 binding, those that prevent co-receptor binding, and those that block gp41 conformational changes.

5.1. CD4-Binding Inhibitors

Soon after the isolation of HIV, it was discovered that soluble forms of the extracellular region of CD4 (sCD4) were able to block the infection of T-lymphocytic and myelomonocytic cell lines by diverse strains of HIV-1, HIV-2, and SIV (Hussey *et al.*, 1988; Clapham *et al.*, 1989). Moreover, it is known that sCD4 mechanism of inhibition involves binding to gp120, an interaction that, for several HIV-1 strains, results in an initial conformational change in the gp120/gp41 complex that exposes the trimeric N-terminal coiled coil of gp41 (allowing the binding of CHR-derived peptides) (Jones *et al.*, 1998). The gp120–CD4 interaction can also be blocked by competition with anti-CD4 antibodies (such as Leu3A), as well as by anti-idiotypic antibodies that mimic the CD4 epitope and therefore bind to gp120 (Dalgleish *et al.*, 1987).

5.2. Co-Receptor-Binding Inhibitors

The common HIV co-receptors are members of the seven TM domain family of integral membrane proteins and, as such, they usually function as receptors of endogenous molecules (Berger *et al.*, 1999). Therefore, it is not surprising that binding of gp120 to the co-receptors can be blocked by the natural co-receptor agonists (such as SDF-1 for CXCR4 (Bleul *et al.*, 1996; Oberlin *et al.*, 1996; Amara *et al.*, 1997) or RANTES and MIP-1α/β for CCR5 (Cocchi *et al.*, 1995; Alkhatib *et al.*, 1996; Deng *et al.*, 1996; Dragic *et al.*, 1996). Furthermore, both peptidic and nonpeptidic small molecule CXCR4 and CCR5 antagonists able to block HIV cell entry have been reported. T22 ([Tyr5,12,Lys7]-polyphemusin II) is an 18-residue synthetic peptide analog of polyphemusin II (a peptide isolated from the american horseshoe crabs, *Limulus polyphemus*) that has been reported to inhibit the cell–cell fusion ability of CXCR4-dependent HIV-1 by specific binding to CXCR4 (Murakami *et al.*, 1997). Furthermore, the bicyclam AMD3100 and the nonapeptide (D-Arg)9 or ALX40-4C are also small-molecule inhibitors that act by selectively antagonizing CXCR4 binding (Doranz *et al.*, 1997; Schols *et al.*, 1997; Donzella *et al.*, 1998). Similarly, TAK-779, a nonpeptide compound with a small molecular weight (Mr 531.13) was shown to bind to CCR5 and inhibit HIV-1 entry (Baba *et al.*, 1999). Indeed, CCR5 antagonists are promising candidates for the development of effective anti-HIV drugs, as blocking of CCR5, contrary to CXCR4, seems to have no serious consequences for the normal functioning of the immune system (Samson *et al.*, 1996).

5.3. Improving the Activity of Inhibitors that Block Conformational Changes

The inhibitory potency of the C-peptide inhibitors (some of which are promising anti-HIV candidates in clinical trials [Kilby *et al.*, 1998]) has been initially correlated with their helical content (Lawless *et al.*, 1996). Indeed, Jin *et al.* (2000) observed that the addition of short helix-capping sequences to both the N- and C-termini of a short peptide, corresponding to the 19 N-terminal residues of C34 or replacing some of its residues by amino acids with stronger helical propensity, significantly enhanced their ability to inhibit HIV-1 gp120/gp41-induced cell–cell fusion. Similarly, it has been reported that replacement of external amino acids in C34 analogs by charged residues that would form matching ion pairs in a helical conformation, resulted in C34 analogs with higher helical content and stronger inhibitory abilities (Otaka *et al.*, 2002). Similarly, stabilization of the helical structure of a 14-residue peptide that partially overlaps the N-terminus of C34, by chemical cross-linking and substitutions with unnatural helix-favoring amino acids has been shown to augment the peptide's inhibitory potency (Sia *et al.*, 2002). Furthermore, Judice *et al.* (1997) found a correlation between the helicity of short constrained T20 analogs designed to adopt α-helical structures and their inhibitory potencies. These studies strongly indicate that helical structure is crucial for the inhibitory activity of the CHR-derived peptides and that the NHR regions (their target of inhibition) are indeed forming an exposed trimeric coiled coil in the pre-hairpin intermediate.

Rational design also has been used to improve the activity of NHR-derived inhibitory peptides. Specifically, Clore and coworkers (Bewley *et al.*, 2002) designed a 36-residue peptide, N36(Mut(e,g)) based on the parent NHR-derived N36 peptide. N36(Mut(e,g)) contained nine substitutions planned to disrupt interactions with the CHR region of gp41, while preserving the interactions required for the formation of the trimeric N-terminal coiled-coil. They observed that N36(Mut(e,g)) folds into a monodisperse, helical trimer (contrary to the tendency of the parent wild-type N36 to form higher order oligomers in aqueous solution) and was about 50-fold more active that the N36. Based on these results, the authors postulated that N36(Mut(e,g)) acts by disrupting the homotrimeric coiled coil of N-terminal helices in the pre-hairpin intermediate to form heterotrimers and, therefore, represents a new class of anti-HIV peptide inhibitors.

The design of inhibitors targeted against the CHR regions, which are presumably exposed in the pre-hairpin intermediate state (Koshiba and Chan, 2002) has been based on constructs that present a trimeric N-terminal coiled coil with exposed surfaces for the interaction of the CHR segments. Clore and coworkers have designed a chimeric protein, termed N(CCG)-gp41, derived from the ectodomain of HIV-1 gp41 (Louis *et al.*, 2001). N(CCG)-gp41 comprises an exposed trimeric coiled coil formed by three NHR helices. The structure is stabilized in two ways: (a) by fusion to a minimal thermostable ectodomain of gp41 and (b) by engineered intersubunit disulfide bonds. N(CCG)-gp41 was shown to inhibit gp41-mediated cell–cell fusion at nanomolar concentrations (Louis *et al.*, 2001).

A different strategy to sequester an exposed CHR region has been used by Kim and his coworkers (Root *et al.*, 2001). They designed a small protein taking advantage of the binding properties of the N-terminal coiled coil while minimizing the tendency of the NHR-derived peptides to aggregate when exposed. The protein, termed 5-Helix, is composed of five of the six helices that make up gp41's trimeric helical hairpin, which are now connected with short peptide linkers. The 5-Helix protein lacks the third CHR helix, and this vacancy created a high-affinity binding site for a CHR segment. The 5-Helix protein was shown to inhibit several HIV-1 variants at nanomolar concentrations (Root *et al.*, 2001).

6. Final Remarks

Over the last years, our understanding of the mechanism by which enveloped viruses (particularly HIV) enter their host cells has greatly improved. We were able to dissect this process in many of its fundamental steps as well as to design potent inhibitors that will eventually lead to new classes of antiviral compounds. However, there are still serious gaps in our knowledge; in particular, we do not know yet the structure of the native or pre-fusion conformation of the gp120/gp41 complex as well as the detailed structural changes triggered by CD4 and co-receptor interactions. Furthermore, the way in which the interactions between the N-terminal fusion peptide, the pre-TM region, and the cellular and viral membranes lead to membrane destabilization and merging still needs to be fully addressed. Moreover, the roles of gp41 intraviral domain in the fusion process and in the structural organization of the mature virion are also poorly understood. New and exciting advances along these lines are likely to come in the next years.

Acknowledgment

Sergio G. Peisajovich is supported by fellowships from The Mifal Ha'paiys Foundation and The Feinberg Graduate School.

References

Alkhatib, G., Combadiere, C., Broder, C.C., Feng, Y., Kennedy, P.E., Murphy, P.M. *et al.* (1996). CC CKR5: A RANTES, MIP-1alpha, MIP-1beta receptor as a fusion cofactor for macrophage-tropic HIV-1. *Science* **272**, 1955–1958.

Amara, A., Gall, S.L., Schwartz, O., Salamero, J., Montes, M., Loetscher, P. *et al.* (1997). HIV coreceptor downregulation as antiviral principle: SDF-1alpha-dependent internalization of the chemokine receptor CXCR4 contributes to inhibition of HIV replication. *J. Exp. Med.* **186**, 139–146.

Baba, M., Nishimura, O., Kanzaki, N., Okamoto, M., Sawada, H., Iizawa, Y. *et al.* (1999). A small-molecule, nonpeptide CCR5 antagonist with highly potent and selective anti-HIV-1 activity. *Proc. Natl. Acad. Sci. USA* **96**, 5698–5703.

Berger, E.A., Murphy, P.M., and Farber, J.M. (1999). Chemokine receptors as HIV-1 coreceptors: Roles in viral entry, tropism, and disease. *Annu. Rev. Immunol.* **17**, 657–700.

Bewley, C.A., Louis, J.M., Ghirlando, R., and Clore, G.M., (2002). Design of a novel peptide inhibitor of HIV fusion that disrupts the internal trimeric coiled-coil of gp41. *J. Biol. Chem.* **277**, 14238–14245.

Bleul, C.C., Farzan, M., Choe, H., Parolin, C., Clark-Lewis, I., Sodroski, J. *et al.* (1996). The lymphocyte chemoattractant SDF-1 is a ligand for LESTR/fusin and blocks HIV-1 entry. *Nature* **382**, 829–833.

Blumenthal, R., Clague, M.J., Durell, S.R., and Epand, R.M. (2003). Membrane fusion. *Chem. Rev.* **103**, 53–70.

Brasseur, R., Vandenbranden, M., Cornet, B., Burny, A., and Ruysschaert, J.M. (1990). Orientation into the lipid bilayer of an asymmetric amphipathic helical peptide located at the N-terminus of viral fusion proteins. *Biochim. Biophys. Acta* **1029**, 267–273.

Bullough, P.A., Hughson, F.M., Skehel, J.J., and Wiley, D.C. (1994). Structure of influenza haemagglutinin at the pH of membrane fusion. *Nature* **371**, 37–43.

Caffrey, M., Cai, M., Kaufman, J., Stahl, S.J., Wingfield, P.T., Covell, D.G. *et al.* (1998). Three-dimensional solution structure of the 44 kDa ectodomain of SIV gp41. *EMBO J.* **17**, 4572–4584.

Carr, C.M. and Kim, P.S. (1993). A spring-loaded mechanism for the conformational change of influenza hemagglutinin. *Cell* **73**, 823–832.

Center, R.J., Schuck, P., Leapman, R.D., Arthur, L.O., Earl, P.L., Moss, B., and Lebowitz, J. (2001). Oligomeric structure of virion-associated and soluble forms of the simian immunodeficiency virus envelope protein in the prefusion activated conformation. *Proc. Natl. Acad. Sci. USA* **98**, 14877–14882.

Chan, D.C., Chutkowski, C.T., and Kim, P.S. (1998). Evidence that a prominent cavity in the coiled coil of HIV type 1 gp41 is an attractive drug target. *Proc. Natl. Acad. Sci. USA* **95**, 15613–15617.

Chan, D.C., Fass, D., Berger, J.M., and Kim, P.S. (1997). Core structure of gp41 from the HIV envelope glycoprotein. *Cell* **89**, 263–273.

Chen, B., Zhou, G., Kim, M., Chishti, Y., Hussey, R.E., Ely, B. *et al.* (2000). Expression, purification, and characterization of gp160e, the soluble, trimeric ectodomain of the simian immunodeficiency virus envelope glycoprotein, gp160. *J. Biol. Chem.* **275**, 34946–34953.

Chernomordik, L., Kozlov, M.M., and Zimmerberg, J. (1995). Lipids in biological membrane fusion. *J. Membr. Biol.* **146**, 1–14.

Chertova, E.J., Bess, W., Jr., Crise, B.J., Sowder, I.R., Schaden, T.M., Hilburn, J.M. *et al.* (2002). Envelope glycoprotein incorporation, not shedding of surface envelope glycoprotein (gp120/SU), is the primary determinant of SU content of purified human immunodeficiency virus type 1 and simian immunodeficiency virus. *J. Virol.* **76**, 5315–5325.

Clapham, P.R., Weber, J.N., Whitby, D., McIntosh, K., Dalgleish, A.G., Maddon, P.J. *et al.* (1989). Soluble CD4 blocks the infectivity of diverse strains of HIV and SIV for T cells and monocytes but not for brain and muscle cells. *Nature* **337**, 368–370.

Cocchi, F., DeVico, A.L., Garzino-Demo, A., Arya, S.K., Gallo, R.C., and Lusso, P. (1995). Identification of RANTES, MIP-1 alpha, and MIP-1 beta as the major HIV-suppressive factors produced by CD8+ T cells. *Science* **270**, 1811–1815.

Dalgleish, A.G., Thomson, B.J., Chanh, T.C., Malkovsky, M., and Kennedy, R.C. (1987). Neutralisation of HIV isolates by anti-idiotypic antibodies which mimic the T4 (CD4) epitope: A potential AIDS vaccine. *Lancet* **2**, 1047–1050.

Deng, H., Liu, R., Ellmeier, W., Choe, S., Unutmaz, D., Burkhart, M. *et al.* (1996). Identification of a major co-receptor for primary isolates of HIV-1. *Nature* **381**, 661–666.

Donzella, G.A., Schols, D., Lin, S.W., Este, J.A., Nagashima, K.A., Maddon, P.J. *et al.* (1998). AMD3100, a small molecule inhibitor of HIV-1 entry via the CXCR4 co-receptor. *Nat. Med.* **4**, 72–77.

Doranz, B.J., Grovit-Ferbas, K., Sharron, M.P., Mao, S.H., Goetz, M.B., Daar, E.S. *et al.* (1997). A small-molecule inhibitor directed against the chemokine receptor CXCR4 prevents its use as an HIV-1 coreceptor. *J. Exp. Med.* **186**, 1395–1400.

Dragic, T., Litwin, V., Allaway, G.P., Martin, S.R., Huang, Y., Nagashima, K.A. *et al.* (1996). HIV-1 entry into CD4+ cells is mediated by the chemokine receptor CC-CKR-5. *Nature* **381**, 667–673.

Dubay, J.W., Roberts, S.J., Brody, B., and Hunter, E. (1992). Mutations in the leucine zipper of the human immunodeficiency virus type 1 transmembrane glycoprotein affect fusion and infectivity. *J. Virol.* **66**, 4748–4756.

Durell, S.R., Martin, I., Ruysschaert, J.M., Shai, Y., and Blumenthal, R. (1997). What studies of fusion peptides tell us about viral envelope glycoprotein-mediated membrane fusion (review). *Mol. Membr. Biol.* **14**, 97–112.

Epand, R.M. (1998). Lipid polymorphism and protein–lipid interactions. *Biochim. Biophys. Acta* **1376**, 353–368.

Furuta, R.A., Wild, C.T., Weng, Y., and Weiss, C.D. (1998). Capture of an early fusion-active conformation of HIV-1 gp41. *Nat. Struct. Biol.* **5**, 276–279.

Gallaher, W.R. (1987). Detection of a fusion peptide sequence in the transmembrane protein of human immunodeficiency virus. *Cell* **50**, 327–328.

Gallaher, W.R., Ball, J.M., Garry, R.F., Griffin, M.C., and Montelaro, R.C. (1989). A general model for the transmembrane proteins of HIV and other retroviruses. *AIDS Res. Hum. Retroviruses* **5**, 431–440.

Gallo, S.A., Puri, A., and Blumenthal, R. (2001). HIV-1 gp41 six-helix bundle formation occurs rapidly after the engagement of gp120 by CXCR4 in the HIV-1 Env-mediated fusion process. *Biochemistry* **40**, 12231–12236.

Ghiara, J.B., Stura, E.A., Stanfield, R.L., Profy, A.T., and Wilson, I.A. (1994). Crystal structure of the principal neutralization site of HIV-1. *Science* **264**, 82–85.

Golding, H., Zaitseva, M., de Rosny, E., King, L.R., Manischewitz, J., Sidorov, I. *et al.* (2002). Dissection of human immunodeficiency virus type 1 entry with neutralizing antibodies to gp41 fusion intermediates. *J. Virol.* **76**, 6780–6790.

Hussey, R.E., Richardson, N.E., Kowalski, M., Brown, N.R., Chang, H.C., Siliciano, R.F. *et al.* (1988). A soluble CD4 protein selectively inhibits HIV replication and syncytium formation. *Nature* **331**, 78–81.

Jin, B.S., Ryu, J.R., Ahn, K., and Yu, Y.G. (2000). Design of a peptide inhibitor that blocks the cell fusion mediated by glycoprotein 41 of human immunodeficiency virus type 1. *AIDS Res. Hum. Retroviruses* **16**, 1797–1804.

Jones, P.L., Korte, T., and Blumenthal, R. (1998). Conformational changes in cell surface HIV-1 envelope glycoproteins are triggered by cooperation between cell surface CD4 and co-receptors. *J. Biol. Chem.* **273**, 404–409.

Judice, J.K., Tom, J.Y., Huang, W., Wrin, T., Vennari, J., Petropoulos, C.J. *et al.* (1997). Inhibition of HIV type 1 infectivity by constrained alpha-helical peptides: Implications for the viral fusion mechanism. *Proc. Natl. Acad. Sci. USA* **94**, 13426–13430.

Kilby, J.M., Hopkins, S., Venetta, T.M., DiMassimo, B., Cloud, G.A., Lee, J.Y. *et al.* (1998). Potent suppression of HIV-1 replication in humans by T-20, a peptide inhibitor of gp41-mediated virus entry. *Nat. Med.* **4**, 1302–1307.

Kliger, Y. and Shai, Y. (2000). Inhibition of HIV-1 entry before gp41 folds into its fusion-active conformation. *J. Mol. Biol.* **295**, 163–168.

Kliger, Y., Aharoni, A., Rapaport, D., Jones, P., Blumenthal, R., and Shai, Y. (1997). Fusion peptides derived from the HIV type 1 glycoprotein 41 associate within phospholipid membranes and inhibit cell–cell fusion. Structure-function study. *J. Biol. Chem.* **272**, 13496–13505.

Koshiba, T. and Chan D.C. (2002). The prefusogenic intermediate of HIV-1 gp41 contains exposed C-peptide regions. *J. Biol. Chem.* **13**, 13.

Kozlovsky, Y. and Kozlov, M.M. (2002). Stalk model of membrane fusion: Solution of energy crisis. *Biophys. J.* **82**, 882–895.

Kwong, P.D., Wyatt, R., Robinson J., Sweet, R.W., Sodroski, J., and Hendrickson, W.A. (1998). Structure of an HIV gp120 envelope glycoprotein in complex with the CD4 receptor and a neutralizing human antibody. *Nature* **393**, 648–659.

Lawless, M.K., Barney, S., Guthrie, K.I., Bucy, T.B., Petteway, S.R., and Merutka, G. (1996). HIV-1 membrane fusion mechanism: Structural studies of the interactions between biologically-active peptides from gp41. *Biochemistry* **35**, 13697–13708.

Louis, J.M., Bewley, C.A., and Clore G.M. (2001). Design and properties of N(CCG)-gp41, a chimeric gp41 molecule with nanomolar HIV fusion inhibitory activity. *J. Biol. Chem.* **276**, 29485–29489.

Markin, V.S., Kozlov, M.M., and Borovjagin, V.L. (1984). On the theory of membrane fusion. The stalk mechanism. *Gen. Physiol. Biophys.* **5**, 361–377.

Melikyan, G.B., Markosyan, R.M., Hemmati, H., Delmedico, M.K., Lambert, D.M., and Cohen, F.S. (2000). Evidence that the transition of HIV-1 gp41 into a six-helix bundle, not the bundle configuration, induces membrane fusion. *J. Cell Biol.* **151**, 413–423.

Munoz-Barroso, I., Salzwedel, K., Hunter, E., and Blumenthal, R. (1999). Role of the membrane-proximal domain in the initial stages of human immunodeficiency virus type 1 envelope glycoprotein-mediated membrane fusion. *J. Virol.* **73**, 6089–6092.

Murakami, T., Nakajima, T., Koyanagi, Y., Tachibana, K., Fujii, N., Tamamura, H. *et al.* (1997). A small molecule CXCR4 inhibitor that blocks T cell line-tropic HIV-1 infection. *J. Exp. Med.* **186**, 1389–1393.

Oberlin, E., Amara, A., Bachelerie, F., Bessia, C., Virelizier, J.L., Arenzana-Seisdedos, F. *et al.* (1996). The CXC chemokine SDF-1 is the ligand for LESTR/fusin and prevents infection by T-cell-line-adapted HIV-1. *Nature* **382**, 833–835.

Otaka, A., Nakamura, M., Nameki, D., Kodama E., Uchiyama S., Nakamura S. *et al.* (2002). Remodeling of gp41-C34 peptide leads to highly effective inhibitors of the fusion of HIV-1 with target cells. *Angew. Chem. Int. Ed. Engl.* **41**, 2937–2940.

Peisajovich, S.G., Epand, R.F., Pritsker, M., Shai, Y., and Epand, R.M. (2000). The polar region consecutive to HIV-1 fusion peptide participates in membrane fusion. *Biochemistry* **39**, 1826–1833.

Pereira, F.B., Goni, F.M., and Nieva, J.L. (1995). Liposome destabilization induced by the HIV-1 fusion peptide effect of a single amino acid substitution. *FEBS Lett.* **362**, 243–246.

Pereira, F.B., Goni, F.M., Muga, A., and Nieva, J.L. (1997a). Permeabilization and fusion of uncharged lipid vesicles induced by the HIV-1 fusion peptide adopting an extended conformation: Dose and sequence effects. *Biophys. J.* **73**, 1977–1986.

Pereira, F.B., Goni, F.M., and Nieva, J.L. (1997b). Membrane fusion induced by the HIV type 1 fusion peptide: Modulation by factors affecting glycoprotein 41 activity and potential anti-HIV compounds. *AIDS Res. Hum. Retroviruses* **13**, 1203–1211.

Pritsker, M., Jones, P., Blumenthal, R., and Shai, Y. (1998). A synthetic all D-amino acid peptide corresponding to the N-terminal sequence of HIV-1 gp41 recognizes the wild-type fusion peptide in the membrane and inhibits HIV-1 envelope glycoprotein-mediated cell fusion. *Proc. Natl. Acad. Sci. USA* **95**, 7287–7292.

Rey, M.A., Laurent, A.G., McClure, J., Krust, B., Montagnier, L., and Hovanessian, A.G. (1990). Transmembrane envelope glycoproteins of human immunodeficiency virus type 2 and simian immunodeficiency virus SIV-mac exist as homodimers. *J. Virol.* **64**, 922–926.

Rhodes, A.D., Spitali, M., Hutchinson, G., Rud, E.W., and Stephens, P.E. (1994). Expression, characterization and purification of simian immunodeficiency virus soluble, oligomerized gp160 from mammalian cells. *J. Gen. Virol.* **75**, 207–213.

Root, M.J., Kay, M.S., and Kim, P.S. (2001). Protein design of an HIV-1 entry inhibitor. *Science* **75**, 5.

Sackett, K. and Shai, Y. (2002). The HIV-1 gp41 N-terminal heptad repeat plays an essential role in membrane fusion. *Biochemistry* **41**, 4678–4685.

Salzwedel, K., Johnston, P.B., Roberts, S.J., Dubay, J.W., and Hunter, E. (1993). Expression and characterization of glycophospholipid-anchored human immunodeficiency virus type 1 envelope glycoproteins. *J. Virol.* **67**, 5279–5288.

Salzwedel, K., West, J.T., and Hunter, E. (1999). A conserved tryptophan-rich motif in the membrane-proximal region of the human immunodeficiency virus type 1 gp41 ectodomain is important for Env-mediated fusion and virus infectivity. *J. Virol.* **73**, 2469–2480.

Samson, M., Libert, F., Doranz B.J., Rucker, J., Liesnard, C., Farber, C.M. *et al.* (1996). Resistance to HIV-1 infection in caucasian individuals bearing mutant alleles of the CCR-5 chemokine receptor gene. *Nature* **382**, 722–725.

Sattentau, Q.J. and Moore, J.P. (1991). Conformational changes induced in the human immunodeficiency virus envelope glycoprotein by soluble CD4 binding. *J. Exp. Med.* **174**, 407–415.

Sattentau, Q.J., Zolla-Pazner, S., and Poignard, P. (1995). Epitope exposure on functional, oligomeric HIV-1 gp41 molecules. *Virology* **206**, 713–717.

Schols, D., Struyf, S., Van Damme, J., Este, J.A., Henson, G., and De Clercq, E. (1997). Inhibition of T-tropic HIV strains by selective antagonization of the chemokine receptor CXCR4. *J. Exp. Med.* **186**, 1383–1388.

Sia, S.K., Carr, P.A., Cochran, A.G., Malashkevich, V.N., and Kim, P.S. (2002). Short constrained peptides that inhibit HIV-1 entry. *Proc. Natl. Acad. Sci. USA* **99**, 14664–14669.

Siegel, D.P. (1999). The modified stalk mechanism of lamellar/inverted phase transitions and its implications for membrane fusion. *Biophys. J.* **76**, 291–313.

Skehel, J.J. and Wiley, D.C. (1998). Coiled coils in both intracellular vesicle and viral membrane fusion. *Cell* **95**, 871–874.

Suarez, T., Gallaher, W.R., Agirre, A., Goni, F.M., and Nieva, J.L. (2000). Membrane interface-interacting sequences within the ectodomain of the human immunodeficiency virus type 1 envelope glycoprotein: Putative role during viral fusion. *J. Virol.* **74**, 8038–8047.

Sullivan, N., Sun, Y., Sattentau, Q., Thali, M., Wu, D., Denisova, G. *et al.* (1998). CD4-induced conformational changes in the human immunodeficiency virus type 1 gp120 glycoprotein: Consequences for virus entry and neutralization. *J. Virol.* **72**, 4694–4703.

Weissenhorn, W., Dessen, A., Harrison, S.C., Skehel, J.J., and Wiley, D.C. (1997). Atomic structure of the ectodomain from HIV-1 gp41. *Nature* **387**, 426–430.

Wild, C., Oas, T., McDanal, C., Bolognesi, D., and Matthews, T. (1992). A synthetic peptide inhibitor of human immunodeficiency virus replication: Correlation between solution structure and viral inhibition. *Proc. Natl. Acad. Sci. USA* **89**, 10537–10541.

Wild, C.T., Shugars, D.C., Greenwell, T.K., McDanal, C.B., and Matthews, T.J. (1994). Peptides corresponding to a predictive alpha-helical domain of human immunodeficiency virus type 1 gp41 are potent inhibitors of virus infection. *Proc. Natl. Acad. Sci. USA* **91**, 9770–9774.

Wilson, I.A., Skehel, J.J., and Wiley, D.C. (1981). Structure of the haemagglutinin membrane glycoprotein of influenza virus at 3 A resolution. *Nature* **289**, 366–373.

Wimley, W.C. and White, S.H. (1996). Experimentally determined hydrophobicity scale for proteins at membrane interfaces. *Nat. Struct. Biol.* **3**, 842–848.

Wyatt, R., Kwong, P.D., Desjardins, E., Sweet, R.W., Robinson, J., Hendrickson, W.A. *et al.* (1998). The antigenic structure of the HIV gp120 envelope glycoprotein. *Nature* **393**, 705–711.

Wyatt, R., Moore, J., Accola, M., Desjardin, E., Robinson, J., and Sodroski, J. (1995). Involvement of the V1/V2 variable loop structure in the exposure of human immunodeficiency virus type 1 gp120 epitopes induced by receptor binding. *J. Virol.* **69**, 5723–5733.

Yang, L. and Huang, H.W. (2002). Observation of a membrane fusion intermediate structure. *Science* **297**, 1877–1879.

Zhang, C.W., Chishti, Y., Hussey, R.E., and Reinherz, E.L. (2001). Expression, purification, and characterization of recombinant HIV gp140. The gp41 ectodomain of HIV or simian immunodeficiency virus is sufficient to maintain the retroviral envelope glycoprotein as a trimer. *J. Biol. Chem.* **276**, 39577–39585.

4

Diversity of Coronavirus Spikes: Relationship to Pathogen Entry and Dissemination

Edward B. Thorp and Thomas M. Gallagher

Coronaviruses are widespread in the environment, infecting humans, domesticated and wild mammals, and birds. Infections cause a variety of diseases including bronchitis, gastroenteritis, hepatitis, and encephalitis, with symptoms ranging from being nearly undetectable to rapidly fatal. A combination of interacting variables determine the pattern and severity of coronavirus-induced disease, including the infecting virus strain, its transmission strategy, and the age and immune status of the infected host. Coronavirus pathogenesis is best understood by discerning how each of these variables dictates clinical outcomes. This chapter focuses on variabilities amongst the spike (S) proteins of infecting virus strains. Diversity of coronavirus surface proteins likely contributes to epidemic disease, an important and timely topic given the recent emergence of the human SARS coronavirus.

1. Introduction

Coronaviruses circulating in nature exhibit considerable genetic variability, and ongoing virus evolution can generate novel variants capable of epidemic diseases such as the recent SARS-CoV. A central goal in coronavirus research is to pinpoint virus strain variations and relate the differences to epidemiologic and pathogenic potentials. Identifying variations correlating with properties such as viral transmission from animals to humans leads toward mechanistic understanding of epidemics and also points to the relevant targets for antiviral therapeutics. For the coronaviruses, it is clear that pronounced variations are accommodated in the spike (S) genes, which encode the "corona" of virion protrusions (Figure 4.1). Each protrusion is a complex, oligomeric assembly of extremely large. ~1,300 amino acid S protein monomers that are integrated into virion membranes by their C-terminal transmembrane anchors. The spikes are essential for virus binding to cell-surface receptors and for virus–cell membrane fusion. During infection, the spikes also accumulate on cell surfaces,

---

Edward B. Thorp and Thomas M. Gallagher • Department of Microbiology and Immunology, Loyola University Medical Center, Maywood, IL 60153, Illinois.

Viral Membrane Proteins: Structure, Function, and Drug Design, edited by Wolfgang Fischer.
Kluwer Academic / Plenum Publishers, New York, 2005.

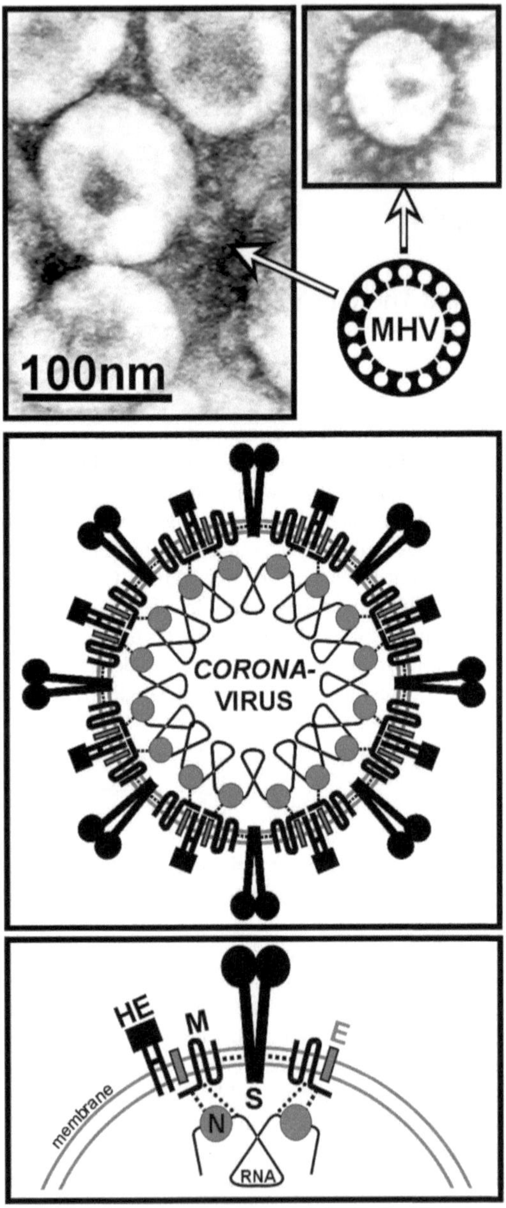

Figure 4.1. Electron micrographs and model of coronavirus structure. Top: Negatively stained (uranyl acetate) micrographs of murine coronavirus *m*ouse *h*epatitis *v*irus (strain A59). Bottom boxes: Schematic of coronavirus particles. *S*: spike glycoprotein, *HE*: hemagglutinin-esterase glycoprotein, *M*: triple-membrane-spanning membrane glycoprotein, *E*: small envelope glycoprotein, *N*: nucleocapsid phosphoprotein. Dotted lines indicate noncovalent protein–protein or protein–RNA interactions.

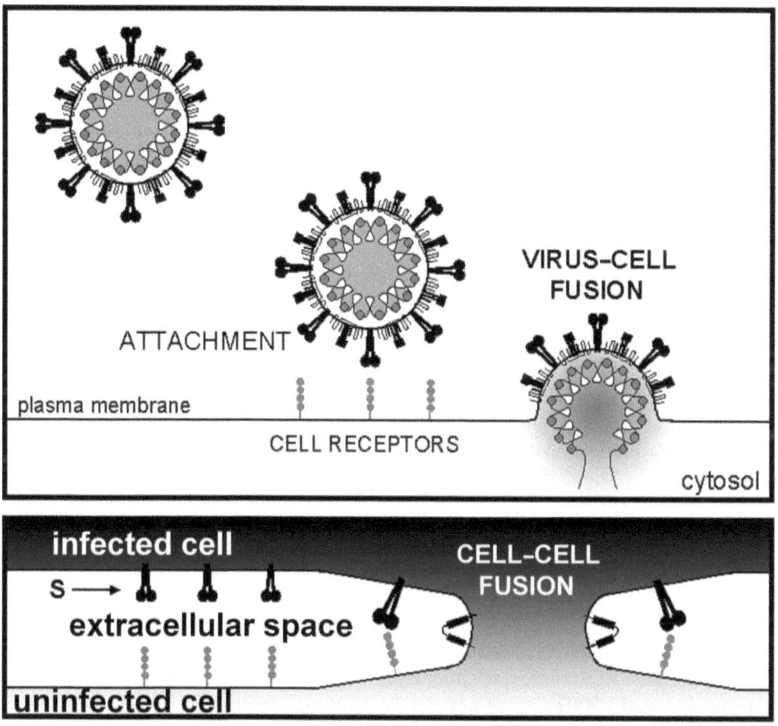

Figure 4.2. Virus entry and dissemination. Top: Depiction of virion attachment to cellular receptors and subsequent fusion of viral envelope with the host cell plasma membrane. Bottom: Depiction of spikes (S) on the infected cell surface recognizing opposing cellular receptors and promoting intercellular fusion and cell–cell spread.

bind to receptors on neighboring uninfected cells, and mediate cell–cell membrane fusion, a process that creates syncytia and causes rapid expansion of infections (Figure 4.2). Thus, the unique characteristics of S proteins from different coronavirus isolates correlate with distinct patterns of virus entry, virus dissemination via syncytia, virus tropism, and pathogenesis.

2. S Functions During Coronavirus Entry

To appreciate the unique characteristics of these S proteins, one must visualize their activities in the context of the infection cycle. We begin with virion binding to susceptible host cells. As mediators of virus attachment to cells, S proteins are set apart by their ability to evolve remarkably varied attachment specificities. Sialic acid, a ubiquitous component of cell-surface carbohydrate complexes, is a documented low-affinity ligand for porcine and bovine coronavirus spikes (Schultze *et al.*, 1991). Aminopeptidase N (APN), a type II-oriented membrane glycoprotein found in abundance on respiratory epithelia, is a receptor for antigenic "group 1" coronaviruses (Delmas *et al.*, 1992; Tresnan *et al.*, 1996). Members of this antigenic cluster include human respiratory viruses such as human CoV 229E, as well as several devastating animal pathogens such as transmissible gastroenteritis virus of swine and infectious peritonitis virus of cats. *C*arcino*E*mbryonic *A*ntigen-related *C*ell *A*dhesion *M*olecules (CEACAMs), immunoglobulin-like type I-oriented membrane glycoproteins that are prevalent in the liver and gastrointestinal tract, serve as receptors for the prototype member of the antigenic "group 2" coronavirus mouse hepatitis virus (Dveksler *et al.*, 1991; Godfraind *et al.*, 1995). Receptors for group 3 coronaviruses, which include several bird viruses causing severe bronchitis in chickens and turkeys, are currently unknown.

High-resolution structures are predicted for APN (Sjostrom *et al.*, 2000; Firla *et al.*, 2002), and are actually known for CEACAM (Tan *et al.*, 2002). Structural homologies between these two proteins are not readily apparent. Thus, the adaptation of coronaviruses to either receptor likely involves substantial remodeling of binding sites on S proteins. In this regard, it is important to remember that Apn or Ceacam receptor usage correlates with the antigenic and genetic relationships used to divide coronaviruses into groups (Siddell, 1995). Therefore, one can reasonably infer that S variations adapt viruses to particular receptor usage, that receptor usage dictates the ecological niche of infection, and that coronaviruses in distinct niches then evolve somewhat independently to create recognizable antigenic/phylogenetic groups. Suggestions that the SARS-CoV constitutes the first member of a fourth coronavirus group (Marra *et al.*, 2003; Rota *et al.*, 2003) may imply that this pathogen has adapted some time ago to bind a novel receptor set apart from either Apn or CEACAM.

High-resolution crystallographic structures for coronavirus S proteins are not yet known. Therefore, one can only speculate about the detailed architecture of their receptor-binding sites. The S proteins are moderately amenable to protein dissection techniques in which expressed fragments are assayed for receptor-binding potential, and these studies have roughly localized the sites of receptor interaction on primary sequences ((Suzuki and Taguchi, 1996; Bonavia *et al.*, 2003), see Figure 4.3). Current hypotheses suggest that, as coronaviruses diverge into types with particular receptor specificities, amino acid changes are fixed into S proteins at putative receptor-binding sites (Baric *et al.*, 1999). This may be the case; however, S protein variabilities are relatively complex, and while many strain differences cluster in amino-terminal regions where receptors are thought to bind (Matsuyama and Taguchi, 2002a; see Figure 4.3), several changes are also found outside of this area. This complex variability can be appreciated by recalling the multifunctional properties of the S proteins, which contain receptor-binding sites as well as the machinery necessary to fuse opposing membranes (Figure 4.2). For S proteins, this membrane fusion activity is not constitutive, but is (with few exceptions) manifest only after receptor binding. In part, complex variability in S proteins may reflect the fact that receptor-binding and membrane fusion processes are coupled during virus entry. Put another way, the S-receptor interaction releases energy that is then used to create the conformational changes leading to S-induced membrane fusions. This coupling of receptor binding with membrane fusion activity suggests that subtle strain-specific polymorphisms virtually anywhere in the large S proteins might affect either or both of these essential functions, and by doing so, alter the course of coronavirus entry into cells.

The mechanism by which coronavirus S proteins mediate membrane fusion, recently clarified in studies by Bosch *et al.* (2003), involves a process in which the proteins respond to target cell receptor binding by undergoing conformational change (Gallagher, 1997; Matsuyama and Taguchi, 2002; Zelus *et al.*, 2003) (Figure 4.4A). Next, a currently unidentified hydrophobic portion of the protein termed the fusion peptide (FP) harpoons target cell membranes (Figure 4.4B). This is followed by irreversible conformational changes in which alpha-helical portions of the protein condense into helical bundles (Figure 4.4C), ultimately bringing opposing membranes into sufficient proximity to coalesce them together (Figure 4.4D). This is a well-documented mechanism by which several viral and cellular proteins catalyze membrane coalescence (Weissenhorn *et al.*, 1999; Russell *et al.*, 2001; Jahn, *et al.*, 2003) and is now classified as a "class-1" type fusion reaction. Algorithms predicting

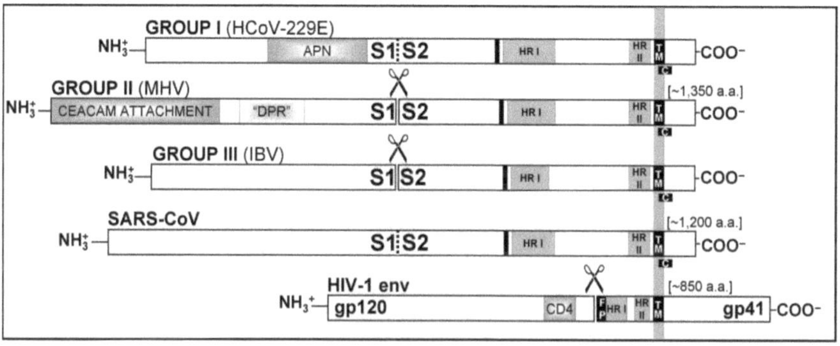

Figure 4.3. Linear depictions of coronavirus spike glycoproteins. Spikes representing groups I–III and the SARS coronavirus are depicted. Scissors indicate proteolytic cleavage between S1 and S2 of spikes. *H*eptad *r*epeat regions (HR) as indicated by *Learn Coil-VMF* and *MultiCoil* are indicated in S2 and are upstream of transmembrane (TM) spans and conserved cysteine-rich stretches (C). Group I spikes recognize aminopeptidase N (CD13) metalloprotease. Group II spikes bind to *c*arcino*e*mbryonic *a*ntigen-related *c*ell *a*dhesion *m*olecule (CEACAM) receptors. Downstream of the CEACAM-binding domain for group II MHV lies a deletion prone region (DPR). Amino acid (a.a.) lengths are drawn approximately to scale and relative to the size of the HIV type I envelope (env) fusion glycoprotein. FP is the hydrophobic fusion peptide for HIV-I. Hydrophobic residues are present N-terminal to HR1 regions in coronaviruses.

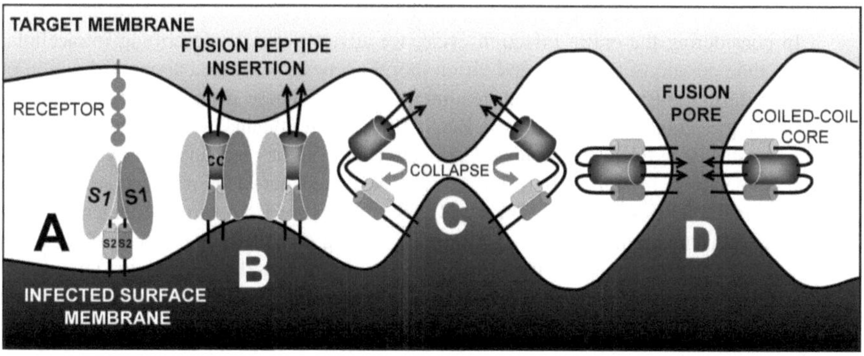

Figure 4.4. Proposed mechanism for S-mediated membrane fusion based on models of class 1-driven viral fusion. (A) Attachment of oligomeric spike S1 domains to receptor. Cylinders represent alpha-helical secondary structure. (B) Arrows depict exposure and insertion of hydrophobic fusion peptides into target membrane, subsequent to displacement of S1. CC indicates formation of alpha-helical *c*oiled *c*oil. (C) Fold-back or collapse of S2 leading to membrane coalescence. S1 domains have been removed for clarity and may in fact be absent as S1 sheds from S2 during fusion activation (see text). (D) Formation of end-stage coiled-coil bundle, fusion pore, and subsequent expansion of pore.

secondary protein structure (Singh *et al.*, 1999) suggest regions of alpha helicity (designated heptad repeat, or HR 1 and 2, see Figure 4.3) in all coronavirus S proteins, including SARS-CoV. Thus, it is generally agreed that the core fusion machinery for all coronaviruses is built in such a way as to catalyze a conserved class-1-type fusion reaction.

It is notable that the fusion module (FP, HR1, HR2, TM span) occupies only about 20% of the inordinately large coronavirus S proteins, and the functional relevance of all but a portion of the remainder is largely unknown. Given that receptor-binding sites are distant from membrane fusion machinery in the primary structures (Figure 4.3), a sensible speculation is that much of the S protein structure is involved in linking receptor binding to the activation of membrane fusion. This is, after all, a crucial coupling that controls the timing and location of virus entry; that is, viral S proteins undergo conformational changes and proceed irreversibly through the class-1-type membrane fusion reaction only when engaged by cellular receptors embedded into the target cell membrane. In considering the activation mechanism, presently available genetic data point toward noncovalent linkages between receptor-binding regions and the fusion machinery (Grosse and Siddell, 1994; Matsuyama and Taguchi, 2002). In the best-studied MHV system, CEACAM binding does cause N-terminal S regions to separate from C-terminal, integral-membrane fragments (Gallagher, 1997), in all likelihood revealing the fusion apparatus (Matsuyama and Taguchi, 2002). This is a process that is augmented by cellular protease(s) that cleave the MHV S proteins at a site between receptor-binding and fusion-inducing domains ((Stauber *et al.*, 1993; Bos *et al.*, 1995; see also Figure 4.3). Proteolytic cleavage likely increases overall S protein conformational flexibility and eases the constraints on exposure of the fusion module, allowing it to advance more readily through the "class-1" pathway (Figure 4.4). These findings are beginning to point toward therapeutic targets interfering with coronavirus entry, and further breakthroughs will likely come from detailed S protein structure determinations.

3. S Functions During Dissemination of Coronavirus Infections

In considering the entire infection cycle, we advance now to describing intracellular events as they pertain to S protein and virion morphogenesis. As stated above, the action of S proteins during entry delivers viral genomes into cells. These genomes are monopartite 27–32 kb single-stranded, positive-sense RNAs (Lai and Stohlman, 1978). The organization of genes on this large RNA has been well characterized, and the mechanisms of gene expression are understood in some detail and are not described here (for reviews, see Lai and Cavanagh, 1997; Sawicki and Sawicki, 1998). As is typical of RNA virus genomes, the vast majority is translated, with the 5′ ~two thirds encoding so-called "nonstructural" proteins that are not found in virions and the remaining ~one third encoding primarily "structural," that is, virion proteins (Figure 4.5). On eclipse, the nonstructural proteins are synthesized without delay, thereby generating RNA-dependent RNA replicase activities that subsequently transcribe antigenomic (negative-sense) RNAs, as well as several subgenomic viral mRNAs. The essential virion proteins S, E (envelope), M (matrix), and N (nucleocapsid) are translated from the newly created set of 3′ proximal subgenomic mRNAs, whose abundance in infected cells is far greater than genomic (virion) RNA, thereby permitting accumulation of virion proteins to the levels required for particle assembly.

A defining characteristic of coronavirion morphogenesis is its intracellular assembly (Figure 4.6), known for some time to take place in the ER-Golgi intermediate compartment

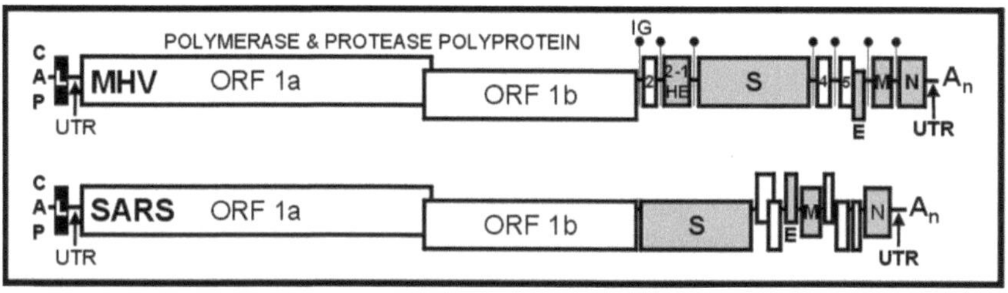

Figure 4.5. Coronavirus genome organization. Depictions of the murine coronavirus MHV (31.2 kb: GenBank accession number NC 001846) and human SARS coronavirus (29.7 kb: # AY278741) positive strand RNA genome. The 5′ end is capped, followed by a leader (L) sequence and *un*translated *r*egion (UTR). The polymerase and protease polyprotein complex is encoded along two open reading frames (1a and 1b) by a ribosomal frame-shifting mechanism and subsequently proteolytically processed into smaller fragments. Vertical lines with globular heads indicate intergenic (IG) sequences. Shaded boxes are structural proteins (sequentially: HE, S, E, M, N) that incorporate into virion particles. Genomes are polyadenylated. Drawn approximately to scale.

(Krijnske-Locker *et al.*, 1994). Infectious virus production requires newly synthesized genome RNA and its associated N proteins, as well as the three integral-membrane proteins S, E, and M. The assembly process, described in greater detail in the legend to Figure 4.6, involves a series of noncovalent interactions; S associating with M (de Haan *et al.*, 1999), M with N (Kuo and Masters, 2002), and N or M with virion RNA (Nelson and Stohlman, 1993; Narayanan *et al.*, 2003). Interestingly, assembly and secretion of intracellular vesicles requires only M and E (Vennema *et al.*, 1996); S and the ribonucleocapsids are dispensable and must be considered to be passive participants in particle morphogenesis. Thus, the S proteins, depending on their affinity for M and their abundance relative to M, may or may not engage in the virion assembly process. In model coronavirus infections, S–M affinities and molar ratios are such that only a portion of S proteins assemble into virions and significant proportions of the population advance as free proteins through the exocytic pathway and on to infected-cell surfaces (Figure 4.6). As there is no evidence that any coronavirus budding takes place at plasma membrane locations, these cell-surface S proteins likely function solely to mediate the cell–cell fusions that facilitate rapid spread of infection. Little is presently known about the relative efficiencies of S–M interactions among the coronaviruses, although the interacting portions of these proteins do indeed differ among virus strains. One might speculate that relatively low S–M affinities reflect adaptations to growth under conditions where syncytial spread of infection provides selective advantages that are greater than those afforded by high S–M affinities favoring efficient infective virion morphogenesis.

4. S Polymorphisms Affect Coronavirus Pathogenesis

It has been known for nearly 20 years that S gene differences correlate with *in vivo* pathogenic potential (Dalziel *et al.*, 1986; Fleming *et al.*, 1986), but only with the recent advent of facile reverse genetics approaches have these S mutations been definitively linked to coronavirus virulence. Definitive links required greater genetic control over the large and heterogeneous coronavirus RNA genome (Figure 4.5), so that one could construct and then characterize panels of recombinant viruses that harbor differences in S genes and nowhere else.

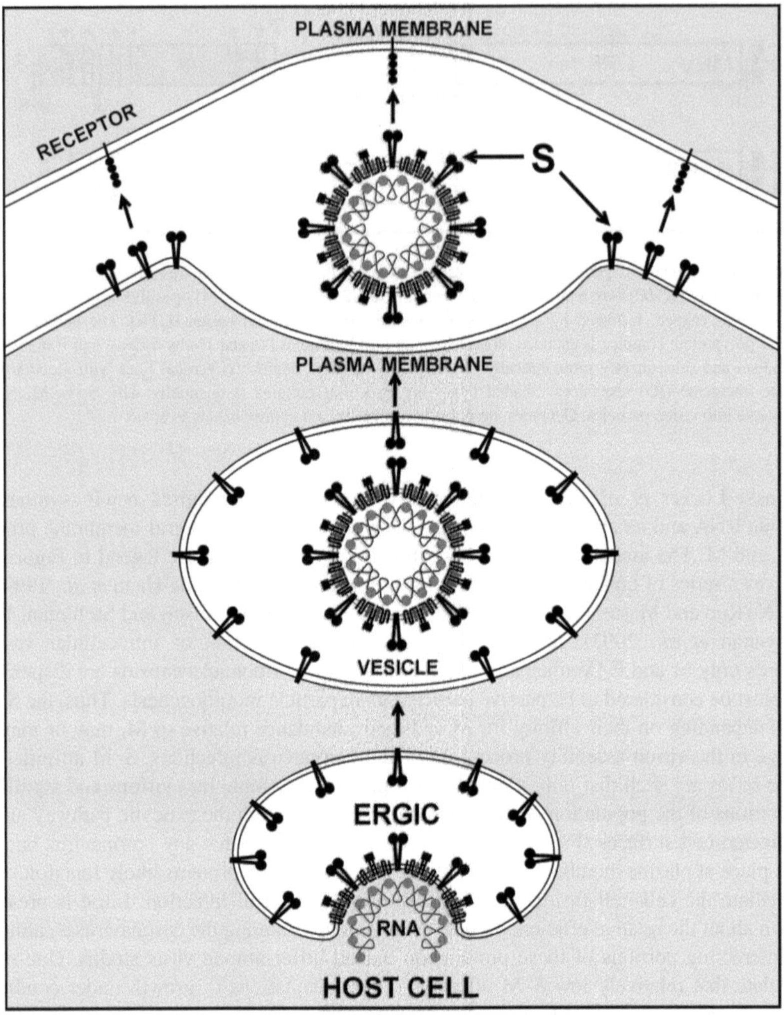

Figure 4.6. Coronavirus assembly. Depiction of accreting coronavirus structural proteins (HE, S, E, M, and N) and RNA at the *e*ndoplasmic *r*eticulum *g*olgi *i*ntermediate *c*ompartment (ERGIC) and the secretion of viral particles along the secretory pathway. Spikes (S) that do not incorporate into particles continue to traffic to the cell surface and promote intercellular spread with neighboring cells.

Studies pioneered in the Masters and Rottier laboratories have yielded creative approaches that are now widely employed to manipulate the 3′ genomic region encompassing "structural" genes (Koetzner *et al.*, 1992; Kuo *et al.*, 2000). The process takes advantage of the fact that coronavirus RNAs tend to recombine, most likely by a copy-choice mechanism (Lai, 1992). Thus, defined site-directed mutant RNAs derived by *in vitro* transcription will recombine with endogenous viral RNAs within infected cells to create site-specific recombinants (Figure 4.7). Isolation of rare recombinants from the far greater parental (nonrecombinant) population depends on the incorporation of a positive-selection marker on the *in vitro* transcript RNAs, which in current approaches amounts to an S ectodomain with strict specificity for APN or CEACAM receptors. Thus, a mutant RNA encoding a group 2 (CEACAM-specific) S ectodomain will recombine in APN+ cells infected by a group 1 (APN-specific) virus, and desired recombinants can be isolated by plaque assays on CEACAM+ cells (Figure 4.7). A subsequent recombination of *in vitro* transcripts encoding group 1 (APN-specific) S ectodomains with the first-generation CEACAM-specific recombinants can generate additionally mutated second-generation recombinants that can be isolated by plaque assay on APN+ cells. This remains a powerful way to manipulate the 3′ portion of the coronavirus genome despite the recent construction of full-length 27–32 kbp infectious coronavirus cDNAs providing for complete genetic control (Almazan *et al.*, 2000; Casais *et al.*, 2001; Thiel *et al.*, 2001; Yount *et al.*, 2002).

The RNA recombination system has been frequently used to specifically incorporate S gene changes, and the general findings indicate that relatively subtle S alterations strongly influence coronavirus virulence and tropism (Sanchez *et al.*, 1999; Phillips *et al.*, 2001;

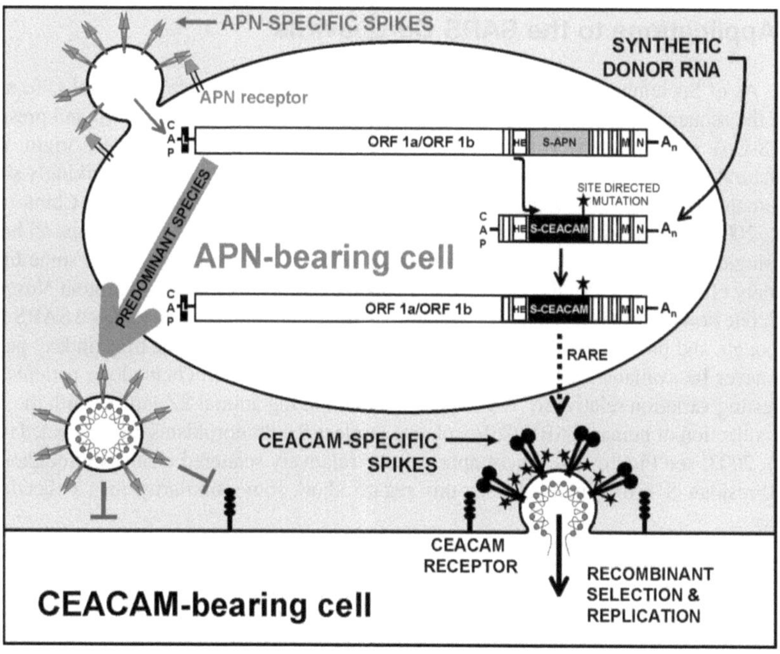

Figure 4.7. Targeted recombination approach to coronavirus reverse genetics. Schematic depicts an input synthetic RNA that harbors a site-directed mutation in the open reading frame (ORF) of a CEACAM-tropic spike (S). The subgenomic transcript recombines during viral replication with the genome of an engineered coronavirus that alternatively harbors APN-tropic spikes. Recombinant full-length genomes encoding CEACAM-tropic spike ORFs are packaged into particles that have incorporated newly synthesized CEACAM-specific S glycoproteins. This small percentage of progeny recombinant virions can enter and replicate in CEACAM-bearing cells. Non-recombinants do not switch tropism and are not selected. (Based on findings by PS Masters & PJ Rottier laboratories.)

Casais *et al.*, 2003; Tsai *et al.*, 2003). Correlating these alterations in virulence with the specific receptor-binding and membrane fusion functions of S proteins has just begun. Recent findings made in the Perlman laboratory have begun to establish these important relationships (Ontiveros *et al.*, 2003). In this laboratory, two variants of the group 2 mouse hepatitis coronavirus were identified with striking differences in neurovirulence, and using the established reverse genetics system, relative virulence was traced to a single S amino acid change, glycine 310 in virulent isolates, serine 310 associated with attenuation. This single difference had global effects on the overall stability of the S proteins, with gly310ser dramatically increasing the stable association of S1 and S2. Concomitant with this stabilization, membrane fusion activities were diminished. In particular, S proteins with the gly310 could mediate cell–cell fusion without the requirement for CEACAM triggering, while those with ser310 could not. These findings indicate that a subtle mutation outside of the core fusion machinery can powerfully influence S-mediated fusion, in this instance affecting its requirements for activation by receptor binding. These findings also point toward S-mediated cell–cell fusion activity as a core agent of coronavirus pathogenicity.

5. Applications to the SARS Coronavirus

As of September 2003, over 30 S_{SARS} sequences were posted in gene banks. To appreciate the sequence variations, one must view the data in the context of known and presumed SARS-CoV epidemiology. This virus is generally considered to be of zoonotic origin. While the natural wild or domestic animal reservoir is probably unknown, isolates strikingly similar to human SARS-CoV has been isolated from exotic animals in Guangdong, China (Guan *et al.*, 2003). These animals included asymptomatic palm civits and raccoon dogs, all housed in a single live animal market. The collection of animal CoV sequences shows some limited diversity (18 nt differences in the 29,709 nt genomes). Speculation is that around November 2002, one or more of these zoonotic "SZ" viruses infected humans and generated SARS fever, dry cough, and pneumonia. Virus from the initially infected human (the true "index" patient) may never be available, but viruses that have been isolated from Guangdong patients have interesting variation relative to "SZ" isolates. In comparing animal SZ viruses with the available collection of human SARS-CoV isolates, 11 clear S polymorphisms were detected (Guan *et al.*, 2003; see Figure 4.8). These appear to be relatively scattered changes throughout the 1,200-residue S ectodomain, and in this regard show some similarity to a collection of

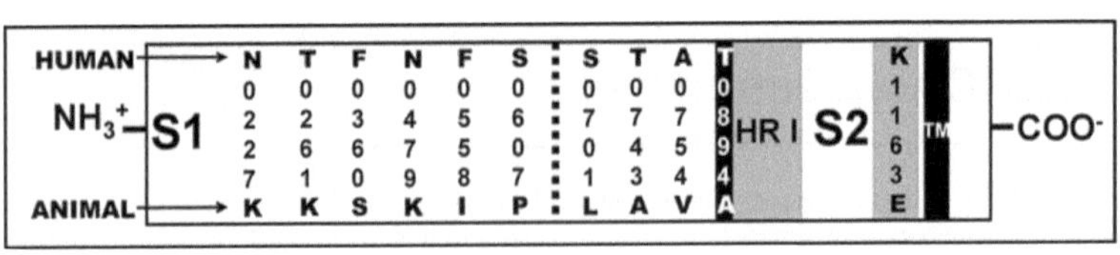

Figure 4.8. Human vs animal spike sequences. Shown are amino acid differences between spikes of human SARS and animal isolates. Numbers indicate location of spike residue. Dashed line demarcates boundary between S1 and S2 regions. Residue 894 resides within a candidate fusion peptide upstream of *h*eptad *r*epeat 1. Residue 1163 is within *h*eptad *r*epeat 2. TM indicates transmembrane region.

16 scattered differences between murine-specific and laboratory-generated zoonotic forms of murine hepatitis coronavirus (Baric *et al.,* 1997, 1999).

Assigning xenotropic potential to a particular combination of these 11 mutations is a challenging but important undertaking. This might be accomplished by employing the approaches used successfully to identify correlates of murine hepatitis virus virulence. S cDNAs encoding SZ or SARS isoforms, as well as SZ/SARS chimeras, can be easily constructed and then used to create recombinant coronaviruses. Tropism of the recombinants for human or animal cells can then be assessed using traditional virological methods. The next challenge will be to correlate S variations to alterations in receptor-binding or membrane fusion potentials. In all likelihood, the successful approaches will again be relatively traditional ones in which soluble S fragments—SZ, SARS, and SZ/SARS chimeras—are developed as mimics of authentic coronaviruses and then used as ligands for binding to human or animal cells, or once identified, SARS cellular receptors and their homologs in animal cells. By titrating soluble S ligands, relative affinities might be obtained. Questions concerning whether SARS polymorphisms specifically affect the membrane fusion reaction can then be addressed by relatively straightforward assays in which S-induced syncytia are measured (Nussbaum *et al.*, 1994). Among the murine coronaviruses, there are S polymorphisms that have no effect on S binding to CEACAM receptors, but yet dramatically impact membrane fusion (Krueger *et al.*, 2001). It will be important to determine whether there are similar variabilities in the SARS S proteins, and whether the membrane fusion process is central to SARS-CoV species transfer and human pathogenicity.

6. Relevance to Antiviral Drug Developments

At present, there are no clinically useful anti-coronavirus drugs, however, the targets for such drugs are clearly in sight. One obvious target is the coronavirus-encoded 3CL protease, as it is essential for the post-translational processing of gene 1 polyproteins into functional subunits ((Ziebuhr *et al.*, 1995; see Figure 4.5). Structure-based, rational anti-3CL protease drug design is at a relatively advanced stage (Anand *et al.*, 2003) and protease inhibitors roughly analogous to those used to combat HIV infection may be forthcoming. A second target, one that is far more relevant to the topic of this chapter, is the S protein. S proteins cause a characteristic syncytial cytopathology in the lung epithelia of SARS patients (Kuiken *et al.*, 2003), and should the S protein dissections described above link syncytial activities with pathogenicity in animal models, investigations would reasonably focus on drugs designed to block S function.

Therapeutics designed to block S-receptor interactions constitute one strategy. Recent structure determinations for a group 2 coronavirus receptor (Tan *et al.*, 2002), and delineation of relevant peptide loops interacting with S proteins (Rao *et al.*, 1997), bring promise to the hypothesis that S-binding receptor fragments might be constructed and used to interfere with virus entry. Such a peptide drug might block infection by inducing the premature triggering of the fusion reaction (Figure 4.4). One must, however, be cautious about advocating this approach because, while many soluble receptors will drive S proteins unproductively into denatured states, some will clearly trigger productive fusion reactions (Matsuyama and Taguchi, 2002). As was found in studies with HIV and soluble CD4 (Moore *et al.*, 1992; Arthos *et al.*, 2002), virus infectivities may be enhanced rather than neutralized.

A second strategy might extend from current hypotheses concerning coronavirus neutralization by antibodies. Several potently neutralizing monoclonal antibodies bind S in regions between CEACAM-binding and fusion-inducing domains (Dalziel *et al.*, 1986). While the mechanisms of neutralization are far from clear, one hypothesis is that the antibodies interfere with conformational transitions linking receptor interaction with fusion activation. High-resolution images of these antibody–S interactions could serve as a guide to construct smaller peptide ligands that neutralize infection by restricting global S conformational change.

Finally, recent convincing evidence that the S proteins of the group 2 mouse hepatitis coronavirus carry out a "class-1" fusion reaction (Bosch *et al.*, 2003) make it probable that several coronaviruses including SARS-CoV will be sensitive to a HR peptide-based fusion inhibition. Peptides derived from the HR regions of structurally similar retroviruses and paramyxoviruses interfere with fusion by associating with complete spikes during the activation reaction, preventing the appropriate collapse into a coiled-coil bundle (Wild *et al.*, 1994; Yao and Compans, 1996; see Figure 4.4). Similarly, a small 38-residue peptide representing mouse hepatitis virus HR2 powerfully inhibited both virus–cell and cell–cell fusion, reducing these activities by several logs when present at 10 μM concentration (Bosch *et al.*, 2003). These HR2 peptides block entry by binding transient intermediate conformations of the fusion protein, depicted in Figure 4.4B, C. It will therefore be important to know whether the genetic variabilities inherent in the coronavirus S proteins alter receptor affinities or fusion kinetics, as these parameters determine the lifespan of the drug-sensitive intermediate structures (Reeves *et al.*, 2002), and by extension they determine whether HR2-based peptides will be effective antiviral agents. By combining comparative studies on S protein receptor binding and membrane fusion with investigations of HR peptide-based antiviral activities, a mechanistic understanding of antiviral action will develop that can lay the groundwork required to develop therapies for human and animal diseases caused by the coronaviruses.

References

Almazan, F., Gonzalez, J.M., Penzes, Z., Izeta, A., Calvo, E., and Plana-Duran, J. *et al.* (2000). Engineering the largest RNA virus genome as an infectious bacterial artificial chromosome. *Proc. Natl. Acad. Sci. USA* **97**(10), 5516–5521.

Anand, K., Ziebuhr, J., Wadhwani, P., Mesters, J.R., and Hilgenfeld, R. (2003). Coronavirus main proteinase (3CLpro) structure: Basis for design of anti-SARS drugs. *Science* **300**(5626), 1763–1767.

Arthos, J., Cicala, C., Steenbeke, T.D., Chun, T.W., Dela Cruz, C., Hanback, D.B. *et al.* (2002). Biochemical and biological characterization of a dodecameric CD4-Ig fusion protein: Implications for therapeutic and vaccine strategies. *J. Biol. Chem.* **277**(13), 11456–11464.

Baric, R.S., Sullivan, E., Hensley, L., Yount, B., and Chen, W. (1999). Persistent infection promotes cross-species transmissibility of mouse hepatitis virus. *J. Virol.* **73**, 638–649.

Baric, R.S., Yount, B., Hensley, L., Peel, S.A., and Chen, W. (1997). Episodic evolution mediates interspecies transfer of a murine coronavirus. *J. Virol.* **71**(3), 1946–1955.

Bonavia, A., Zelus, B.D., Wentworth, D.E., Talbot, P.J., and Holmes, K.V. (2003). Identification of a receptor-binding domain of the spike glycoprotein of human coronavirus HCoV-229E. *J. Virol.* **77**(4), 2530–2538.

Bos, E.C.W., Heijnen, L., Luytjes, W., and Spaan, W.J.M. (1995). Mutational analysis of the murine coronavirus spike protein: Effect on cell-to-cell fusion. *Virology* **214**, 453–463.

Bosch, B.J., van der Zee, R., de Haan, C.A., and Rottier, P.J. (2003). The coronavirus spike protein is a class I virus fusion protein: Structural and functional characterization of the fusion core complex. *J. Virol.* **77**(16), 8801–8811.

Casais, R., Dove, B., Cavanagh, D., and Britton, P. (2003). Recombinant avian infectious bronchitis virus expressing a heterologous spike gene demonstrates that the spike protein is a determinant of cell tropism. *J. Virol.* **77**(16), 9084–9089.

Casais, R., Thiel, V., Siddell, S.G., Cavanagh, D., and Britton, P. (2001). Reverse genetics system for the avian coronavirus infectious bronchitis virus. *J. Virol.* **75**(24), 12359–12369.

Dalziel, R.G., Lampert, P.W., Talbot, P.J., and Buchmeier, M.J. (1986). Site-specific alteration of murine hepatitis virus type 4 peplomer glycoprotein E2 results in reduced neurovirulence. *J. Virol.* **59**, 463–471.

de Haan, C.A.M., Smeets, M., Vernooij, F., Vennema, H., and Rottier, P.J.M. (1999). Mapping of the coronavirus membrane protein domains involved in interaction with the spike protein. *J. Virol.* **73**, 7441–7452.

Delmas, B., Gelfi, J., L'Haridon, R., Vogel, L.K., Sjostrom, H., Noren O. *et al.* (1992). Aminopeptidase N is a major receptor for the entero-pathogenic coronavirus TGEV. *Nature* **357**(6377), 417–420.

Dveksler, G.S., Pensiero, M.N., Cardellichio, C.B., Williams, R.K,.Jiang, G.S., Holmes, K.V., *et al.* (1991). Cloning of the mouse hepatitis virus (MHV) receptor: Expression in human and hamster cell lines confers susceptibility to MHV. *J. Virol.* **65**(12), 6881–6891.

Firla, B., Arndt, M., Frank, K., Thiel, U., Ansorge, S., Tager, M. *et al.* (2002). Extracellular cysteines define ectopeptidase (APN, CD13) expression and function. *Free. Radic. Biol. Med.* **32**(7), 584–595.

Fleming, J.O., Trousdale, M.D., el-Zaatari, F.A., Stohlman, S.A., and Weiner, L.P. (1986). Pathogenicity of antigenic variants of murine coronavirus JHM selected with monoclonal antibodies. *J. Virol.* **58**(3), 869–875.

Gallagher, T.M. (1997). A role for naturally occurring variation of the murine coronavirus spike protein in stabilizing association with the cellular receptor. *J. Virol.* **71**, 3129–3137.

Godfraind, C., Langreth, S.G., Cardellichio, C.B., Knobler, R., Coutelier, J.P., Dubois-Dalcq, M. *et al.* (1995). Tissue and cellular distribution of an adhesion molecule in the carcinoembryonic antigen family that serves as a receptor for mouse hepatitis virus. *Lab. Invest.* **73**(5), 615–627.

Grosse, B. and Siddell, S.G. (1994). Single amino acid changes in the S2 subunit of the MHV surface glycoprotein confer resistance to neutralization by S1 subunit-specific monoclonal antibody. *Virology* **202**, 814–824.

Guan, Y., Zheng, B.J., He, Y.Q., Liu, X.L., Zhuang, Z.X., Cheung, C.L. *et al.* (2003). Isolation and characterization of viruses related to the SARS coronavirus from animals in Southern China. *Science* **302**, 276–278.

Jahn, R., Lang, T., and Sudhof, T.C. (2003). Membrane fusion. *Cell* **112**(4), 519–533.

Koetzner, C.A., Parker, M.M., Ricard, C.S., Sturman, L.S., and Masters, P.S. (1992). Repair and mutagenesis of the genome of a deletion mutant of the coronavirus mouse hepatitis virus by targeted RNA recombination. *J. Virol.* **66**(4), 1841–1848.

Krijnske-Locker, J., Ericsson, M., Rottier, P.J., and Griffiths, G. (1994). Characterization of the budding compartment of mouse hepatitis virus: Evidence that transport from the RER to the Golgi complex requires only one vesicular transport step. *J. Cell Biol.* **124**, 55–70.

Krueger, D.K., Kelly, S.M., Lewicki, D.N., Ruffolo, R., and Gallagher, T.M. (2001). Variations in disparate regions of the murine coronavirus spike protein impact the initiation of membrane fusion. *J. Virol.* **75**(6), 2792–2802.

Kuiken, T., Fouchier, R.A., Schutten, M., Rimmelzwaan, G.F., van Amerongen, G., van Riel, D. *et al.* (2003). Newly discovered coronavirus as the primary cause of severe acute respiratory syndrome. *Lancet* **362**(9380), 263–270.

Kuo, L., Godeke, G.J., Raamsman, M.J., Masters, P.S., and Rottier, P.J. (2000). Retargeting of coronavirus by substitution of the spike glycoprotein ectodomain: Crossing the host cell species barrier. *J. Virol.* **74**, 1396–1406.

Kuo, L. and Masters, P.S. (2002). Genetic evidence for a structural interaction between the carboxy termini of the membrane and nucleocapsid proteins of mouse hepatitis virus. *J. Virol.* **76**, 4987–4999.

Lai, M.M.C. (1992). Genetic recombination in RNA viruses. *Curr. Top. Microbiol. Immunol.* **176**, 21–32.

Lai, M.M. and Stohlman, S.A. (1978). RNA of mouse hepatitis virus. *J. Virol.* **26**(2), 236–242.

Lai, M.M.C. and Cavanagh, D. (1997). The molecular biology of coronaviruses. *Adv. Virus Res.* **48**, 2–100.

Marra, M.A., Jones, S.J., Astell, C.R., Holt, R.A., Brooks-Wilson, A., Butterfield, Y.S. *et al.* (2003). The genome sequence of the SARS-associated coronavirus. *Science* **300**(5624), 1399–1404.

Matsuyama, S. and Taguchi, F. (2002a). Communication between S1N330 and a region in S2 of murine coronavirus spike protein is important for virus entry into cells expressing CEACAM1b receptor. *Virology* **295**, 160–171.

Matsuyama, S. and Taguchi, F. (2002b). Receptor-induced conformational changes of murine coronavirus spike protein. *J. Virol.* **76**(23): 11819–11826.

Moore, J.P., McKeating, J.A., Huang, Y.X., Ashkenazi, A., and Ho, D.D. (1992). Virions of primary human immunodeficiency virus type 1 isolates resistant to soluble CD4 (sCD4) neutralization differ in sCD4 binding and glycoprotein gp120 retention from sCD4-sensitive isolates. *J. Virol.* **66**(1), 235–243.

Narayanan, K., Chen, C.J., Maeda, J., and Makino, S. (2003). Nucleocapsid-independent specific viral RNA packaging via viral envelope protein and viral RNA signal. *J. Virol.* **77**(5), 2922–2927.

Nelson, G.W. and Stohlman, S.A. (1993). Localization of the RNA-binding domain of mouse hepatitis virus nucleocapsid protein. *J. Gen. Virol.* **74**(Pt 9), 1975–1979.

Nussbaum, O., Broder, C.C., and Berger, E.A. (1994). Fusogenic mechanisms of enveloped-virus glycoproteins analyzed by a novel recombinant vaccinia virus-based assay quantitating cell fusion-dependent reporter gene activation. *J. Virol.* **68**, 5411–5422.

Ontiveros, E., Kim, T.S., Gallagher, T.M., and Perlman, S. (2003). Enhanced virulence mediated by the murine coronavirus, mouse hepatitis virus strain JHM, is associated with a glycine at residue 310 of the spike glycoprotein. *J. Virol.* **77**(19), 10260–10269.

Phillips, J.J., Chua, M., Seo, S.H., and Weiss, S.R. (2001). Multiple regions of the murine coronavirus spike glycoprotein influence neurovirulence. *J. Neurovirol.* **7**, 421–431.

Rao, P.V., Kumari, S., and Gallagher, T.M. (1997). Identification of a contiguous 6-residue determinant in the MHV receptor that controls the level of virion binding to cells. *Virology* **229**, 336–348.

Reeves, J.D., Gallo, S.A., Ahmad, N., Miamidian, J.L., Harvey, P. E., Sharron, M. *et al.* (2002). Sensitivity of HIV-1 to entry inhibitors correlates with envelope/coreceptor affinity, receptor density, and fusion kinetics. *Proc. Natl. Acad. Sci. USA* **99**(25), 16249–16254.

Rota, P.A., Oberste, M.S., Monroe, S.S., Nix, W.A., Campagnoli, R., Icenogle, J.P. *et al.*, (2003). Characterization of a novel coronavirus associated with severe acute respiratory syndrome. *Science* **300**(5624), 1394–1399.

Russell, C.J., Jardetzky, T.S., and Lamb, R.A. (2001). Membrane fusion machines of paramyxoviruses: Capture of intermediates of fusion. *EMBO J.* **20**(15), 4024–4034.

Sanchez, C.M., Izeta, A., Sanchez-Morgado, J.M., Alonso, S., Sola, I., Balasch, M. *et al.* (1999). Targeted recombination demonstrates that the spike gene of transmissible gastroenteritis coronavirus is a determinant of its enteric tropism and virulence. *J. Virol.* **73**, 7607–7618.

Sawicki, S.G., and Sawicki, D.L. (1998). A new model for coronavirus transcription. *Adv. Exp. Med. Biol.* **440**, 215–219.

Schultze, B., Gross, H.J., Brossmer, R., and Herrler, G. (1991). The S protein of bovine coronavirus is a hemagglutinin recognizing 9-O-acetylated sialic acid as a receptor determinant. *J. Virol.* **65**(11), 6232–6237.

Siddell, S.G. (1995). The coronaviridae: An introduction. In S.G. Siddell (ed.), *The Coronaviridae*, Plenum Press, New York and London, pp. 1–10.

Singh, M., Berger, B., and Kim, P.S. (1999). LearnCoil-VMF: Computational evidence for coiled-coil-like motifs in many viral membrane fusion proteins. *J. Mol. Biol.* **290**, 1031–1041.

Sjostrom, H., Noren, O., and Olsen, J. (2000). Structure and function of aminopeptidase N. *Adv. Exp. Med. Biol.* **477**, 25–34.

Stauber, R., Pfleiderara, M., and Siddell, S. (1993). Proteolytic cleavage of the murine coronavirus surface glycoprotein is not required for fusion activity. *J. Gen. Virol.* **74**, 183–191.

Suzuki, H., and Taguchi, F. (1996). Analysis of the receptor-binding site of murine coronavirus spike protein. *J. Virol.* **70**(4), 2632–2636.

Tan, K., Zelus, B.D., Meijers, R., Liu, J.-H., Bergelson, J.M., Duke, N. *et al.* (2002). Crystal structure of murine sCEACAM1a[1,4]: A coronavirus receptor in the CEA family. *EMBO J.* **21**, 2076–2086.

Thiel, V., Herold, J., Schelle, B., and Siddell, S.G. (2001). Infectious RNA transcribed in vitro from a cDNA copy of the human coronavirus genome cloned in vaccinia virus. *J. Gen. Virol.* **82**(Pt 6), 1273–1281.

Tresnan, D.B., Levis, R., and Holmes, K.V. (1996). Feline aminopeptidase N serves as a receptor for feline, canine, porcine, and human coronaviruses in serogroup I. *J. Virol.* **70**(12), 8669–8674.

Tsai, J.C., de Groot, L., Pinon, J.D., Iacono, K.T., Phillips, J.J., Seo, S.H. *et al.* Weiss (2003). Amino acid substitutions within the heptad repeat domain 1 of murine coronavirus spike protein restrict viral antigen spread in the central nervous system. *Virology* **312**(2), 369–380.

Vennema, H., Godeke, G.J., Rossen, J.W., Voorhout, W.F., Horzinek, M.C., Opstelten, D.J. *et al.* (1996). Nucleocapsid-independent assembly of coronavirus-like particles by co-expression of viral envelope protein genes. *EMBO J.* **15**, 2020–2028.

Weissenhorn, W., Dessen, A., Calder, L.J., Harrison, S.C., Skehel, J.J., and Wiley, D.C. (1999). Structural basis for membrane fusion by enveloped viruses. *Mol. Membr. Biol.* **16**(1), 3–9.

Wild, C.T., Shugars, D.C., Greenwell, T.K., McDanal, C.B., and Matthews, T.J. (1994). Peptides corresponding to a predictive alpha-helical domain of human immunodeficiency virus type 1 gp41 are potent inhibitors of virus infection. *Proc. Natl. Acad. Sci. USA* **91**(21), 9770–9774.

Yao, Q., and Compans, R.W. (1996). Peptides corresponding to the heptad repeat sequence of human parainfluenza virus fusion protein are potent inhibitors of virus infection. *Virology* **223**(1), 103–112.

Yount, B., Denison, M.R., Weiss, S.R., and Baric, R.S. (2002). Systematic assembly of a full-length infectious cDNA of mouse hepatitis virus strain A59. *J. Virol.* **76**(21), 11065–11078.

Zelus, B.D., Schickli, J.H., Blau, D.M., Weiss, S.R., and Holmes, K.V. (2003). Conformational changes in the spike glycoprotein of murine coronavirus are induced at 37 degrees C either by soluble murine CEACAM1 receptors or by pH 8. *J. Virol.* **77**(2), 830–840.

Ziebuhr, J., Herold, J., and Siddell, S.G. (1995). Characterization of a human coronavirus (strain 229E) 3C-like proteinase activity. *J. Virol.* **69**(7), 4331–4338.

5

Aspects of the Fusogenic Activity of Influenza Hemagglutinin Peptides by Molecular Dynamics Simulations

L. Vaccaro, K.J. Cross, S.A. Wharton, J.J. Skehel, and F. Fraternali

The interactions of the fusion domain of influenza hemagglutinin (HA) (N-terminal 20 residues, the fusion peptide) with a POPC lipid bilayer have been studied by molecular dynamics (MD) simulations. This domain is considered a model system for studying processes occurring during viral membrane fusion. Synthetic peptides corresponding to this domain are able to cause red blood cell hemolysis and leakage of liposomal content. Mutations in their sequences lead to homologs with fusion properties similar to whole HA molecules with the corresponding sequence changes. Detailed information about the molecular interactions leading to insertion into the core of the lipid bilayer can be obtained by the analysis of our simulations. We observe that the N-terminal 11 residues of the fusion peptides are helical and insert with a tilt angle with respect to the membrane plane of about 30°, in very good agreement with experimental data. The tilt angle stabilizes around the final value only after 4 ns. Residues Glu11, Glu15, and Asp19 are positioned at the level of the lipid phosphate groups, and the last peptide segment can either be helical or unfolded without altering dramatically the tilting of the first N-terminal 11 residues. The membrane bilayer experiences a thinning caused by the presence of the peptides and the calculated order parameters show larger disorder of the alkyl chains. These results indicate a perturbed lipid packing upon peptide insertion that could facilitate membrane fusion.

1. Introduction

Membrane fusion is one of the most frequently occurring mechanisms in biology and yet one of the most poorly understood. Amongst the known fusion events, a fusion caused by viral infection represents a relatively simple system because of the small number of viral proteins involved. Many viral fusion proteins fold into coiled coil bundles, suggesting that a common structural element is necessary for the bridging of membranes as a preliminary step

---

L. Vaccaro, K.J. Cross, S.A. Wharton, J.J. Skehel, and F. Fraternali • National Institute for Medical Research, Mill Hill, London NW7 1AA, United Kingdom.

Viral Membrane Proteins: Structure, Function, and Drug Design, edited by Wolfgang Fischer.
Kluwer Academic / Plenum Publishers, New York, 2005.

to fusion (Skehel and Wiley, 2000). Moreover, most viral proteins contain a conserved stretch of about 20 amino acids at the N-terminus, called fusion peptide FP, which provides the attachment point of the virus to the target membrane. One of the best structurally studied membrane fusion processes is that induced by the Influenza HA protein. Native HA is a homotrimer, with each subunit comprising two glycopolypeptides linked by a single disulfide bond, HA1 and HA2, produced during infection by cleavage of the biosynthetic precursor HA0. At neutral pH, the fusion peptides are buried in the interior of the trimer (Wilson *et al.*, 1981), but endosomal acidification induces substantial conformational changes in the HA structure, the most important being the relocation of the fusion peptide from its buried position to form the membrane-distal tip of a 100 Å long triple stranded coiled-coil (Skehel *et al.*, 1982; Bullough *et al.*, 1994; Bizebard *et al.*, 1995).

Synthetic FPs have been assayed for their capacity to perturb natural membranes and to fuse liposomes at neutral pH (Lear and De Grado, 1987). Peptides with specific amino acid substitutions have fusion properties similar to whole HA molecules with the corresponding mutations (Wharton *et al.*, 1988; Steinhauer *et al.*, 1995; Han *et al.*, 1999). The incorporation of mutants of HA into infectious influenza viruses by reverse genetics (Cross *et al.*, 2001) allowed for investigations on the importance of the conserved spacing of glycines in the FP sequences. These studies demonstrate that FPs are good models to study the membrane-associated structures and the processes that cause viral fusion. The most supported hypothesis is that the bilayer perturbations caused by these peptides are necessary but not sufficient for the occurrence of viral fusion in the physiological context (Nieva and Agirre, 2003).

One of the key features of the fusogenic sequence is that very few mutations are tolerated when replacing Gly1, especially polar residues are prohibitive. In order to maintain fusogenic properties, the first three N-terminal residues have to be of hydrophobic nature. Moreover, deletion of Gly1 leads to a non-fusogenic mutant (Gething *et al.*, 1986; Wharton *et al.*, 1988). Recently, the pKa of the N-terminal amino group of Gly1 was measured by ^{15}N NMR (Zhou *et al.*, 2000) in the presence of DOPC. The measured high value (8.69) is indicative of the stabilization of the protonated form of the amino group by means of non-covalent interactions. In general, peptides with protected amino terminus are found to inhibit viral fusion, and by deprotecting the amino group promotion of negative curvature strain of membranes has been observed (Epand *et al.*, 1993). All these observations underline the specificity of fusion peptide sequences and hint to an important role of this segment in the interaction with the membrane. The protonation state of the N-terminus, the orientation of the peptide with respect to the membrane bilayer and to the phosphate groups seem all to be key factors of the exhibited fusogenicity. Molecular simulations have been demonstrated to be very useful in analyzing in atomic detail processes not directly accessible to experimental techniques. It is very difficult to obtain detailed experimental structural data on non-bilayer and intermediate structures in the cell fusion process; therefore, theoretical models can help in outlining key features and in directing new experimental work.

The first atomic MD simulation of spontaneous membrane fusion has been recently published (Marrink and Tieleman, 2002), demonstrating that force field parametrization is accurate enough to produce atomic details of processes like membrane destabilization and binding of peptides to lipid bilayers (Tieleman *et al.*, 1997; Sansom *et al.*, 1998). We, therefore, decided to investigate the aforementioned aspects of the FP insertion and membrane perturbation by MD simulations in a preformed POPC bilayer. The main aspects that have been investigated are the conformational stability of the modeled structures in lipid bilayers, their orientation and depth of insertion, and the disorder created to the membrane's alkyl chains.

2. Methods

The initial structures for the simulations are the pdb files 1IBN (pH 5.0) and 1IBO (pH 7.4), hereafter mentioned as NMR5 and NMR7 (Han *et al.*, 2001). The peptides were inserted in a fully equilibrated 128 POPC lipid bilayer (Tieleman *et al.*, 1999). The Glu11 residue was placed at the level of the lipid phosphate groups (Han *et al.*, 2001) in the bilayer. Each system was then solvated, resulting in a total of about 19.000 atoms (box dimensions were $6.2 \times 6.7 \times 6.2$ nm^3), subjected to 500 ps of solute restrained MD simulations in order to let the lipid equilibrate. The final structures were submitted to 5 ns simulation runs. Simulations were run using GROMACS (Berendsen *et al.*, 1995). The LINCS algorithm was used to constrain all bond lengths within the lipids. A cutoff of 0.9 nm for Coulomb and Lennard-Jones interactions was used and Particle Mesh Ewald (Essmann *et al.*, 1995) to calculate the remaining electrostatic contributions on a grid with 0.12 nm spacing.

NPT conditions (i.e., constant number of particles, pressure, and temperature) were used in the simulations. A constant pressure of 1 bar in all three directions was used, with a coupling constant of 1.0 ps (Berendsen *et al.*, 1984). Water, lipids, and protein were coupled separately to a temperature bath at 300 K with a coupling constant of 0.1 ps. The lipid parameters were as in previous MD studies of lipid bilayers (Tieleman *et al.*, 1999) and the GROMOS96 force field (43a1) was used for the peptide (van Gunsteren *et al.*, 1996). The SPC water model was used for the solvent (Berendsen *et al.*, 1981). Analysis programs from GROMACS were used, together with self-written programs (tilt angle and secondary structure calculations). The bilayer thickness has been measured as the average difference in the Y coordinates (the membrane plane is in XZ orientation) between the phosphorous atoms of the upper and lower leaflets. The depth of residues in the bilayer has been measured as the average difference in the Y coordinates of each residue's C^{α} atom and the phosphorous atoms of the upper leaflet.

3. Results

3.1. Comparison with Experimental Structures

For several years, numerous biophysical studies have been performed to determine the structure of fusion peptides within a membrane-like environment and peptide binding to the lipidic bilayer. The high hydrophobicity of these peptides forbids measurements of their partitioning between aqueous and membrane phases. To overcome this problem, the insertion of hydrophilic residues into the sequence has allowed the NMR structure determination of a fusogenic homolog of the HA fusion peptide (GLFGAIAGFIENGWEGMIDG), called E5 (GLFEAIAEFIEGGWEGLIEG), in DPC and in SDS micelles (Dubovskii *et al.*, 2000; Hsu *et al.*, 2002). In order to study the native sequence in an *in vivo*-like environment, a new host–guest fusion peptide system has been engineered so as to render the HA fusion peptide completely water soluble and with high affinity for biological lipid model membranes. The fusion peptide is tethered with a flexible linker to a host peptide that solubilizes the entire system (Han and Tamm, 2000). This innovative design has allowed not only partition experiments of fusion peptides to lipid bilayers (Li *et al.*, 2003), but also NMR studies of the structure of the fusion domain in detergent micelles at pH 5.0 and 7.4. Two different conformations for the two pH environments have been proposed in this study (we will refer to these structures as NMR5 and NMR7) (Han *et al.*, 2001). It is supposed by these authors that the secondary structure of the peptide changes to some extent by changing the pH. In particular the C-terminal segment GWEGMIDG (residues 13–20) is quite disordered in the NMR7 structure. The first segment is mostly α-helical in both cases, but the pH 5.0 conformer presents a short stretch of 3_{10} helix from residues 13 to 18. The NMR study in SDS micelles of the E5 peptide (Hsu *et al.*, 2002) also proposes two different structures at pH 4.3 and 7.3. All the low pH structures are very similar, with most of the peptide in helical structure, with a hinge in the region around residues G12 and G13. In Figure 5.1, the NMR5 structure is compared with the structure of the E5 peptide (Dubovskii *et al.*, 2000) in SDS, determined at the same pH (5.0). The "V"-shape adopted by the NMR5 structure is smoothed for E5 and the peptide assumes a boomerang-shaped conformation with an oblique orientation of the first 11 residues. No hydrophobic pocket due to interaction between Phe9 and Trp12 is formed in the E5 structure, although there is a clear hydrophobic side facing the membrane. The average rmsd for the first 11 C^{α} atoms is 0.96 Å between NMR5 and E5. This striking similarity outlines the importance of these residues in the fusogenic activity of these peptides. The Glu11 and Glu15 residues of both peptides are pointing to the same direction (toward the phosphate groups of the bilayer interface) and the Phe residues are on the opposite side pointing toward the lipidic tails of the membrane.

All these considerations seem to imply that the N-terminal 11 residues of the FPs are helical and that residues 11 and 15 should be located at the phosphate–water interface. We, therefore, constructed two structures starting from the coordinates of NMR5 and NMR7, inserted into a preformed POPC lipid bilayer (see Methods section) and we simulated each system for 5 ns (we will refer to simulated conformers as HA5 and HA7, respectively). We did not take into account differences in the protonation state of charged residues (except for the N- and C-termini) depending on the pH values at which the structures were determined. The two conformers are simply different starting points for the simulations of which we want to study the time evolution and stability. With the purpose of studying in more detail the location of the N-terminus at the interface between membrane phosphate groups and water together with the effects of the protonation state of the N-terminal amino group, we constructed two additional starting structures with charged N- and C-termini (referred to as HA5_*c* and HA7_*c*). In Figure 5.2, the final snapshots of all the simulated systems are

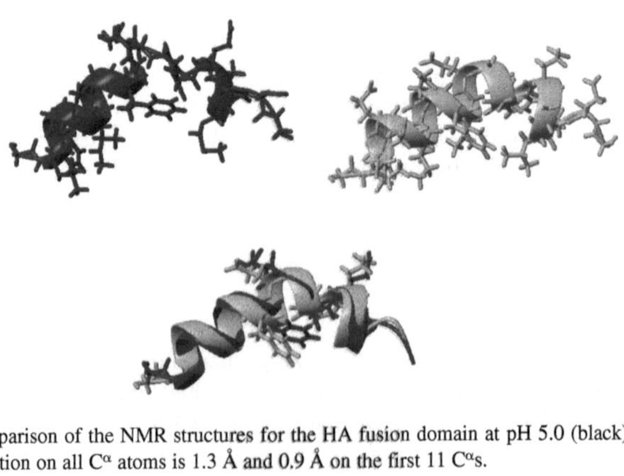

Figure 5.1. Comparison of the NMR structures for the HA fusion domain at pH 5.0 (black) and the E5 peptide (grey). Superimposition on all C^{α} atoms is 1.3 Å and 0.9 Å on the first 11 C^{α}s.

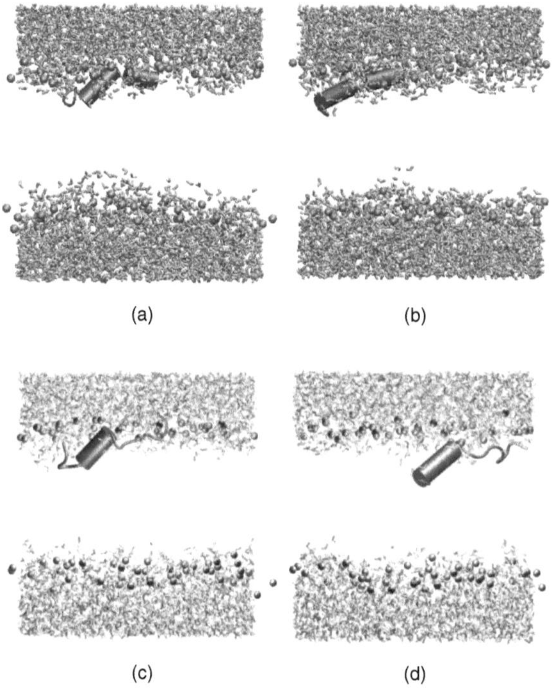

Figure 5.2. Final snapshots after 5 ns of MD simulations. (a) HA5_c; (b) HA5; (c) HA7_c; (d) HA7. Helical segments are represented by cylinders. Only the water molecules and the phosphorous atoms (cpk spheres) have been dispayed for clarity.

reported. Both HA5 and HA5_c structures retained their helical conformation during most of the simulated time, but the charged termini peptides adopted a "hook"-shaped turn at the N-terminal three residues that allowed the charged amino group to reach the aqueous phase (*vide infra* Figure 5.4). The "V"-shaped form was kept in the HA5_c simulation, while it was slightly widened in HA5 [Figure 5.2(b)]. All peptides adopted a tilted insertion angle at the end of the simulation (~30° with respect to the membrane surface). The HA7 starting structure remained disordered in the segment 12–20, and remained tilted, although the tilting was more pronounced for the HA7_c conformer [Figure 5.2(c)].

In Figure 5.3(a) the NMR5 (black) and NMR7 (grey) structure bundles are reported and in Figure 5.3(b), the superimposition of the final conformers of HA5 (black) and HA5_c (grey) with the NMR5 structure is shown. The negatively charged side chain groups are all pointing toward the aqueous phase, and the hydrophobic groups are directed toward the hydrophobic lipid chains. In all MD structures, we observed large mobility of Phe3, while the relative position of Phe9 and Trp14 is preserved. The already mentioned hook-shaped conformation of residues 1–3 is clearly visible in the HA5_c structure. The relative position of Glu11 and Glu15 is maintained and their side-chain carboxylate groups are both reaching the polar phase of the membrane, acting as anchoring points at the membrane interface. For the NMR7 structures, one can observe that, apart from the previously discussed differences,

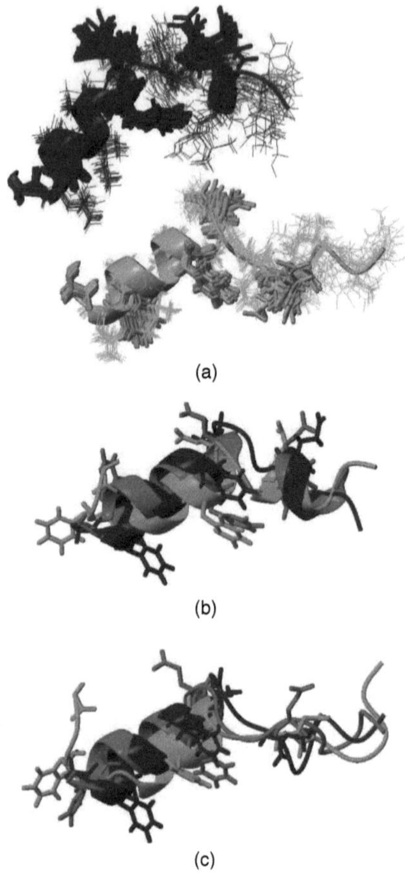

Figure 5.3. (a) NMR structures for HA5 (black) and HA7 (grey); (b) superimposition of the HA5_*c* structure (light grey) and of the HA5 structure (dark grey) to the experimental structure (black); (c) superimposition of the HA7_*c* structure (light grey) and of the HA7 structure (dark grey) to the experimental structure (black).

Glu15 is pointing in a different direction with respect to Glu11, and it is directed toward the hydrophobic phase of the membrane. In our simulations, the carboxylate groups are unprotonated and, therefore, they tend to move toward the aqueous phase [Figure 5.3(c)]. The fact that this segment remains unstructured during the simulated time implies that if the peptide would enter the membrane in an already partially disordered conformation, the process of refolding would be prevented by the competitive favorable interactions with the polar head groups at the membrane interface.

Several measured properties for the studied peptides are reported in Table 5.1. The helical content derived from the simulations is often higher than measured ones, but it should be considered that folding–refolding events occur on larger time scales than the ones sampled here. HA5 and HA5_*c* show higher helical content than the HA7 structures, and neutral termini conformers are in both cases more helical. The mechanism of anchoring to the

Table 5.1. Relevant Structural and Insertion Parameters for the Studied Peptides

Peptide	H^a_{calc}(%)	H^b_{CD} (%)	H^c_{NMR} (%)	Tilt angle (°)	Thickness[d] (Å)
HA5	71.0	33.0	66.0	32.7	31.7
HA5_*c*	51.1	—	—	28.4	28.2
HA7	47.6	—	40.2	31.0	26.9
HA7_*c*	39.3	—	—	34.3	27.3

Notes:
[a]Helical content (in % over all residues) calculated from the 5 ns trajectory.
[b]Helical content measured from CD experiments in POPC.
[c]Helical content calculated from the pdb structures in SDS micelles.
[d]Membrane thickness measured as the average difference in the Y coordinates of the phosphorous atoms of the upper and lower leaflets (membrane is in XZ plane).

membrane of the charged N-termini, with consequent disruption of helical conformation at residues 1–3, will be discussed in the next paragraph.

3.2. Membrane Anchoring

The charged termini conformers present the four N-terminal residues in a curled conformation, caused by the amino group reaching out towards the polar head groups of the membrane and the aqueous phase. The remaining helical segments between residues 5 and 11 and residues 15 and 18 are not affected by this curling, and remain stable during the simulated time. In order to show the behavior of the amino group during the simulated time in more detail, snapshots of its time evolution for the HA5_*c* structure are shown in Figure 5.4. The charged amino group is initially deeply inserted into the membraneous hydrophobic phase. Within the first 60 ps of simulations, a shell of water solvation has already formed around the amino terminus and remains stable for the rest of the simulation. The calculated Radial Distribution Function (RDF) for this group with water shows an average number of two/three molecules of water solvating this group (data not shown). Analogous behavior is observed for the HA7_*c* conformer. This simulation result could explain at a molecular level the experimentally observed high pKa value for this amino group (Zhou *et al.*, 2000), because the base strength increases when the protonated form is stabilized. In this study, the authors conclude that the N-terminus of the peptide is close to the aqueous phase and protonated. Our simulations, starting from a situation in which the peptide is deeply inserted into the membrane, allow us to follow the inverse process. The peptide inserting into the membrane will pass through the aqueous phase, then the polar phase, and, finally, reaches the hydrophobic phase of the bilayer. In our simulations, we observe the amino terminus reaching out to the polar phase, yielding a stable solvated state. We decided to perform additional simulations with a neutral amino terminus in order to rule out artificial perturbations of the membrane generated by the charged amino group, deeply inserted into the hydrophobic phase. The insertion of these peptides at a tilted angle into the membrane bilayer seems to be one of the common features of all the viral fusion peptides (Brasseur, 1991; Gray *et al.*, 1996; Lins *et al.*, 2001). Therefore, we analyzed this parameter for all the performed simulations. In Table 5.1, the average value of the tilt angle, defined here as the angle formed with the membrane plane, is reported. For all the studied conformers, this angle is around 30°, in very good agreement with experimental data measured by spin-labeling electron paramagnetic resonance (EPR)

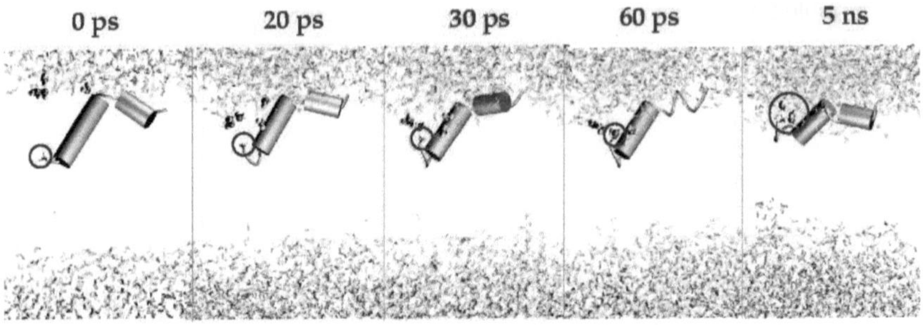

Figure 5.4. Time evolution of the solvation process of the charged N-terminus (circled) for the HA5_*c* simulation. After 60 ps, two molecules of water are solvating the charged group throughout the resting simulated time.

techniques (Macosko *et al.*, 1997; Zhou *et al.*, 2000; Han *et al.*, 2001). The values are not dependent on the chosen charged state of the N- and C-termini; as we will see later for the HA5_*c* and HA7_*c* simulations, the immersion depth of the first two residues will be affected but not the average tilt angle. Recently, the fusion domain of the human immunodeficiency virus gp41 has been studied by MD for 1 ns (Kamath and Wong, 2002). In our case, longer simulation times were necessary to distinguish clearly between the slow process of migration to the interface and the stabilization of the tilt angle.

The simulated bilayer thickness, measured as the average distance between phosphorous atoms of the upper and lower leaflets of POPC, is for all the systems smaller than the one for the simulated POPC bilayer alone (36 Å, data not shown) and of the experimentally measured value (40 Å) (Table 5.1) (Kinoshita *et al.*, 1998). There is a thinning effect of about 8 Å on average by comparison to the value of the equilibrated POPC bilayer. The smallest thickness values are observed for HA7 and HA7_*c* simulations; therefore, we cannot discriminate between HA5 and HA7 peptides on the basis of their insertion mode and bilayer perturbation. The effect of reducing the membrane thickness could be related to two main causes (in relation to the presence of the peptide): (a) the interaction of the polar residues with the polar interface that tends to "pull down" the phosphate groups of the upper leaflet; (b) the disorder in the hydrophobic tails generated by the peptide, which results in lower order parameters especially for the palmitic chains (data not shown). In the case of charged termini peptides, the first cause has a stronger effect. In order to analyze in detail the insertion mode of the peptide, we report in Figure 5.5, the depth of insertion of the 11 N-terminal residues. We do not observe a substantially different behavior with respect to the tilted orientation of the peptide and to the thinning of the membrane between the two conformers HA5 and HA7, and the partially unfolded state of the C-terminal segment in HA7 does not affect the behavior of the 11 N-terminal residues.

Therefore, we concentrated our analysis only on those residues that determine the tilted insertion into the membrane. The charged termini peptides are less deeply inserted, with strong differences at positions 1–2. The tilted insertion is mainly due to residues 3, 5, and 6 in agreement with previous experimental observations (Macosko *et al.*, 1997; Zhou *et al.*, 2000; Han *et al.*, 2001). For neutral termini peptides, the first three N-terminal residues remain more deeply inserted and no migration to the interface is observed during the

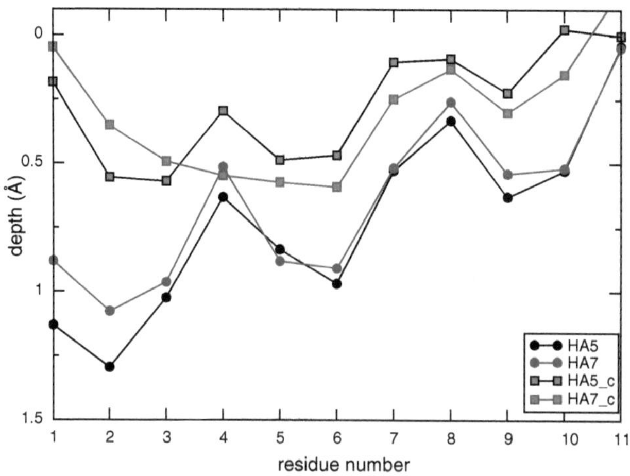

Figure 5.5. Calculated depth of insertion for the first 11 residues of the fusion peptides. Circles refer to neutral termini peptides, squares to charged termini peptides.

simulated time. For the remaining residues, the most deeply inserted Ile6 is located at about 10 Å depth (for HA5 and HA7).

4. Conclusions

This work describes, at a molecular level, the mechanism of oblique insertion of the fusion peptide of influenza HA into lipid bilayers. The amphiphilic character of the sequence and the typical glycine pattern favor helical structures in hydrophobic environments. Our work supports previous structural observations of highly amphiphatic conformations adopting an inverse "V"-shaped structure, with a bend at residue 12 that forms a hydrophobic pocket. We believe that one of the most critical parameters to exhibit fusogenicity is the insertion with a tilted angle in the membrane in order to perturb the bilayer. In particular, the presence of a fairly stable helix spanning the first 11 N-terminal residues is deemed necessary to stabilize the tilted angle to a value of about ~30°. This tilted insertion is mainly due to residues 3, 5, and 6 in agreement with experimental EPR measurements, and especially good agreement is found for residue 6, which is inserted at 10 Å. The C-terminal segment can be partially unfolded, without modifying the tilted orientation of the first segment. The lipid bilayer is perturbed by the presence of the peptides and a thinning of about 8 Å is observed after 5 ns of simulation.

Acknowledgments

We wish to thank Dr. J. Kleinjung and Dr. P. Temussi for critical reading of the manuscript.

References

Berendsen, H.J.C., Postma, J.P.M., van Gunsteren, W.F., DiNola A., and Haak, J.R. (1984). Molecular dynamics with coupling to an external bath. *J. Chem. Phys.* **81**, 3684–3690.

Berendsen, H.J.C., Postma, J.P.M., van Gunsteren, W.F., and Hermans, J. (1981). Interaction models for water in relation to protein hydration. In B. Pullman (ed.), pp. 331–342. *Intermolecular Forces*. Reidel, Dordrecht.

Berendsen, H.J., van der Spoel, C.D., and van Drunen, R. (1995). GROMACS: A message-passing parallel molecular dynamics implementation. *Comp. Phys. Comm.* **95**, 43–56.

Bizebard, T.B., Gigant, P., Rigolet, B., Rasmussen, O., Diat, P., Boseke *et al.* (1995). Structure of influenza haemagglutinin complexed with a neutralizing antibody. *Nature* **376**, 92–94.

Brasseur, R. (1991). Differentiation of lipid-associating helices by use of 3-dimensional molecular hydrophobicity potential calculation. *J. Biol. Chem.* **266**, 16120–16127.

Bullough, P.A., Hughson, F.M., Skehel, J.J., and Wiley, D.C. (1994). Structure of influenza haemagglutinin at the pH of membrane fusion. *Nature* **371**, 37–43.

Cross, K.J., Wharton, S.A., Skehel, J.J., Wiley, D.C., and Steinhauer, D.A. (2001). Studies on influenza haemagglutinin fusion peptide mutants generated by reverse genetics. *EMBO J.* **20**, 4432–4442.

Dubovskii, P.V., Li, H., Takahashi, S., Arseniev, A.S., and Akasaka, K. (2000). Structure of an analog of fusion peptide from haemagglutinin. *Protein Sci.* **9**, 786–798.

Epand, R.M., Epand, R.F., Richardson, C.D., and Yeagle, P.L. (1993). Structural requirements for the inhibition of membrane fusion by carbobenzoxy-D-Phe–Phe–Gly. *Biochim. Biophys. Acta* **1152**, 128–134.

Essmann, U., Perera, L., and Berkowitz, M.L. (1995). A smooth particle mesh Ewald method. *J. Chem. Phys.* **103**, 8577–8593.

Gething, M.J., Doms, R.W., York, D., and White, J.M. (1986). Studies on the mechanism of membrane fusion: Site-specific mutagenesis of the haemagglutinin of influenza virus. *J. Cell Biol.* **102**, 11–23.

Gray, C., Tatulian, S.A., Wharton, S.A., and Tamm, L.K. (1996). Effect of the N-terminal glycine on the secondary structure, orientation, and interaction of the influenza haemagglutinin fusion peptide with lipid bilayers. *Biophys. J.* **70**, 2275–2286.

Han, X. and Tamm, L. (2000). A host–guest system to study structure–function relationships of membrane fusion peptides. *Proc. Natl. Acad. Sci.* **97**, 13097–13102.

Han, X., Bushweller, J.H., Cafiso, D.S., and Tamm, L. (2001). Membrane structure and fusion-triggering conformational change of the fusion domain from influenza haemagglutinin. *Nature Struct. Biol.* **8**, 715–720.

Han, X., Steinhauer, D.A., Wharton, S.A., and Tamm, L.K. (1999). Interaction of mutant influenza virus haemagglutinin fusion peptides with lipid bilayers: Probing the role of hydrophobic residue size in the central region of the fusion peptide. *Biochemistry* **38**, 15052–15059.

Hsu, C.H., Wu, S.H., Chang, D.K., and Chen, C. (2002). Structural characterizations of fusion peptide analogs of influenza virus haemagglutinin. *J. Biol. Chem.* **25**, 22725–22733.

Kamath, S. and Wong, T.C. (2002). Membrane structure of the human immunodeficiency virus gp41 fusion domain by molecular dynamics simulations. *Biophys. J.* **83**, 135–143.

Kinoshita, K., Furuike, S., and Yamazaki, M. (1998). Intermembrane distance in multilamellar vesicles of phosphatidylcholine depends on the interaction free energy between solvents and the hydrophilic segments of the membrane surface. *Biophys. Chem.* **74**, 237–249.

Lear, J.D. and De Grado, W.F. (1987). Membrane binding and conformational properties of peptide representing the amino terminus of influenza virus HA2. *J. Biol. Chem.* **262**, 6500–6505.

Li, Y., Han, X., and Tamm, L. (2003). Thermodynamics of fusion peptide–membrane interactions. *Biochemistry* **42**, 7245–7251.

Lins, L., Charloteaux, B., Thomas, A., and Brasseur, R. (2001). Computational study of lipid-destabilising protein fragments: Towards a comprehensive view of tilted peptides. *Proteins: Struct. Funct. Gen.* **44**, 435–447.

Macosko, J.C., Kim, C., and Shin, Y. (1997). The membrane topology of the fusion peptide region of influenza haemagglutinin determined by spin-labeling EPR. *J. Mol. Biol.* **267**, 1139–1148.

Marrink, S.-J. and Tieleman, D.P. (2002). Molecular dynamics simulations of spontaneous membrane fusion during a cubic–hexagonal phase transition. *Biophys. J.* **83**, 2386–2392.

Nieva, J.L. and Agirre, A. (2003). Are fusion peptides a good model to study viral cell fusion? *Biochim. Biophys. Acta* **1614**, 104–115.

Sansom, M.S., Tieleman, D.P., Forrest, L.R., and Berendsen, H.J. (1998). Molecular dynamics simulations of membranes with embedded proteins and peptides: Porin, alamethicin and influenza virus M2. *Biochem. Soc. Trans.* **26**, 438–443.

Skehel, J.J. and Wiley, D.C. (2000). Receptor binding and membrane fusion in virus entry: The influenza haemagglutinin. *Annu. Rev. Biochem.* **69**, 531–569.

Skehel, J.J., Bayley, P.M., Brown, E.B., Martin, S.R., Waterfield, M.D., White, J.M. *et al.* (1982). Changes in the conformation of influenza virus haemagglutinin at the pH optimum of virus-mediated membrane fusion. *Proc. Natl. Acad. Sci. USA* **79**, 968–972.

Steinhauer, D.A., Wharton, S.A., Skehel, J.J., and Wiley, D.C. (1995). Studies of the membrane fusion activities of fusion peptide mutants of influenza virus haemagglutinin. *J. Virol.* **69**, 6643–6651.

Tieleman, D.P., Sansom, M., and Berendsen, H. (1999). Alamethicin helices in a bilayer and in solution: Molecular dynamics simulations. *Biophys. J.* **76**, 40–49.

Tieleman, D.P., Marrink, S.-J., and Berendsen, H.J. (1997). A computer perspective of membranes: Molecular dynamics studies of lipid bilayer systems. *Biochim. Biophys. Acta* **1331**, 235–270.

van Gunsteren, W.F., Billeter, S.R., Eising, A.A., Hünenberger, P.H., Krüger, P., Mark, A.E. *et al.* (1996). Biomolecular Simulations: The GROMOS96 Manual and User Guide, 1st edn. BIOMOS b.v. Laboratory of Physical Chemistry, Zürich.

Wharton, S.A., Martin, S.R., Ruigrok, R.W., Skehel, J.J., and Wiley, D.C. (1988). Membrane fusion by peptide analogues of influenza virus haemagglutinin fusion. *J. Gen. Virol.* **69**, 1847–1857.

Wilson, I.A., Skehel, J.J., and Wiley, D.C. (1981). Structure of the haemagglutinin membrane glycoprotein of influenza virus at 3.0 A resolution. *Nature* **289**, 366–373.

Zhou, Z., Macosko, J.C., Hughes, D.W., Sayer, B.G., Hawes J., and Epand, R.M. (2000). NMR study of the ionization properties of the influenza virus fusion peptide in zwitterionic phospholipid dispersions. *Biophys. J.* **78**, 2418–2425.

Part III

Viral Ion Channels/viroporins

6

Viral Proteins that Enhance Membrane Permeability

María Eugenia González and Luis Carrasco

1. Introduction

During the infection of cells by animal viruses, membrane permeability is modified at two different steps of the virus life cycle (Carrasco, 1995) (Figure 6.1). Initially, when the virion enters cells, a number of different-sized molecules are able to co-enter the cytoplasm with the virus particles (Fernandez-Puentes and Carrasco, 1980; Otero and Carrasco, 1987). Membrane potential is reversibly destroyed, being restored several minutes later. Endosomes are involved in the co-entry process, since inhibitors of the proton ATPase block early permeabilization even with viruses that do not require endosomal function. A chemiosmotic model has been advanced to explain the molecular basis of early membrane modification by virus particles (Carrasco, 1994). The viral molecules involved are components of virions: glycoproteins when enveloped particles are analyzed or, still unidentified, domains of the structural proteins in the case of naked viruses. Attachment of the particle to the cell surface receptor does not alter membrane permeability by itself. Inhibitors that hamper virus decapsidation, still allowing virus attachment to the cell surface, block early membrane permeabilization (Almela *et al.*, 1991).

At late times of infection, when there is active translation of late viral mRNAs, the plasma membrane becomes permeable to small molecules and ions (Carrasco, 1978) (Figure 6.1). Different viral molecules may be responsible for this late enhancement of membrane permeability, including viroporins (Gonzalez and Carrasco, 2003), glycoproteins, and even proteases (Chang *et al.*, 1999; Blanco *et al.*, 2003). This chapter is devoted to reviewing some characteristics of membrane permeabilization by viral proteins. In addition, the methodology used to assay enhanced permeability in animal cells is described. Finally, the design of selective viral inhibitors based on the modification of cellular membranes during virus entry or at late times of infection is also discussed.

María Eugenia González • Unidad de Expresión Viral, Centro Nacional de Microbiologia, Instituto de Salud Carlos III, Carretera de Majadahonda-Pozuelo, Km 2, Majadahonda 28220, Madrid, Spain.
Luis Carrasco • Centro de Biología Molecular Severo Ochoa, Facultad de Ciencias, Universidad Autónoma, Cantoblanco 28049, Madrid, Spain.

Viral Membrane Proteins: Structure, Function, and Drug Design, edited by Wolfgang Fischer.
Kluwer Academic / Plenum Publishers, New York, 2005.

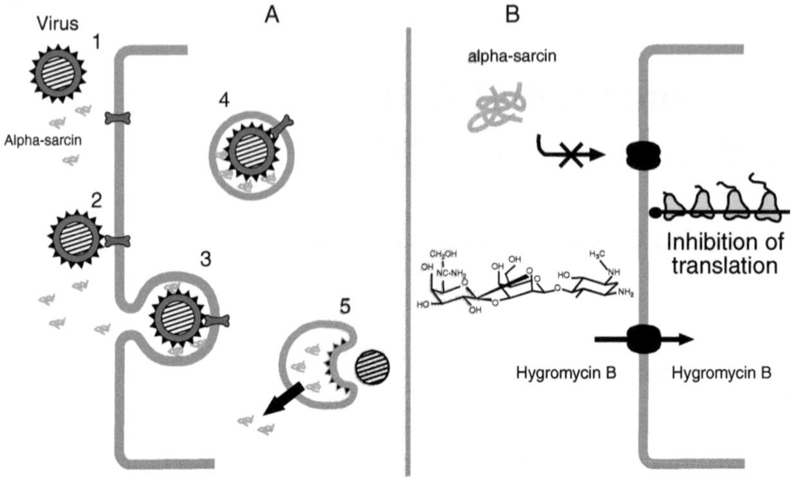

Figure 6.1. (A) Schematic representation of early membrane permeability. The different steps of virus attachment, entry, and fusion events are shown. The protein toxin alpha-sarcin co-enters endosomes in conjunction with animal virus particles. The fusion of the viral membrane with the endosome membrane induces the release of alpha-sarcin to the cytoplasm. (B) Permeabilization of the plasma membrane at late times of infection. The figure depicts the entry of the low molecular weight translation inhibitor hygromycin B through pores created at the plasma membrane. Alpha-sarcin is unable to pass through these pores.

2. Measuring Alterations in Membrane Permeability

2.1. The Hygromycin B Test

A number of hydrophilic molecules, including some antibiotics, poorly permeate through cellular membranes (Contreras *et al.*, 1978; Lacal *et al.*, 1980). This is the case of hygromycin B, anthelmycin, blasticidin S, destomycin A, gougerotin, and edein complex. The aminoglycoside antibiotic hygromycin B (MW 527) is produced by *Streptomyces hygroscopicus*. This hydrophilic molecule is an efficient inhibitor of protein synthesis in cell-free systems but interferes very poorly with translation in intact cells. However, the modification of the plasma membrane by viruses or by other means leads to a rapid blockade of translation (Carrasco, 1995). Concentrations of the antibiotic ranging from about 0.1–1 mM are added to the culture medium and protein synthesis is estimated by incubation with radioactive methionine for 1 hr (see Figure 6.2) (Gonzalez and Carrasco, 2001). In addition to its simplicity, the hygromycin B test has a number of advantages for assaying changes in membrane permeability. One is its great sensitivity, and another is that this test measures membrane modifications only in cells that are metabolically active. Moreover, in cultures where some cells are uninfected, hygromycin B would only enter virus-infected cells that are synthesizing proteins. The hygromycin B test has been applied with success to prokaryotic (Lama and Carrasco, 1992) and eukaryotic cells, including yeast (Barco and Carrasco, 1995, 1998) and mammalian cells (Gatti *et al.*, 1998; Gonzalez and Carrasco, 2001).

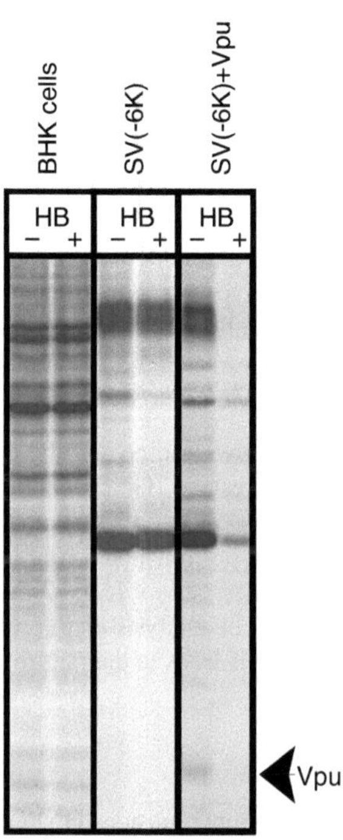

Figure 6.2. Entry of hygromycin B into BHK cells promoted by the expression of HIV-1 Vpu. Control BHK cells (left), cells infected with Sindbis virus lacking the 6K gene (SV(-6K)) (center), or with (SV(-6K)) containing the HIV-1 *vpu* gene (SV(-6K)+Vpu) (right) were treated or not with 0.5 mM hygromycin B for 10 min and protein synthesis was assayed by extended incubation with [^{35}S] Met/Cys for 1 hr. The labeled proteins were analyzed by autoradiography after SDS PAGE. For additional details see Gonzalez and Carrasco (2001).

2.2. Entry of Macromolecules into Virus-Infected Cells

Alpha-sarcin is a protein of 150 amino acid residues, which is produced by *Aspergillus giganteus* (Oka *et al.*, 1990). This protein inhibits translation by modifying ribosomes in an enzymatic manner. Thus, a molecule of alpha-sarcin is able to inactivate a great number of ribosomes by hydrolysis of the A4324-G4325 phosphodiester bond in the 28S rRNA (Chan *et al.*, 1983). This toxin does not enter mammalian cells because it does not attach to the cell surface and is therefore unable to cross the plasma membrane. However, alpha-sarcin efficiently interferes with protein synthesis in cell-free systems or in cells where the permeability barrier has been destroyed (Fernandez-Puentes and Carrasco, 1980). Alpha-sarcin co-enters cells in conjunction with virus particles, and is liberated to the cytoplasm (Otero and Carrasco, 1987; Liprandi *et al.*, 1997). In this manner, this toxin irreversibly blocks translation several minutes after virus entry. The molecular basis of the co-entry of macromolecules with virus particles has been analyzed in detail elsewhere (Carrasco, 1994, 1995). Apart from alpha-sarcin, a number of proteins that interfere with translation, many of them of plant origin, have been described (Fernandez-Puentes and Carrasco, 1980; Lee *et al.*, 1990). The release of all these toxins into cells is enhanced by virus particles. Not only proteins, but also other macromolecules, including nucleic acids efficiently co-enter with virus particles (Cotten *et al.*, 1992). However, none of these macromolecules passes into cells at late times of infection.

2.3. Other Assays to Test the Entry or Exit of Molecules from Virus-Infected Cells

Apart from the use of translation inhibitors that do not easily permeate into intact cells, a number of assays can be employed to assess modifications in membrane permeability. Amongst these assays, we can list the following.

2.3.1. Entry or Exit of Radioactive Molecules

Cells are preloaded with radioactive uridine and the exit of nucleotides can be monitored after induction of viroporin expression (Gonzalez and Carrasco, 1998). Unlike with most amino acids, the pool of uridine nucleotides is abundant in the cell interior, thus providing a convenient and sensitive assay for monitoring the exit of molecules from cells. Other tests use radioactive glucose derivatives (e.g., 2-deoxyglucose), which cannot be metabolized and accumulate in cells. The analysis of the release of radioactive compounds that are not actively transported into cells leads to the failure to measure enhanced membrane permeability.

2.3.2. Entry of ONPG and Dyes

Entry of *o*-nitrophenyl-β-D-galactopyranoside (ONPG) into the bacterial cells can be determined very simply. This β-galactosidase substrate can be incubated with bacterial cells and production of the resulting compound can be followed by determining the absorbance at 420 nm (Lama and Carrasco, 1992). There are a number of non-vital dyes employed to characterize cell mortality. It should be noted that the entry of these compounds, in fact, determines a modification in cell membranes, which in some cases does not directly correlate with cell death. Trypan blue is a dye widely used for monitoring enhanced membrane permeability. However, this assay is not very sensitive. In addition, trypan blue staining does not discriminate between metabolically active or dead cells, as the hygromycin B tests does. Another assay employs the polyamine neurobiotin that needs specific connexin channels to enter mammalian cells (Elfgang *et al.*, 1995). Permeabilization of cell membrane increases uptake of this cationic molecule. Internalized neurobiotin can be detected in paraformaldehyde-fixed cells, by fluorescence microscopy, using fluorescein isothiocyanate-conjugated streptavidin (Gonzalez and Carrasco, 1998).

2.3.3. Entry of Propidium Iodide

Propidium iodide (PI) is a DNA-intercalating compound that does not enter intact cells. However, those cells that exhibit increased membrane permeability are able to take up PI, which can be assayed by cell fluorometric analysis (Arroyo *et al.*, 1995).

2.3.4. Release of Cellular Enzymes to the Culture Medium

The appearance in the culture medium of cellular enzymes is a clear indicator of cell mortality. This is the case for lactic dehydrogenase and bacterial β-galactosidase, present outside the cells (Sanderson *et al.*, 1996). Commercial kits to measure this enzymatic activity are available. The release of cellular proteins to the medium occurs at very late times of viral infection, when cells have already died. This alteration takes place at much later times after hygromycin B entry can be detected (Blanco *et al.*, 2003).

3. Viral Proteins that Modify Permeability

3.1. Viroporins

Viroporins are small proteins encoded by viruses that contain a stretch of hydrophobic amino acids (Gonzalez and Carrasco, 2003). Typically, viroporins are comprised of some 60–120 amino acids. The hydrophobic domain is able to form an amphipathic α-helix. The insertion of these proteins into membranes followed by their oligomerization creates a hydrophilic pore. The architecture of this channel is such that the hydrophobic amino acid residues face the phospholipid bilayer while the hydrophilic residues form part of the pore. In addition to this domain, there are other features of viroporin structure, including a second hydrophobic region in some viroporins that also interacts with membranes. This second interaction may further disturb the organization of the lipid bilayer. These proteins may also contain a stretch of basic amino acids that acts in a detergent-like fashion. All these structural features contribute to membrane destabilization. More recently, another domain has been described in some glycoproteins and also viroporins that has the capacity to interact with membranes. This domain is rich in aromatic amino acids and is usually inserted at the interface of the phospholipid bilayer (Suarez *et al.*, 2000; Sanz *et al.*, 2003). This type of interaction also leads to membrane destabilization, further enhancing membrane permeability.

A number of viroporins from different viruses that infect eukaryotic cells have been reported. This group of proteins includes picornavirus 2B and 3A, alphavirus 6K, retrovirus Vpu, paramyxovirus SH, orthomyxovirus M2, reovirus p10, flavivirus p7, phycodnavirus Kcv, coronavirus E, and rhabdovirus alpha 10p. A recent review devoted to viroporins discusses the structure and function of a number of proteins of this group (Gonzalez and Carrasco, 2003), and so the details of each particular viroporin will not be reviewed in this chapter.

The main activity of viroporins is to create pores at biological membranes to permit the passage of ions and small molecules. The cloning and individual expression of viroporin genes has allowed their effects in bacterial and animal cells to be analyzed. Thus, the expression of this type of viral gene enhances the permeation of ions and several hydrophilic molecules in or out of cells (Carrasco, 1995). In addition, the purified viroporin molecules open pores in model membranes, providing a system that is amenable to biophysical analysis (Fischer and Sansom, 2002). The pore size created by viroporins allows the diffusion of different molecules with a molecular weight below about 1,000 Da.

The main step affected in animal viruses containing a deleted viroporin gene is the assembly and exit of virions from the infected cells (Klimkait *et al.*, 1990; Liljestrom *et al.*, 1991; Loewy *et al.*, 1995; Betakova *et al.*, 2000; Watanabe *et al.*, 2001; Kuo and Masters, 2003). These genes are not essential for virus replication in culture cells, but the plaque size is much smaller in viroporin-defective viruses. Notably, virus entry and gene expression in viroporin-deleted viruses occur as in their wild-type counterparts. An aspect of viroporin function at the molecular level that is still not understood is the link between pore activity and virus budding.

3.2. Viral Glycoproteins that Modify Membrane Permeability

In addition to small hydrophobic viral proteins, there are other virus products that promote membrane permeabilization. This occurs with a number of virus glycoproteins (GP) that are known to increase cell membrane permeability, such as the human immunodeficiency virus gp41 (Chernomordik *et al.*, 1994; Arroyo *et al.*, 1995), the Ebola virus GP (Yang *et al.*, 2000), the cytomegalovirus US9 protein (Maidji *et al.*, 1996), the Vaccinia virus A38L protein (Sanderson *et al.*, 1996), rotavirus VP7 and NS4 proteins (Charpilienne *et al.*, 1997; Newton *et al.*, 1997), the hepatitis C virus E1 protein (Ciccaglione *et al.*, 1998), and the alphavirus E1 protein (Nyfeler *et al.*, 2001; Wengler *et al.*, 2003).

The architecture of some viral glycoproteins is such that upon oligomerization, the transmembrane (TM) domains may form a physical pore. In principle, two different regions of a viral fusion glycoprotein could form pores. One such region contains the fusion peptide that would create a pore in the cell membranes upon insertion (Skehel and Wiley, 1998), while the TM domain would form a pore in the virion membrane (Wild *et al.*, 1994). Moreover, sequences adjacent to the TM region could have motifs designed to destabilize membrane structure (Suarez *et al.*, 2000). Entry of enveloped animal viruses leads to early membrane permeabilization, which is mediated by the formation of the two pores (fusion and TM) formed by viral fusion glycoproteins. This early permeabilization induced during the entry of virions requires conformational changes of the fusion glycoproteins. By contrast, after virus replication, newly synthesized glycoproteins may affect membrane permeability when they reach the plasma membrane (Figure 6.3). This modification is achieved only by the TM domain, while the fusion peptide does not participate in this late modification. In viruses that lack the typical viroporin, its function could be replaced by these pore-forming glycoproteins, while for other viruses viroporin activity may be redundant (Bour and Strebel, 1996). In the latter case, pore formation may be generated by viral glycoproteins and viroporins (Figure 6.3). We would like to propose the possibility that pore-forming glycoproteins play a key role mainly during virus entry and, in some cases, also during virus budding, while viroporins come into action when viruses need to exit the cell.

Early membrane permeabilization is always carried out by a virion component. In the case of enveloped viruses, this early event is executed by a structural glycoprotein, which is coupled to the fusion process. An understanding of fusion at the molecular level also requires an explanation of the phenomenon of early membrane permeabilization. We have advanced the idea that viral glycoproteins involved in membrane fusion participate in the dissipation of the chemiosmotic gradient, thus providing the energy to push the nucleocapsid and neighboring macromolecules to the cell interior (Carrasco, 1994; Irurzun *et al.*, 1997). Fusion glycoproteins do not simply serve to bridge the cellular and the viral membrane, but instead are designed to open pores in both membranes. This pore-opening activity may be necessary to lower membrane potential and to dissipate ionic gradients. Several chapters of this book are devoted to the detailed description of the structure and function of these glycoproteins, so we will focus our attention on viral glycoproteins that permeabilize membranes when individually expressed in cells. These membrane active proteins may exhibit this activity later on in the virus life cycle.

3.2.1. Rotavirus Glycoprotein

Rotavirus infection provokes a number of alterations in cellular membranes during infection (del Castillo *et al.*, 1991). Amongst these alterations, there is an increase in the

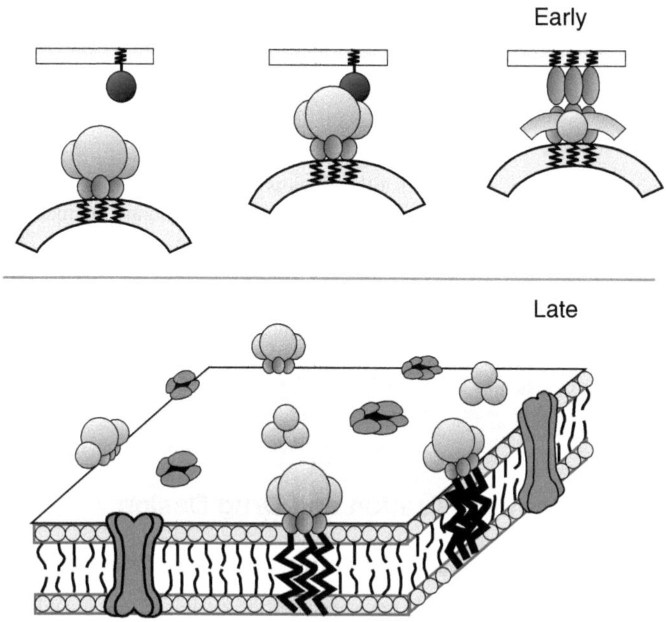

Figure 6.3. Participation of pore formation by viral glycoproteins and viroporins in membrane permeability. Early membrane permeabilization is coupled to the fusion activity of the corresponding viral glycoprotein. This fusion glycoprotein may create two pores. One is located at the target cell membrane and the other is formed by the TM domain. Late membrane permeabilization may be carried out by viroporins or by the TM domains of viral glycoproteins.

concentration of cytoplasmic calcium (Michelangeli *et al.*, 1991). Several rotavirus proteins exhibit membrane-destabilizing activity. The enterotoxin NSP4 induces alterations in membrane permeability (Tian *et al.*, 1994). The individual expression of the non-structural glycoprotein NSP4 has the ability to increase the concentration of cytoplasmic calcium. This increase may be mediated by activation of phospholipase C activity (Dong *et al.*, 1997).

Rotavirus particles induce the co-entry of protein toxins into cells (Cuadras *et al.*, 1997). At least two structural components possess the ability to permeabilize cells, including VP5 protein and VP7 glycoprotein (Charpilienne *et al.*, 1997; Irurzun *et al.*, 1997).

3.2.2. The HIV-1 gp41

Infection of lymphocytic human cells by HIV-1 enhances membrane permeability to ions and several compounds (Voss *et al.*, 1996; Gatti *et al.*, 1998). There are at least three different HIV-encoded proteins responsible for these alterations: Vpu protein, the retroviral protease, and the fusion glycoprotein gp41. Apart from the fusion peptide, there are two regions of gp41 that exhibit membrane permeability; one is located at the carboxy terminus (Arroyo *et al.*, 1995; Comardelle *et al.*, 1997) and another corresponds to the membrane-spanning domain (Arroyo *et al.*, 1995). The C-terminus of gp41 includes two 20–30 residues, which may form cationic amphipathic α-helices, designated as lentivirus lytic peptides 1 and 2 (LLP-1 and LLP-2). Synthetic LLP-1 peptide forms pores in planar phospholipid bilayers

(Chernomordik *et al.*, 1994), permeabilizes HIV-1 virions to deoxyribonucleoside triphosphates (Zhang *et al.*, 1996), and induces alterations in ion permeability of *Xenopus* oocytes (Comardelle *et al.*, 1997).

3.2.3. Other Viral Glycoproteins

Inducible expression of the hepatitis C virus E1 glycoprotein increases membrane permeability in bacterial cells. The ability of E1 to modify membrane permeability has been mapped to the carboxy terminus of the protein (Ciccaglione *et al.*, 1998, 2001). Similar permeabilization was found with *Escherichia coli* cells that synthesize Semliki forest virus E1 glycoprotein after exposure to low pH (Nyfeler *et al.*, 2001). Finally, overexpression of Vaccinia virus A38L glycoprotein produces changes in the morphology, permeability, and adhesion of mammalian cells. The potential capacity of A38L protein to form pores at the plasma membrane promotes the entry of calcium ions and PI and the release of lactic dehydrogenase into the culture medium (Sanderson *et al.*, 1996).

4. Membrane Permeabilization and Drug Design

4.1. Antibiotics and Toxins that Selectively Enter Virus-Infected Cells

Different approaches have been envisaged for the design of compounds that interfere with virus replication based on modifications in membrane permeability. One such approach makes use of inhibitors of cellular or viral functions that do not permeate easily into intact animal cells. Notably, these agents selectively enter into virus-infected cells (Carrasco, 1978; Benedetto *et al.*, 1980). Most of the inhibitors used thus far interfere with protein synthesis, although compounds that affect other functions could also be employed. Entry of these agents leads to the inhibition of translation specifically in virus-infected cells, leading to a profound inhibition of virus growth (Contreras *et al.*, 1978; Carrasco, 1995; Gatti *et al.*, 1998). Although this approach discriminates well between uninfected or virus-infected cells in culture, the high toxicity of the agents thus far assayed has hampered its use in whole animals. Perhaps future searches for less toxic compounds would make this approach amenable to application in therapy. In fact, some of the plant toxins that co-enter with virus particles have been described as being antiviral agents (Lee *et al.*, 1990). Even compounds such as hygromycin B, which has been used in the veterinary field as an antibacterial agent, could also be used as an antiviral compound for rotavirus infections (Liprandi *et al.*, 1997).

4.2. Viroporin Inhibitors

The paradigm of an inhibitor of a viral ion channel is amantadine (Hay, 1992) (Figure 6.4). This compound has been used as an anti-influenza agent in humans (De Clercq, 2001). The target of amantadine is the influenza-encoded protein M2. Residues 27, 30, 31, and 34 of M2 determine the amantadine sensitivity of this ion channel. A drawback of amantadine is the high doses necessary to affect influenza. The search for more effective compounds may provide a more efficacious treatment for this illness.

Compounds that interfere with the functioning of other viroporins have also been described. This is the case of amiloride derivatives that block HIV-1 Vpu activity (Ewart *et al.*,

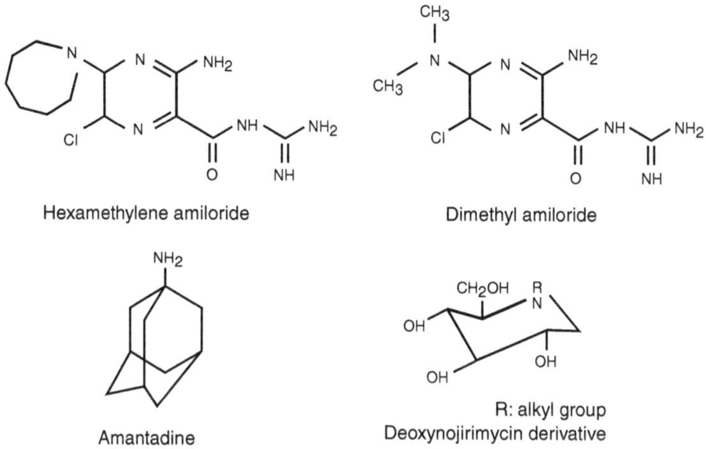

Figure 6.4. Chemical structures of several inhibitors of viroporin activity.

2002). In this manner, the production of infectious HIV-1 is reduced in the presence of this agent. Recently, long alkyl-chain iminosugar derivatives have been found to interfere with the function of the hepatitis C virus p7 protein as an ion channel (Griffin *et al.*, 2003; Pavlovic *et al.*, 2003). These compounds exhibited antiviral activity with bovine viral diarrhea virus, which is closely related to the hepatitis C virus (Durantel *et al.*, 2001).

4.3. Antiviral Agents that Interfere with Viral Glycoproteins

Much effort has been concentrated recently on the development of antiviral agents that inhibit the fusion step of HIV. Binding of HIV gp120 to the CD4 receptor is followed by further interaction of this viral glycoprotein with the coreceptor molecules CXCR4 and CCR5. After this initial interaction, the conformation of the ectodomain of the TM glycoprotein gp41 is profoundly modified. Exposure of the fusion peptide at the amino terminus of gp41 triggers its insertion into the target cellular membrane, leading to the fusion of the viral and the cellular plasma membranes. All these steps have been used as targets for anti-HIV therapy (Cooley and Lewin, 2003). As regards the fusion step, a variety of peptide mimetic inhibitors have been developed. The pioneering work on peptide T20 has demonstrated that this compound is a potent inhibitor of gp41-induced membrane fusion. T20 exhibits antiviral activity in HIV-infected patients. The detailed mechanism of action of T20 at the molecular level is known. This peptide is homologous with 36 amino acids within the C-terminal heptad repeat region (HR2) of HIV-1 gp41. T20 competitively binds to HR1 and interferes with the formation of the six helix HR1–HR2 bundle complex necessary for membrane fusion. At present there are a great number of peptides that interfere with binding of gp120 or with gp41-induced membrane fusion and that have been tested for their anti-HIV activity and clinical efficacy. In this regard, T1249 is one of the second generation of HR-2 peptide mimetic inhibitors that consists of 39 amino acids. PRO 542 is a soluble CD4 receptor (CD4-IgG2) that binds to and neutralizes gp120 before virus binding occurs. SCH-C is an oxime–piperidine compound that is a coreceptor antagonist. This small molecule acts as an inhibitor of CCR5. MD3100 is a non-peptidic, low molecular weight bicyclam compound that prevents interactions between CXCR4 and gp120, blocking signal transduction from CXCR4. Future research in this field will provide us with additional antiviral compounds to add to the anti-HIV armory.

Acknowledgments

This work was supported by grants from the Fundación para la Investigación y Prevención del SIDA en España (24291), Instituto de Salud Carlos III (01/0042), and the DGICYT PM99-0002. The authors also acknowledge the Fundación Ramón Areces for an institutional grant awarded to the Centro de Biología Molecular "Severo Ochoa."

References

Almela, M.J., Gonzalez, M.E., and Carrasco, L. (1991). Inhibitors of poliovirus uncoating efficiently block the early membrane permeabilization induced by virus particles. *J. Virol.* **65**, 2572–2577.

Arroyo, J., Boceta, M., Gonzalez, M.E., Michel, M., and Carrasco, L. (1995). Membrane permeabilization by different regions of the human immunodeficiency virus type 1 transmembrane glycoprotein gp41. *J. Virol.* **69**, 4095–4102.

Barco, A. and Carrasco, L. (1995). A human virus protein, poliovirus protein 2BC, induces membrane proliferation and blocks the exocytic pathway in the yeast *Saccharomyces cerevisiae*. *EMBO J.* **14**, 3349–3364.

Barco, A. and Carrasco, L. (1998). Identification of regions of poliovirus 2BC protein that are involved in cytotoxicity. *J. Virol.* **72**, 3560–3570.

Benedetto, A., Rossi, G.B., Amici, C., Belardelli, F., Cioe, L., Carruba, G. *et al.* (1980). Inhibition of animal virus production by means of translation inhibitors unable to penetrate normal cells. *Virology* **106**, 123–132.

Betakova, T., Wolffe, E.J., and Moss, B. (2000). The vaccinia virus A14.5L gene encodes a hydrophobic 53-amino-acid virion membrane protein that enhances virulence in mice and is conserved among vertebrate poxviruses. *J. Virol.* **74**, 4085–4092.

Blanco, R., Carrasco, L., and Ventoso, I. (2003). Cell killing by HIV-1 protease. *J. Biol. Chem.* **278**, 1086–1093.

Bour, S. and Strebel, K. (1996). The human immunodeficiency virus (HIV) type 2 envelope protein is a functional complement to HIV type 1 Vpu that enhances particle release of heterologous retroviruses. *J. Virol.* **70**, 8285–8300.

Carrasco, L. (1978). Membrane leakiness after viral infection and a new approach to the development of antiviral agents. *Nature* **272**, 694–699.

Carrasco, L. (1994). Entry of animal viruses and macromolecules into cells. *FEBS Lett.* **350**, 151–154.

Carrasco, L. (1995). Modification of membrane permeability by animal viruses. *Adv. Virus Res.* **45**, 61–112.

Chan, Y.L., Endo, Y., and Wool, I.G. (1983). The sequence of the nucleotides at the alpha-sarcin cleavage site in rat 28S ribosomal ribonucleic acid. *J. Biol. Chem.* **258**, 12768–12770.

Chang, Y.S., Liao, C.L., Tsao, C.H., Chen, M.C., Liu, C.I., Chen, L.K. *et al.* (1999). Membrane permeabilization by small hydrophobic nonstructural proteins of Japanese encephalitis virus. *J. Virol.* **73**, 6257–6264.

Charpilienne, A., Abad, M.J., Michelangeli, F., Alvarado, F., Vasseur, M., Cohen, J. *et al.* (1997). Solubilized and cleaved VP7, the outer glycoprotein of rotavirus, induces permeabilization of cell membrane vesicles. *J. Gen. Virol.* **78**, 1367–1371.

Chernomordik, L., Chanturiya, A.N., Suss-Toby, E., Nora, E., and Zimmerberg, J. (1994). An amphipathic peptide from the C-terminal region of the human immunodeficiency virus envelope glycoprotein causes pore formation in membranes. *J. Virol.* **68**, 7115–7123.

Ciccaglione, A.R., Costantino, A., Marcantonio, C., Equestre, M., Geraci, A., and Rapicetta, M. (2001). Mutagenesis of hepatitis C virus E1 protein affects its membrane-permeabilizing activity. *J. Gen. Virol.* **82**, 2243–2250.

Ciccaglione, A.R., Marcantonio, C., Costantino, A., Equestre, M., Geraci, A., and Rapicetta, M. (1998). Hepatitis C virus E1 protein induces modification of membrane permeability in *E. coli* cells. *Virology* **250**, 1–8.

Comardelle, A.M., Norris, C.H., Plymale, D.R., Gatti, P.J., Choi, B., Fermin, C.D. *et al.* (1997). A synthetic peptide corresponding to the carboxy terminus of human immunodeficiency virus type 1 transmembrane glycoprotein induces alterations in the ionic permeability of *Xenopus laevis* oocytes. *AIDS Res. Hum. Retroviruses* **13**, 1525–1532.

Contreras, A., Vazquez, D., and Carrasco, L. (1978). Inhibition, by selected antibiotics, of protein synthesis in cells growing in tissue cultures. *J. Antibiot. (Tokyo)* **31**, 598–602.

Cooley, L.A. and Lewin, S.R. (2003). HIV-1 cell entry and advances in viral entry inhibitor therapy. *J. Clin. Virol.* **26**, 121–132.

Cotten, M., Wagner, E., Zatloukal, K., Phillips, S., Curiel, D.T., and Birnstiel, M.L. (1992). High-efficiency receptor-mediated delivery of small and large (48 kilobase) gene constructs using the endosome-disruption activity of defective or chemically inactivated adenovirus particles. *Proc. Natl. Acad. Sci. USA* **89**, 6094–6098.

Cuadras, M.A., Arias, C.F., and Lopez, S. (1997). Rotaviruses induce an early membrane permeabilization of MA104 cells and do not require a low intracellular Ca_2^+ concentration to initiate their replication cycle. *J. Virol.* **71**, 9065–9074.

De Clercq, E. (2001). Antiviral drugs: Current state of the art. *J. Clin. Virol.* **22**, 73–89.

del Castillo, J.R., Ludert, J.E., Sanchez, A., Ruiz, M.C., Michelangeli, F., and Liprandi, F. (1991). Rotavirus infection alters Na^+ and K^+ homeostasis in MA-104 cells. *J. Gen. Virol.* **72**, 541–547.

Dong, Y., Zeng, C.Q., Ball, J.M., Estes, M.K., and Morris, A.P. (1997). The rotavirus enterotoxin NSP4 mobilizes intracellular calcium in human intestinal cells by stimulating phospholipase C-mediated inositol 1,4,5-trisphosphate production. *Proc. Natl. Acad. Sci. USA* **94**, 3960–3965.

Durantel, D., Branza-Nichita, N., Carrouee-Durantel, S., Butters, T.D., Dwek, R.A., and Zitzmann, N. (2001). Study of the mechanism of antiviral action of iminosugar derivatives against bovine viral diarrhea virus. *J. Virol.* **75**, 8987–8998.

Elfgang, C., Eckert, R., Lichtenberg-Frate, H., Butterweck, A., Traub, O., Klein, R.A. *et al.* (1995). Specific permeability and selective formation of gap junction channels in connexin-transfected HeLa cells. *J. Cell Biol.* **129**, 805–817.

Ewart, G.D., Mills, K., Cox, G.B., and Gage, P.W. (2002). Amiloride derivatives block ion channel activity and enhancement of virus-like particle budding caused by HIV-1 protein Vpu. *Eur. Biophys. J.* **31**, 26–35.

Fernandez-Puentes, C. and Carrasco, L. (1980). Viral infection permeabilizes mammalian cells to protein toxins. *Cell* **20**, 769–775.

Fischer, W.B. and Sansom, M.S. (2002). Viral ion channels: Sructure and function. *Biochim. Biophys. Acta* **1561**, 27–45.

Gatti, P.J., Choi, B., Haislip, A.M., Fermin, C.D., and Garry, R.F. (1998). Inhibition of HIV type 1 production by hygromycin B. *AIDS Res. Hum. Retroviruses* **14**, 885–892.

Gonzalez, M.E. and Carrasco, L. (1998). The human immunodeficiency virus type 1 Vpu protein enhances membrane permeability. *Biochemistry* **37**, 13710–13719.

Gonzalez, M.E. and Carrasco, L. (2001). Human immunodeficiency virus type 1 VPU protein affects Sindbis virus glycoprotein processing and enhances membrane permeabilization. *Virology* **279**, 201–209.

Gonzalez, M.E. and Carrasco, L. (2003). Viroporins. *FEBS Lett.* **552**, 28–34.

Griffin, S.D., Beales, L.P., Clarke, D.S., Worsfold, O., Evans, S.D., Jaeger, J. *et al.* (2003). The p7 protein of hepatitis C virus forms an ion channel that is blocked by the antiviral drug, Amantadine. *FEBS Lett.* **535**, 34–38.

Hay, A.J. (1992). The action of adamantanamines against influenza A viruses: Inhibition of the M2 ion channel protein. *Semin. Virol.* **3**, 21–30.

Irurzun, A., Nieva, J.L., and Carrasco, L. (1997). Entry of Semliki forest virus into cells: Effects of concanamycin A and nigericin on viral membrane fusion and infection. *Virology* **227**, 488–492.

Klimkait, T., Strebel, K., Hoggan, M.D., Martin, M.A., and Orenstein, J.M. (1990). The human immunodeficiency virus type 1-specific protein vpu is required for efficient virus maturation and release. *J. Virol.* **64**, 621–629.

Kuo, L. and Masters, P.S. (2003). The small envelope protein E is not essential for murine coronavirus replication. *J. Virol.* **77**, 4597–4608.

Lacal, J.C., Vazquez, J.M., Fernandez-Sousa, D., and Carrasco, L. (1980). Antibiotics that specifically block translation in virus-infected cells. *J. Antibiot. (Tokyo)* **33**, 441–446.

Lama, J. and Carrasco, L. (1992). Expression of poliovirus nonstructural proteins in *Escherichia coli* cells. Modification of membrane permeability induced by 2B and 3A. *J. Biol. Chem.* **267**, 15932–15937.

Lee, T., Crowell, M., Shearer, M.H., Aron, G.M., and Irvin, J.D. (1990). Poliovirus-mediated entry of pokeweed antiviral protein. *Antimicrob. Agents Chemother.* **34**, 2034–2037.

Liljestrom, P., Lusa, S., Huylebroeck, D., and Garoff, H. (1991). *In vitro* mutagenesis of a full-length cDNA clone of Semliki Forest virus: The small 6,000-molecular-weight membrane protein modulates virus release. *J. Virol.* **65**, 4107–4113.

Liprandi, F., Moros, Z., Gerder, M., Ludert, J.E., Pujol, F.H., Ruiz, M.C. *et al.* (1997). Productive penetration of rotavirus in cultured cells induces co-entry of the translation inhibitor alpha-sarcin. *Virology* **237**, 430–438.

Loewy, A., Smyth, J., von Bonsdorff, C.H., Liljestrom, P., and Schlesinger, M.J. (1995). The 6-kilodalton membrane protein of Semliki Forest virus is involved in the budding process. *J. Virol.* **69**, 469–475.

Maidji, E., Tugizov, S., Jones, T., Zheng, Z., and Pereira, L. (1996). Accessory human cytomegalovirus glycoprotein US9 in the unique short component of the viral genome promotes cell-to-cell transmission of virus in polarized epithelial cells. *J. Virol.* **70**, 8402–8410.

Michelangeli, F., Ruiz, M.C., del Castillo, J.R., Ludert, J.E., and Liprandi, F. (1991). Effect of rotavirus infection on intracellular calcium homeostasis in cultured cells. *Virology* **181**, 520–527.

Newton, K., Meyer, J.C., Bellamy, A.R., and Taylor, J.A. (1997). Rotavirus nonstructural glycoprotein NSP4 alters plasma membrane permeability in mammalian cells. *J. Virol.* **71**, 9458–9465.

Nyfeler, S., Senn, K., and Kempf, C. (2001). Expression of Semliki Forest virus E1 protein in *Escherichia coli*. Low pH-induced pore formation. *J. Biol. Chem.* **276**, 15453–15457.

Oka, T., Natori, Y., Tanaka, S., Tsurugi, K., and Endo, Y. (1990). Complete nucleotide sequence of cDNA for the cytotoxin alpha sarcin. *Nucleic Acids Res.* **18**, 1897.

Otero, M.J. and Carrasco, L. (1987). Proteins are cointernalized with virion particles during early infection. *Virology* **160**, 75–80.

Pavlovic, D., Neville, D.C., Argaud, O., Blumberg, B., Dwek, R.A., Fischer, W.B. *et al.* (2003). The hepatitis C virus p7 protein forms an ion channel that is inhibited by long-alkyl-chain iminosugar derivatives. *Proc. Natl. Acad. Sci. USA* **100**, 6104–6108.

Sanderson, C.M., Parkinson, J.E., Hollinshead, M., and Smith, G.L. (1996). Overexpression of the vaccinia virus A38L integral membrane protein promotes Ca_2^+ influx into infected cells. *J. Virol.* **70**, 905–914.

Sanz, M.A., Madan, V., Carrasco, L., and Madan, J.L. (2003). Interfacial domains in Sindbis virus 6K protein. Detection and functional characterization. *J. Biol. Chem.* **278**, 2051–2057.

Skehel, J.J. and Wiley, D.C. (1998). Coiled coils in both intracellular vesicle and viral membrane fusion. *Cell* **95**, 871–874.

Suarez, T., Gallaher, W.R., Agirre, A., Goni, F.M., and Nieva, J.L. (2000). Membrane interface-interacting sequences within the ectodomain of the human immunodeficiency virus type 1 envelope glycoprotein: Putative role during viral fusion. *J Virol.* **74**, 8038–8047.

Tian, P., Hu, Y., Schilling, W.P., Lindsay, D.A., Eiden, J., and Estes, M.K. (1994). The nonstructural glycoprotein of rotavirus affects intracellular calcium levels. *J. Virol.* **68**, 251–257.

Voss, T.G., Fermin, C.D., Levy, J.A., Vigh, S., Choi, B., and Garry, R.F. (1996). Alteration of intracellular potassium and sodium concentrations correlates with induction of cytopathic effects by human immunodeficiency virus. *J. Virol.* **70**, 5447–5454.

Watanabe, T., Watanabe, S., Ito, H., Kida, H., and Kawaoka, Y. (2001). Influenza A virus can undergo multiple cycles of replication without M2 ion channel activity. *J. Virol.* **75**, 5656–5662.

Wengler, G., Koschinski, A., Wengler, G., and Dreyer, F. (2003). Entry of alphaviruses at the plasma membrane converts the viral surface proteins into an ion-permeable pore that can be detected by electrophysiological analyses of whole-cell membrane currents. *J. Gen. Virol.* **84**, 173–181.

Wild, C.T., Shugars, D.C., Greenwell, T.K., McDanal, C.B., and Matthews, T.J. (1994). Peptides corresponding to a predictive alpha-helical domain of human immunodeficiency virus type 1 gp41 are potent inhibitors of virus infection. *Proc. Natl. Acad. Sci. USA* **91**, 9770–9774.

Yang, Z.Y., Duckers, H.J., Sullivan, N.J., Sanchez, A., Nabel, E.G., and Nabel, G.J. (2000). Identification of the Ebola virus glycoprotein as the main viral determinant of vascular cell cytotoxicity and injury. *Nat. Med.* **6**, 886–889.

Zhang, H., Dornadula, G., Alur, P., Laughlin, M.A., and Pomerantz, R.J. (1996). Amphipathic domains in the C terminus of the TM protein (gp41) permeabilize HIV-1 virions: A molecular mechanism underlying natural endogenous reverse transcription. *Proc. Natl. Acad. Sci. USA* **93**, 12519–12524.

7

FTIR Studies of Viral Ion Channels

Itamar Kass and Isaiah T. Arkin

1. Introduction

Obtaining high-resolution structures of membrane proteins using solution NMR spectroscopy or diffraction methods has proven to be a difficult task. As a result, only a few dozen structures of membrane proteins can be found in the PDB (Berman *et al.*, 2000). Any structural insight obtained from other biophysical methods, therefore is particularly useful. One such method is Fourier Transform Infrared Spectroscopy (FTIR) that can provide useful structural information on membrane proteins in general, and viral ion channels in particular. In this chapter, we will begin by briefly describing the principles of infrared spectroscopy in general and site-specific infrared dichroism (SSID) in particular. We will then discuss in detail a few examples in which SSID has been applied to the study of viral ion channels. Finally, we discuss future directions and possibilities in which infrared spectroscopy can enhance our understanding of this important family of proteins.

1.1. Principles of Infrared Spectroscopy

Infrared spectroscopy is the measurement of the wavelength and intensity of absorption of infrared light (*c.*200–5,000 cm^{-1}) due to molecular vibrations. Not all possible vibrations within a molecule will interact with electromagnetic waves in the infrared region. In order to interact with infrared light, the vibration must result in a change of the molecular dipole moment during the vibration. Heteronuclear diatomic molecules, such as carbon monoxide or hydrogen chloride, which possess a permanent dipole moment, have infrared activity because stretching of this bond leads to a change in the dipole moment. In contrast, homonuclear diatomic molecules such as dihydrogen or dioxygen have no infrared absorption, as these molecules have zero dipole moment and vibrations of the bonds will not produce one. It is important to note that it is not necessary for a compound to have a permanent dipole moment to interact with infrared light. As an example, centrosymmetric linear molecules such as carbon dioxide do not have a permanent dipole moment. This means that the symmetric stretch will not be infrared active, because no dipole moment is generated. However, in the case of the asymmetric stretch, a dipole moment will be produced, resulting in infrared activity.

Itamar Kass and Isaiah T. Arkin • The Alexander Silberman Institute of Life Science, Department of Biological Chemistry, The Hebrew University, Givat-Ram, Jerusalem 91904, Israel.

Viral Membrane Proteins: Structure, Function, and Drug Design, edited by Wolfgang Fischer.
Kluwer Academic / Plenum Publishers, New York, 2005.

1.1.1. Amide Group Vibrations

Nine vibrational bands can be characterized as pertaining to the peptide group, named amide A, B, I, II–VII (Krimm, 1983). Amide I and amide II bands are the two major vibrations of the protein infrared spectrum. The amide I band (between 1,600 cm^{-1} and 1,700 cm^{-1}) is mainly associated with C=O group stretching and is directly influenced by the backbone conformation (see below). The amide II mode is derived mainly from the N–H bending vibration and from the C–N stretching vibration (18–40%), and is also conformationally sensitive.

1.1.2. Secondary Structure

Protein infrared spectra are greatly affected by the local microenvironment of the different secondary groups. As a result, infrared spectra can be used to identify secondary structures (Susi and Byler, 1986). A large number of synthetic polypeptides, with a defined secondary structure content, has been used for the measurement and characterization of infrared spectra for secondary structure elements. For example, polylysine which adopts β-sheet or α-helical conformation depending on temperature and pH of the solution. Experimental and theoretical work on a large number of synthetic polypeptides have provided insights into the variability of the frequencies for particular secondary structure conformations (Krimm and Bandekar, 1986).

For α-helical structures, the mean wavenumber of the amide I mode is found to be *c.*1,652 cm^{-1}, whereas the wavenumber of the amide II mode is found to be around 1,548 cm^{-1} (Chirgadze and Nevskaya, 1976). The width of the peak at half of the height for the α-helix band depends on the stability of the helix. Half-widths of about 15 cm^{-1} correspond to an α-helix to random coil free energy of transition of more than 300 cal/mol. Half-widths of 38 cm^{-1} correspond to an α-helix to random coil free energies of transition of about 90 cal/mol (Nevskaya and Chirgadze, 1976; Arrondo *et al.*, 1993).

Spectrum from synthetic polypeptides with an antiparallel β-chain conformation, have been collected (Chirgadze and Nevskaya, 1976). From the data, it follows that, the amide I absorption is primarily determined by the backbone conformation and is independent of the amino acid sequence, the amino acid hydrophilic, and the hydrophobic properties and charges. The average wavenumber of the main absorption is about 1,629 cm^{-1} with a minimum of 1,615 cm^{-1} and a maximum of 1,637 cm^{-1}. The average wavenumber value for the second absorption is 1,696 cm^{-1} (lowest value 1,685 cm^{-1}). Parallel β-sheet structure results in an amide I absorption near 1,640 cm^{-1}.

The β turn structure involves four amino acid residues which form an i to $i + 3$ hydrogen bond. A number of turn structures have been identified from protein structures: type I (42%, nonhelical), type II (15%, nonhelical, requires Gly in position 3), and type III (18%, corresponds to one turn of 3_{10}-helix). It was found that strong overlapping of the different types of turns with the α-helical absorption occurs in the infrared spectrum. However, an absorption near 1,680 cm^{-1} is now clearly assigned to β turns.

2. SSID FTIR

In this part, the theory of SSID and its application to the determination of rotational and orientational constraints for oriented transmembrane helices is presented. Infrared spectroscopy dichroism measurements of single amide I vibrations corresponding to $^{13}C{=}^{18}O$ (Torres *et al.*, 2001) or the symmetric, *ss*, and antisymmetric, *as*, stretching modes of the Gly CD_2 (Torres *et al.*, 2000) or Ala CD_3 (Torres and Arkin, 2002) contain information about the helix tilt and rotation angles. This information can be extracted by analysis of the dichroism of a set of labeled sites along the peptide sequence, and used as an input to molecular dynamics simulations (Kukol *et al.*, 1999). The approach provides rotational and orientational constraints about selectively labeled peptides even under conditions of modest fractional sample order.

The orientation of a transition dipole moment of a single vibrational mode in an α-helix, represented as vector $\vec{P}$ in Figure 7.1, can be expressed as a function of several parameters: (a) the helix tilt angle β, (b) the angle α relating the helix axis and the vibrational mode [in the case of the amide I mode it is measured to be 141° (Tsuboi, 1962)], and (c) the rotation angle ω of the vibration about the helix axis. The orientation of the helix with respect to the axes can, therefore, be readily extracted given a perfect dichroic ratio $\mathcal{R}$, defined as the ratio between absorption of parallel and perpendicular polarized light, from a site-specific FTIR dichroism measurement.

Although possible in theory, calculating the molecular orientation restrains from experimental dichroic ratios is complicated. First, many vibrational modes in a peptide are located at different rotation angles. This results in overlapping of the different dichroism signals, whereby the information about the helix rotational pitch angle is lost. The second problem arises due to less than perfectly ordered sample: while a fraction of the helices are tilted from the membrane normal by the angle β, others are not, resulting in a reduction of the observed sample dichroism (due to vibrational modes located within disordered peptides).

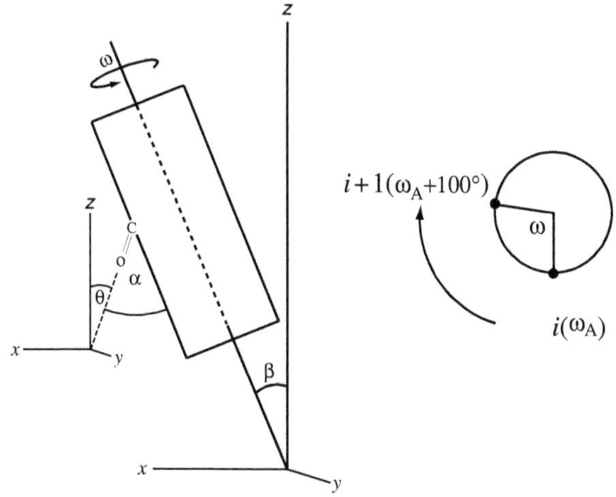

Figure 7.1. Left: Schematic of the geometric parameters that define the amide I (mostly C=O stretch) transition dipole moment orientation in a transmembrane helix. The helix tilt β and the rotational orientation ω are derived from the experimental data. The angle α is 141° (Tsuboi, 1962). Right: Schematic vertical view for the relative positions of labels i and $i + 1$ showing their distribution around the helical axis. The helical segment has been drawn as a perfect cylinder for simplicity.

The first of these problems is addressed by isotope labeling in order to shift the frequency of a specific vibrational mode so that its dichroism may be directly measured. The second problem is overcome by using more than one labeled site along the peptide backbone and the direct determination of the fractional order of each of the samples. This analysis is aided by the fact that the relative rotational pitch angles of the labeled sites are known: in α-helices, consecutive sites are related by a 100° rotation. Alternatively, one can use the fact that symmetric and antisymmetric vibrational modes of the CD_2 group in Gly CD_2 are mutually perpendicular, in order to determine two dichroic ratios, for *ss* and *as* vibrations, in a single residue (Torres *et al.*, 2000; Torres and Arkin, 2002).

2.1. Dichroic Ratio

The experimentally measured dichroic ratio $\mathcal{R}$ is defined as the ratio between the absorption of parallel and perpendicular polarized light, by a chromophore whose dipole vector is $\vec{P}$ (see Figure 7.1):

$$\mathcal{R} \equiv \frac{\mathcal{A}_{\parallel}}{\mathcal{A}_{\perp}} \qquad (1)$$

The absorption of light is equal to the squared scalar product between the electric field components $\mathcal{E}_x$, $\mathcal{E}_y$, and $\mathcal{E}_z$ and the corresponding dimensionless integrated absorption coefficients $\mathcal{K}_x$, $\mathcal{K}_y$, and $\mathcal{K}_z$ (Arkin *et al.*, 1997). In the geometrical configuration of attenuated total internal reflection shown in Figure 7.2 (Harrick, 1967), the dichroic ratio is given by:

$$\mathcal{R} = \frac{\mathcal{E}_z^2\,\mathcal{K}_z^2 + \mathcal{E}_x^2\,\mathcal{K}_x^2}{\mathcal{E}_y^2\,\mathcal{K}_y^2} \qquad (2)$$

2.2. Sample Disorder

In equation 2, sample disorder is not taken into account, in that all helices are assumed to share the same tilt angle (i.e., there is no variation in β). However, in practice this is never the case, as shown schematically in Figure 7.3. Following Fraser (Fraser, 1953), the fraction of perfectly ordered material about a particular helix tilt is denoted by *f*, and thus $1 - f$

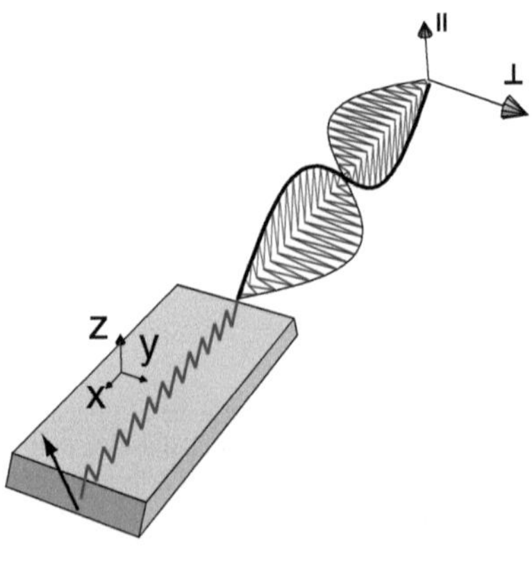

Figure 7.2. Schematic of the optical configuration of attenuated total internal reflection.

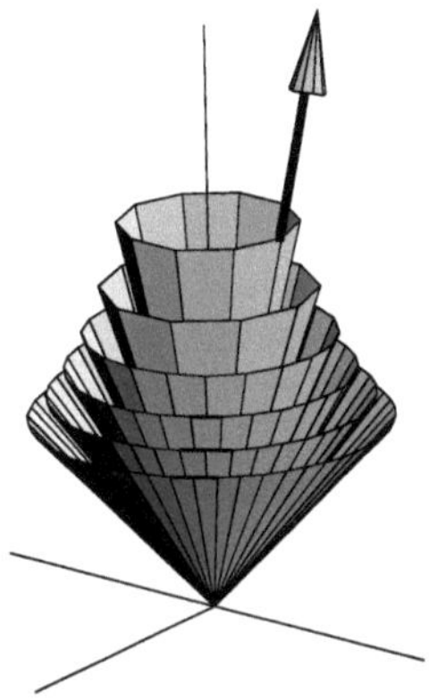

Figure 7.3. Schematic representation of the distribution of a chromophore's transition dipole moment in a uniaxial averaged system.

represents the random fraction. An appropriate and mathematically simple correction for the integrated absorption coefficients yields the desired function describing the dichroic ratio:

$$\mathcal{R} = \frac{\mathcal{E}_z^2\left(f\mathcal{K}_z^2 + \frac{1-f}{3}\right)\mathcal{E}_x^2\left(f\mathcal{K}_x^2 + \frac{1-f}{3}\right)}{\mathcal{E}_y^2\left(f\mathcal{K}_y^2 + \frac{1-f}{3}\right)} \tag{3}$$

2.3. Orientational Parameters Derivation

As stated above, in order to overcome the uncertainty of the sample's disorder we rely on the use of multiple labels, as follows. In the first sample i, a single isotopic label is inserted in a particular position. Dichroism analysis of the sample yields two measure quantities: (a) the dichroism of the helix, $\mathcal{R}_{Helix_i}$, which is due to the randomly rotated transition dipole moment of the peptidic C=O stretching modes, and is a function of the helix tilt β and the sample disorder f_i:

$$\mathcal{R}_{Helix_i} = \mathcal{F}(\beta, f_i) \tag{4}$$

(b) Similarly, the dichroism of the labeled site $\mathcal{R}_{Site_i}$ is obtained, which is a function of the helix tilt β, sample disorder f_i, and the rotational position of the chromophore about the helix axis, ω.

$$\mathcal{R}_{Site_i} = \mathcal{F}(\beta, \omega, f_i) \tag{5}$$

Thus, from a single peptide with one label two equations are obtained with three unknowns. The solution is obtained upon using another peptide, which is identical to the first in all aspects except for the position of the isotopic label, which in this case is adjacent to the first label (i.e., $\omega_{j} = \omega_i + 100°$).

$$\mathcal{R}_{Helix_j} = \mathcal{F}(\beta, f_j) \tag{6}$$

$$\mathcal{R}_{Site_j} = \mathcal{F}(\beta, \omega + 100°, f_j) \tag{7}$$

Taken together, two additional measurables are obtained, leading in total to four coupled equations with four unknowns: the helix tilt angle, the rotational position of the first labeled site, and the two sample disorders.

2.4. Data Utilization

The strength of the orientational restraints is through the objective incorporation in energy refinement procedures. Explicitly, the angle measured experimentally by SSID for different C=O bonds in the helical bundle can be restrained using a harmonic energy penalty function during the modeling process (Kukol *et al.*, 1999). One such modeling process is global searching molecular dynamics in which a large number of starting structures are used to search the conformational space of a transmembrane α-helical bundle (Adams *et al.*, 1995).

3. Examples

3.1. M2 H^+ Channel from Influenza A Virus

The first model for transmembrane segments of a virus channel, based on SSID FTIR method was determined for the M2 H^+ channel from Influenza A virus (Kukol *et al.*, 1999). M2 is a homotetrameric membrane protein, composed of 97 amino acids (Lamb *et al.*, 1985; Holsinger *et al.*, 1995). In a landmark study Pinto, Lamb, and coworkers (Pinto *et al.*, 1992) discovered that the M2 protein has ion channel activity.

FTIR measurements of peptides encompassing the transmembrane domain of M2 labeled at residue Ala29 or residue Ala30 with ^{13}C (Kukol *et al.*, 1999), are shown in Figure 7.4. The values obtained were $\beta = (31.6 \pm 6.2)^\circ$ and $\omega_{Ala29} = (-59.8 \pm 9.9)^\circ$ whereby ω is defined as zero for a residue located in the direction of the helix tilt. Note that ω_{Ala30} is simply $\omega_{Ala29} + 100^\circ$. The results obtained were in excellent agreement with those obtained from solid state NMR spectroscopy (Kovacs and Cross, 1997).

As shown in Figure 7.4, the helices in M2 are significantly tilted from the membrane normal and the Trp and His residues implicated in channel gating (Tang *et al.*, 2002) and pH activation (Wang *et al.*, 1995), respectively, are located in the channel pore.

3.2. *vpu* Channel from HIV

A model for the transmembrane domain of *vpu* viral ion channel from HIV-1, was determined using double labeled residues (Kukol and Arkin, 1999). The 81-residue *vpu* protein belongs to the auxiliary proteins of the human immunodeficiency virus type 1 (HIV-1) (Cohen *et al.*, 1988; Strebel *et al.*, 1988). *vpu* is responsible for: (a) the degradation of one of the HIV-1 coreceptor molecules, CD4 (Schubert *et al.*, 1996; Schubert and Strebel, 1994), allowing the *env* glycoprotein to be transported to the cell surface; (b) virus particle release (Schubert *et al.*, 1996). The molecular basis of these actions is unknown. The transmembrane region of *vpu* is composed of 22 residues that span the membrane (Maldarelli *et al.*, 1993). *vpu* forms homo-oligomers of at least four subunits as detected by gel electrophoresis (Maldarelli *et al.*, 1993). By analogy with the M2 protein of Influenza A virus (Lamb and Pinto, 1997), it has been suggested that the transmembrane domain of *vpu* may act as an ion channel (Strebel *et al.*, 1989; Klimkait *et al.*, 1990; Maldarelli *et al.*, 1993; Schubert and

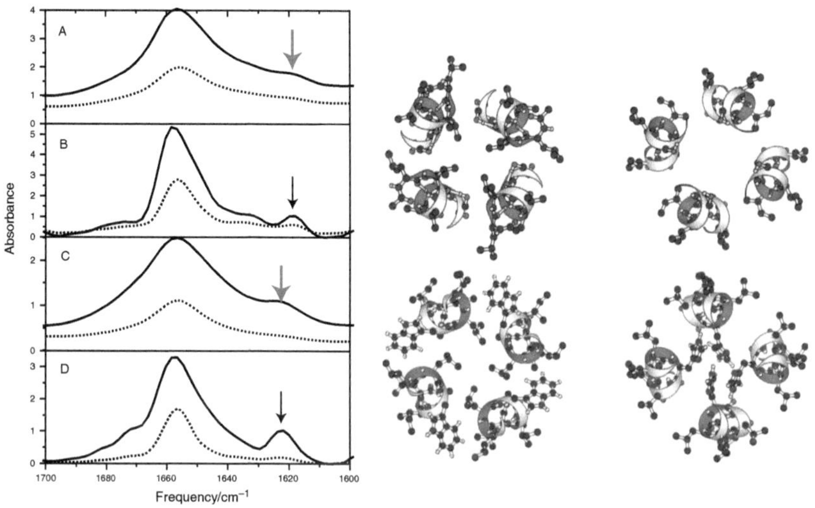

Figure 7.4. Left: ATR-FTIR spectra of the amide I region obtained with parallel (solid line) and perpendicular (dotted line) polarized light the M2 transmembrane domain peptide in lipid vesicles. Panels A and C depict the spectra of 1-^{13}C-labeled Ala29 and Ala30 peptides, respectively. Panels B and D show the Fourier-self deconvoluted spectra of A and C. The position of the arrows indicate the isotope edited peaks in the original (broad arrows) and Fourier-self deconvoluted spectra (narrow arrows). Right: Consecutive slices showing the helix as a ribbon diagram and the conformation of the side chains in a ball and stick representation. Top left, Leu26-Ser31; top right, Ser31-Ile35. Bottom left, Ile35-Ile39; bottom right, Ile39-Leu43. The figure was created with MOLSCRIPT (Kraulis, 1991).

Strebel, 1994). Recently it has been shown that the channel activity of *vpu* is blocked by amiloride derivatives (Ewart *et al.*, 2002). Furthermore, amiloride derivatives inhibited the virus release activity of *vpu*, thereby linking its channel activity with its native function.

In the structural analysis of M2, the main problem that arose from using site-specific labels with ^{13}C, was the spectal overlap of the signal arising from the isotopic labeled residue and the ^{12}C natural abundance. To enhance the ^{13}C amide I mode intensity, two labels were introduced into the peptide, at positions i and $i + 7$. These two labels have approximately the same rotational pitch angle $\omega \approx \omega_{i+7}$ in α-helical geometry (Pauling *et al.*, 1951). The error introduced in this assumption is compensated by the fact that the signal of ^{13}C amide I mode is enhanced.

Two labeled peptides at positions Val13/Val20 and Val14/Ala21 were synthesized and examined in order to find orientational constraints. The helix tilt was determined to be $\beta = (6.5 \pm 1.7)^\circ$, where the rotational pitch angle was calculated to be $\omega = (283 \pm 11)^\circ$ for the ^{13}C-labels Val13/Val20, and $\omega = (23 \pm 11)^\circ$ for the ^{13}C-labels Ala14/Val21. The oligomerization number of *vpu* is not known, hence, a different molecular dynamics search were used for calculating a model based on the FTIR results. The molecular dynamics search was done for tetrameric, pentameric, and hexameric homo-oligomers, as the orientational data are independent of the helix oligomerization. The combined spectroscopic and molecular approach yields, in this case, a single pentameric model of the transmembrane domain of the protein, as shown in Figure 7.5.

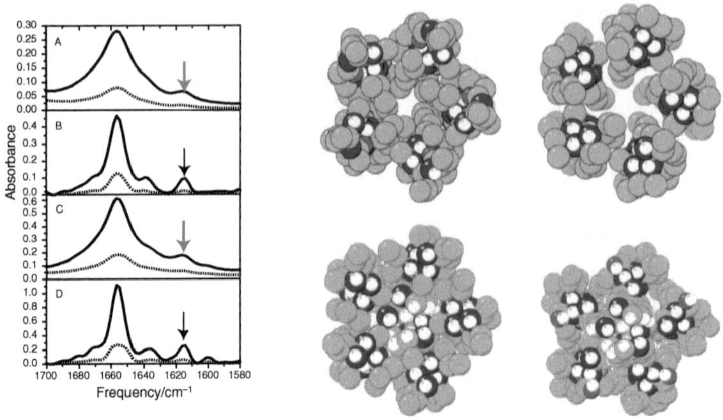

Figure 7.5. Left: ATR-FTIR spectra of the amide I region obtained with parallel (solid line) and perpendicular (dotted line) polarized light of *vpu* transmembrane domain peptide in lipid vesicles. Panels A and C depict the spectra of 1-^{13}C-labeled Val13/Val20 and Val14/Ala21 peptides, respectively. Panels B and D show the Fourier-self deconvoluted spectra of A and C. The position of the arrows indicate the isotope edited peaks in the original (broad arrows) and Fourier-self deconvoluted spectra (narrow arrows). Right: Consecutive slices of the *vpu* structure in a CPK respresentation. Top left, Ile6-Val12; top right, Val12-Ile17. Bottom left, Ile17-Trp22; bottom right Trp22-Ile27. The figure was created with MOLSCRIPT (Kraulis, 1991).

3.3. CM2 from Influenza C Virus

The transmembrane domain of CM2 from Influenza C virus (Hongo *et al.*, 1994), composed of 23 residues, was also subject to SSID analysis (Kukol and Arkin, 2000). CM2 protein of Influenza C virus (Hongo *et al.*, 1994) has been characterized as an integral membrane glycoprotein, which forms disulfide linked dimers and tetramers. CM2 is assumed to be structurally similar to the Influenza A M2 protein and the Influenza B NB protein (Hongo *et al.*, 1997; Pekosz and Lamb, 1997). The transmembrane domains of M2 and NB both form ion channels in lipid membranes (Duff and Ashley, 1992; Sunstrom *et al.*, 1996). Based upon two pairs of labeled residues, Gly59/Leu66 and Gly61/Leu68, the helix tilt was determined to be $\beta = (14.6 \pm 3.0)°$, whereas the rotational pitch angle was calculated to be $\omega = (218 \pm 17)°$ for the ^{13}C-labels Gly59/Leu66 (Figure 7.6).

4. Future Directions

Although ^{13}C labels have been shown to relay site-specific secondary structure and orientational information, the use of this label is limited. The reason for this limitation is the high natural abundance of ^{13}C and the lack of baseline resolution between the main amide I band and the isotope-edited peak. To overcome these problems, the use of new labels, 1-$^{13}C{=}^{18}O$ (Torres *et al.*, 2000, 2001), Gly CD_2 (Torres *et al.*, 2000), and Ala CD_3 (Torres & Arkin, 2002), were found to be valuable. The double-isotope label 1-$^{13}C{=}^{18}O$, virtually eliminates any contribution from natural abundance of ^{13}C. More importantly, the isotope-edited peak is further red-shifted compated with that of 1-^{13}C and is completely baseline resolved from the main amide I band. Recent work on the structure of transmembrane domain of phospholamban (Torres *et al.*, 2000) have illustrated the utility of the new isotopic label.

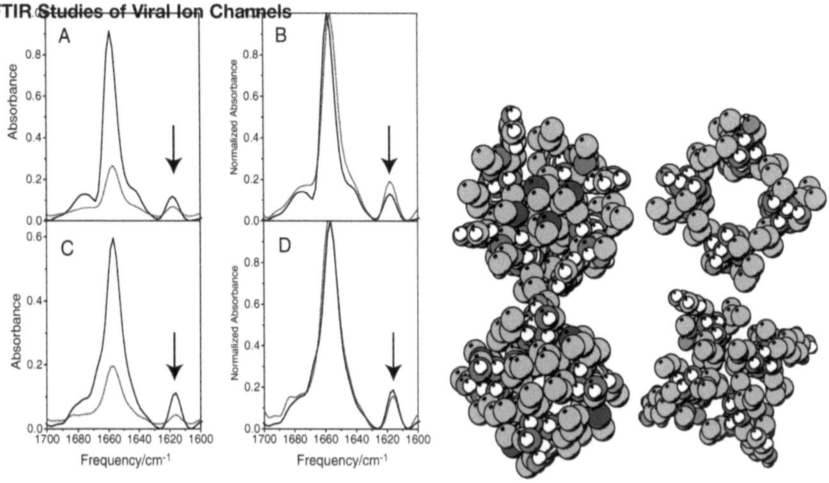

Figure 7.6. Left: Attenuated total reflection deconvoluted FTIR-spectra of the lipid vesicle reconstituted CM2 peptide ^{13}C-labeled at Gly59/Leu66 (A,B) or Gly61/Leu68 (C,D) obtained with parallel (solid line) or perpendicular (dotted line) polarized light. In panels B and D, the spectra of panels A and C are shown whereby the absorbance is normalized to the amide I maximal intensity in order to show the dichroism of the ^{13}C amide I absorption band. The position of the arrows indicate the isotope edited peaks. Right: Consecutive slices of the CM2 structure in a CPK representation. From top to bottom: Tyr51-Ala56, Ala56-Gly61, Gly61-Leu66, and Leu66-Val70. The figure was created with MOLSCRIPT (Kraulis, 1991).

A different advantage arises from the use of labels such as Gly CD_2 or Ala CD_3. The symmetric and antisymmetric vibrational modes of the groups are mutually perpendicular, allowing the determination of two dichroic ratios, *ss* and *as*, in a single residue. This is important because according to the theory of site-directed dichroism, at least three dichroic ratios have to be known to determine the main structural parameters of the helix. In the case of the labels ^{13}C and 1-$^{13}C{=}^{18}O$, the dichroism of the helix and at least two labeled positions are needed, that is, two samples, each of them labeled at a different residue. In the case of (Gly CD_2), *ss* and *as* dichroic ratios can be determined using a single residue.

Finally, an additional approach taken for the T-cell receptor CD3-ζ (Torres *et al.*, 2002a, b) and the transmembrane domain of the trimeric MHC class II-associated invariant chain (Kukol *et al.*, 2002) involves the use of multiple labels (i.e., >2). In both of the above cases, a large series of peptide were studied containing around 10 consecutive labels. Thus, local helix kinks and twists can be observed due to the localized information. This approach is now being employed to refine the structure of the aforementioned viral ion channels as well as new targets.

In conclusion, the structural information obtained from SSID FTIR is capable of defining structural models for transmembrane α-helices in general and viral ion channels in particular. Improvements in the methods, including new isotopic labels and the combination of multiple labels set the stage for increased reliability and higher resolution models in the future.

Acknowledgment

This research was supported in part by a grant from the Israel Science Foundation (784/01) to ITA.

References

Adams, P.D., Arkin, I.T., Engelman, D.M., and Brunger, A.T. (1995). *Nat. Struct. Biol.* **2**(2), 154–162.
Arkin, I.T., MacKenzie, K.R., and Brunger, A.T. (1997). *J. Am. Chem. Soc.* **119**(38), 8973–8980.
Arrondo, J.L., Muga, A., Castresana, J., and Goni, F.M. (1993). *Prog. Biophys. Mol. Biol.* **59**(1), 23–56.
Berman, H.M., Westbrook, J., Feng, Z., Gilliland, G., Bhat, T.N., Weissig, J. *et al.* (2000). *Nucleic Acids Res.* **28**(1), 235–242.
Chirgadze, Y.N. and Nevskaya, N.A. (1976). *Biopolymers* **15**(4), 607–625.
Cohen, E.A., Terwilliger, E.F., Sodroski, J.G., and Haseltine, W.A. (1988). *Nature* **334**(6182), 532–534.
Duff, K.C. and Ashley, R.H. (1992). *Virology* **190**(1), 485–489.
Ewart, G.D., Mills, K., Cox, G.B., and Gage, P.W. (2002). *Eur. Biophys. J.* **31**(1), 26–35.
Fraser, R.D.B. (1953). *J. Chem. Phys.* **70**, 1511–1515.
Harrick, N. (1967). *Internal Reflection Spectroscopy*, 1st edn. Interscience Publishers, New York.
Holsinger, L.J., Shaughnessy, M.A., Micko, A., Pinto, L.H., and Lamb, R.A. (1995). *J. Virol.* **69**(2), 1219–1225.
Hongo, S., Sugawara, K., Muraki, Y., Kitame, F., and Nakamura, K. (1997). *J. Virol.* **71**(4), 2786–2792.
Hongo, S., Sugawara, K., Nishimura, H., Muraki, Y., Kitame, F., and Nakamura, K. (1994). *J. Gen. Virol.* **75**(Pt 12), 3503–3510.
Klimkait, T., Strebel, K., Hoggan, M.D., Martin, M.A., and Orenstein, J.M. (1990). *J.Virol.* **64**(2), 621–629.
Kovacs, F.A. and Cross, T.A. (1997). *Biophys. J.* **73**(5), 2511–2517.
Kraulis, P.J. (1991). *J. Appl. Cryst.* **24**, 946–950.
Krimm, S. (1983). *Biopolymers* **22**(1), 217–225.
Krimm, S. and Bandekar, J. (1986). *Adv. Protein Chem.* **38**, 181–364.
Kukol, A., Adams, P.D., Rice, L.M., Brunger, A.T., and Arkin, I.T. (1999). *J. Mol. Biol.* **288**(3), 951–962.
Kukol, A. and Arkin, I.T. (1999). *Biophys. J*, **77**(3), 1594–1601.
Kukol, A. and Arkin, I.T. (2000). *J. Biol. Chem.* **275**(6), 4225–4229.
Kukol, A., Torres, J., and Arkin, I.T. (2002). *J. Mol. Biol.* **320**(5), 1109–1117.
Lamb, R.A. and Pinto, L.H. (1997). *Virology* **229**(1), 1–11.
Lamb, R.A., Zebedee, S.L., and Richardson, C.D. (1985). *Cell* **40**(3), 627–633.
Maldarelli, F., Chen, M.Y., Willey, R.L., and Strebel, K. (1993). *J. Virol.* **67**(8), 5056–5061.
Nevskaya, N.A. and Chirgadze, Y.N. (1976). *Biopolymers* **15**(4), 637–648.
Pauling, L., Corey, R.B., and Branson, H.R. (1951). *Proc. Natl. Acad. Sci. USA* **37**(), 205–211.
Pekosz, A., and Lamb, R.A. (1997). *Virology* **237**(2), 439–451.
Pinto, L.H., Holsinger, L.J., and Lamb, R.A. (1992). *Cell* **69**(3), 517–528.
Schubert, U., Bour, S., Ferrer-Montiel, A.V., Montal, M., Maldarell, F., and Strebel, K. (1996). *J. Virol.* **70**(2), 809–819.
Schubert, U. and Strebel, K. (1994). *J. Virol.* **68**(4), 2260–2271.
Strebel, K., Klimkait, T., Maldarelli, F., and Martin, M.A. (1989). *J. Virol.* **63**(9), 3784–3791.
Strebel, K., Klimkait, T., and Martin, M.A. (1988). *Science* **241**(4870), 1221–1223.
Sunstrom, N.A., Premkumar, L.S., Premkumar, A., Ewart, G., Cox, G.B., and Gage, P.W. (1996). *J. Membr. Biol.* **150**(2), 127–132.
Susi, H. and Byler, D.M. (1986). *Meth. Enzymol.* **130**, 290–311.
Tang, Y., Zaitseva, F., Lamb, R.A., and Pinto, L.H. (2002). *J. Biol. Chem.* **277**(42), 39880–39886.
Torres, J., Adams, P.D., and Arkin, I.T. (2002). *J. Mol. Biol.* **300**(4), 677–685.
Torres, J. and Arkin, I.T. (2002). *Biophys. J.* **82**(2), 1068–1075.
Torres, J., Briggs, J.A., and Arkin, I.T. (2002a). *J. Mol. Biol.* **316**(2), 365–374.
Torres, J., Briggs, J.A., and Arkin, I.T. (2002b). *J. Mol. Biol.* **316**(2), 375–384.
Torres, J., Kukol, A., and Arkin, I.T. (2000). *Biophys. J.* **79**(6), 3139–3143.
Torres, J., Kukol, A., Goodman, J.M., and Arkin, I.T. (2001). *Biopolymers* **59**(6), 396–401.
Tsuboi, M. (1962). *J. Polym. Sci.* **59**, 139–153.
Wang, C., Lamb, R.A., and Pinto, L.H. (1995). *Biophys. J.* **69**(4), 1363–1371.

8

The M2 Proteins of Influenza A and B Viruses are Single-Pass Proton Channels

Yajun Tang, Padmavati Venkataraman, Jared Knopman, Robert A. Lamb, and Lawrence H. Pinto

This chapter summarizes and evaluates the evidence that the M2 proteins of influenza A and B viruses possess intrinsic ion channel activity that is essential to the life cycle of the virus. Both of these proteins are homotetramers with fewer than 100 residues per subunit, an N-terminal ectodomain and a single transmembrane (TM) domain. There is little amino acid homology between the two proteins other than a H–X–X–X–W motif in the TM domain. The proton selectivity of the A/M2 ion channel depends on the presence of a TM domain histidine residue, which serves as a "selectivity filter." It is possible that this filter functions by binding protons and subsequently releasing them to the opposite side of the TM pore. Protons normally flow from the acidic medium bathing the ectodomain of the protein through the TM pore and then into the virion interior. Protons are prevented from flowing through the pore of the A/M2 channel in the opposite direction by a TM tryptophan residue, which serves as the activation "gate" that is closed if the medium bathing the ectodomain is neutral or alkaline. Thus, protons serve as both the activating stimulus and the conducted ion for the A/M2 channel. The BM2 protein has very little amino acid homology to the A/M2 protein, but it has ion channel activity that depends on TM histidine and tryptophan residues. Together, these proteins constitute the first-known members of the single-pass proton channel family.

1. Introduction

It is difficult to demonstrate the intrinsic ion channel activity of proteins that are encoded by viruses because they often encode proteins for which there is no known cellular

---

Yajun Tang, Padmavati Venkataraman, Jared Knopman, and Lawrence H. Pinto • Department of Neurobiology and Physiology, Northwestern University, Evanston, Illinois. **Robert A. Lamb** • Department of Biochemistry, Molecular Biology and Cell Biology, and Howard Hughes Medical Institute, Northwestern University, Evanston, Illinois.

Viral Membrane Proteins: Structure, Function, and Drug Design, edited by Wolfgang Fischer.
Kluwer Academic / Plenum Publishers, New York, 2005.

homolog. An additional complication stems from the possibilities that virally encoded protein might function either in the virus particle (virion) itself, in the infected cell, or in a manner to upregulate an endogenous cellular ion channel that is normally silent.

Electrical activity has been reported for several proteins encoded by viruses: NB of influenza B virus (Sunstrom *et al.*, 1996), VPU of HIV virus (Schubert *et al.*, 1996), A/M2 protein of influenza A virus (Pinto *et al.*, 1992), BM2 protein of influenza B virus (Mould *et al.*, 2003), and K^+ channel protein of Chlorella virus (Plugge *et al.*, 2000), and for some of these proteins the claim has been made that these proteins serve to provide ion channel activity that is needed at some stage during the life cycle of the virus. Ion channels are distinguished from pores by the criteria that ion channels are selective for a limited range of ions and remain closed until they are activated by a chemical or electrical stimulus that is specific to the ion channel. The flow of ions across an ion channel is driven by the electrochemical gradient of the conducted ion across the membrane, and neither the flux of another ion or molecule nor the hydrolysis of ATP play a direct role.

The principal experimental obstacle to demonstrating ion channel activity of virally encoded proteins is the difficulty in recording ion channel activity from the extremely small viruses or intracellular organelles that contain the presumed ion channel proteins. Even if it were possible to record from these structures or infected cells, it would still be necessary to distinguish the ion channel activity of the virally encoded protein from that of host cell proteins. Instead of recording from a virus, organelle, or host cell, the properties of presumed ion channel proteins have been studied in expression systems and these properties compared with those expected from the life cycle of the virus at the relevant stage proposed for the protein. The greater the number of correlations that can be made between the ion channel function predicted from the biological role and the actual ion channel function in an expression system, the stronger is the case that can be made for intrinsic ion channel activity. We will first review the evidence that the A/M2 protein of influenza functions as an ion channel in the life cycle of the virus and then discuss the mechanism for its ion selectivity and activation.

2. Intrinsic Activity of the A/M2 Protein of Influenza Virus

The evidence that the A/M2 protein has intrinsic ion channel activity does not come only from measurements of its ion channel activity in expression systems, but also from a detailed knowledge of the life cycle of the virus, experiments employing the antiviral drug amantadine, and measurements on mutant A/M2 proteins.

The influenza A virion is bounded by a membrane formed by budding from the plasma membrane of the infected cell and contains three integral membrane proteins, hemagglutinin (HA), neuraminidase (NA), and A/M2 (reviewed in Lamb and Krug, 2001). Influenza A virus HA binds to a sialic acid receptor on the surface membrane of the infected cell and is then endocytosed (reviewed in Lamb and Krug, 2001). While in the acidic endosome, the HA protein undergoes a conformational change to its low pH form that exposes a hydrophobic fusion peptide to the interior of the endosome, so that after fusion with the endosome, the interior of the virus is exposed to the cytoplasm of the infected cell. However, fusion alone is not sufficient for the release of the viral genetic material (uncoating). Experiments with detergent-lysed virions have shown that the ribonuclear proteins (RNPs) are only released from the matrix (M) protein at low pH (Zhirnov, 1992). For this reason, a low pH step is needed for uncoating. Several lines of evidence implicated the A/M2 protein to be responsible for this acidification and led to the postulate that the A/M2 protein might have ion channel activity (Sugrue *et al.*, 1990; Sugrue and Hay, 1991). This ion channel activity was postulated to cause acidification of the virion while it is contained within the acidic endosome and, presumably, to continue to provide acidification after the process of fusion has begun in order to maintain the low pH conditions needed to complete release of the RNPs from the M protein.

The first line of evidence implicating the A/M2 protein in the acidification of the virion came from experiments with amantadine, which inhibits the "early" uncoating step in replication; this evidence has been reviewed (Hay, 1992; Lamb *et al.*, 1994). Amantadine-resistant "escape" mutant viruses contain alterations in amino acids in the A/M2 protein TM domain (Hay *et al.*, 1985), suggesting that the inhibiting molecule and the A/M2 protein interact in some manner. The notion that the A/M2 protein has ion channel activity or is able to accomplish acidification came from experiments involving the "late" phase of the infective cycle, in which the HA protein is transported from the endoplasmic reticulum to the surface of the infected cells. It had been known that the HA protein of certain subtypes of influenza is inserted into the cell membrane in its low pH conformational form when the infected cells were exposed to micromolar concentrations of amantadine (Sugrue *et al.*, 1990), presumably due to the low pH of the trans-Golgi network (Ciampor *et al.*, 1992; Grambas *et al.*, 1992; Grambas and Hay, 1992). When the ionophore monensin, which catalyzes the exchange of Na^+ for H^+ across membranes, was added to the amantadine-treated infected cells, the HA was inserted into the plasma membrane in its high pH form, suggesting that the A/M2 protein and monensin (Sugrue *et al.*, 1990; Grambas and Hay, 1992) both act to shunt the pH gradient across the Golgi membrane.

The A/M2 ion channel protein is a homotetrameric integral membrane protein with each chain of the mature protein containing 96 amino acid residues (Lamb and Choppin, 1981; Lamb *et al.*, 1985; Zebedee *et al.*, 1985; Holsinger and Lamb, 1991; Sugrue and Hay, 1991; Panayotov and Schlesinger, 1992; Sakaguchi *et al.*, 1997; Tobler *et al.*, 1999). The coding regions for the A/M2 protein have been conserved in all known subtypes of avian, swine, equine, and human influenza A viruses, and the amino acid sequence of the A/M2 TM domain has been conserved to a greater extent than the remainder of the protein (Ito *et al.*, 1991). The TM domain consists of 19 residues, including polar residues and a histidine residue, making it possible for the A/M2 protein TM domain to form the channel pore, a proteinaceous core that allows protons to flow across the membrane.

Direct evidence that the A/M2 protein has ion channel activity was obtained in recordings from oocytes of *Xenopus laevis* (Pinto *et al.*, 1992) and mammalian cells (Wang *et al.*, 1994; Chizhmakov *et al.*, 1996; Mould *et al.*, 2000b) that expressed the protein. The oocytes were found to have ion channel activity that was increased when the pH of the solution bathing the N-terminal ectodomain of the protein was lowered (Wang *et al.*, 1995; Chizhmakov *et al.*, 1996; Pinto *et al.*, 1997; Mould *et al.*, 2000). This activity is so strong that the oocytes and mammalian cells can become acidified in a short time (Shimbo *et al.*, 1996; Mould *et al.*, 2000) when bathed in a mildly acidic solution. The ion channel activity and acidification were both inhibited by amantadine with an apparent Ki of *ca.*0.5 μM (Wang *et al.*, 1993). However, these findings, although consistent with intrinsic ion channel activity, were not sufficient to demonstrate that the activity was intrinsic to the A/M2 protein and not merely the result of upregulation of a normally silent, endogenous protein of the expressing cell. Two additional lines of evidence support this conclusion. First, mutations made in the TM domain of the A/M2 protein produce alterations in the functional characteristics of the protein that are predicted from the biology of the virus infection or anticipated properties of the TM domain. A/M2 proteins with mutations corresponding to those found in amantadine-resistant escape mutant viruses have ion channel activity expressed in oocytes that is resistant to amantadine (Holsinger *et al.*, 1994; Pinto *et al.*, 1997). Second, mutations of two amino acids in the TM domain of the protein produce distinct changes in the proton-handling properties of the recorded activity. Proteins in which His_{37} has been mutated to Gly, Ala, or Glu possess ion channel activity different from that of the wild-type (wt) protein, in that their activity is not specific for protons and is relatively independent of the pH of the solution bathing the N-terminal ectodomain than the wt protein (Wang *et al.*, 1995). A/M2 mutant proteins for which Trp_{41} is mutated to Ala, Cys, Phe, or Tyr become activated at higher values of pH than the wt protein and are capable of supporting efflux of protons into external media of high pH, in contrast to the wt protein which "traps" protons in the cytoplasm (Tang *et al.*, 2002). In addition, purified recombinant A/M2 protein, when introduced into artificial lipid bilayers or vesicles, produces ion channel activity that is amantadine sensitive (Tosteson *et al.*, 1994; Lin and Schroeder, 2001) and selective to protons (Lin and Schroeder, 2001). These results, taken together, show that the A/M2 protein has the necessary ion channel activity in expression systems to explain its postulated role in viral uncoating.

It is important to note that no one piece of evidence would be sufficient to make the conclusion that A/M2 protein has intrinsic ion channel activity. For example, the activity recorded from artificial lipid bilayers might not reflect the activity found in the virus or the infected cell. The activity recorded in oocytes from wt A/M2 protein might be the result of upregulation of an endogenous ion channel, as is the case for other virally encoded proteins (Shimbo *et al.*, 1995). However, the body of evidence from recordings of ion channel activity, taken together with knowledge of the viral life cycle and the proposed role of the protein, together with the results of experiments with the inhibitor amantadine and amantadine-resistant escape mutations, help to make a strong case for the conclusion of intrinsic ion channel activity.

It should be noted that alterations in the morphological and functional properties of cells ensue with time when they express large quantities of the A/M2 protein. Cells overexpressing the A/M2 protein display a dilation of the lumen of the Golgi apparatus and delay in transport through the secretory pathway that are ameliorated by amantadine and mimicked by treatment with the ionophore monensin (Sakaguchi *et al.*, 1996). In addition, we have noticed that cells overexpressing the A/M2 protein also show a greater amount of nonspecific "leakage" current (that is not amantadine-sensitive) when studied with the patch clamp method (mammalian cells) or with two-electrode voltage clamp (oocytes) than do control cells that do not express the protein. The amplitude of these leakage currents often exceeded that of the amantadine-sensitive current after expressing the A/M2 protein for a few days. Thus, secondary effects due to the presence of the A/M2 protein have been found, and these effects are a consequence of the intrinsic activity of the A/M2 protein. These secondary effects give rise to a general increase in "leakiness" of the expressing cells and alterations in morphology of the cells (Lamb *et al.*, 1994; Giffin *et al.*, 1995), and it would be incorrect to conclude that this increase in "leakiness" in cells expressing the A/M2 protein represents an intrinsic property of the protein.

Results obtained from reverse genetic experiments on the influenza A virus support the conclusion from the above evidence that the A/M2 protein is essential to the life cycle of the virus. If the gene encoding the A/M2 protein is deleted from the genome of the virus, or if a mutation encoding a functionally defective A/M2 protein replaces the wt gene in the virus, then viral replication is severely impaired (Watanabe *et al.*, 2001; Takeda *et al.*, 2002).

It should be noted that in spite of either deletion or mutation of the gene encoding the A/M2 protein, the virus is still able to replicate in carefully controlled tissue culture conditions, at levels comparable to that found in the presence of the inhibitor amantadine (Takeda *et al.*, 2002). However, this ability to replicate at a very reduced rate does not mean that the A/M2 protein is "nonessential" because the fitness of the mutant viruses is so low that when placed in competition with the wt virus they would become extinct within a few generations (Takeda *et al.*, 2002).

Recently, the case for the A/M2 protein has been further strengthened by the finding that the B/M2 protein of influenza B (see below) virus is an integral membrane protein (Paterson *et al.*, 2003) that has ion channel activity (Mould *et al.*, 2003).

3. Mechanisms for Ion Selectivity and Activation of the A/M2 Ion Channel

Ion channels differ from simple TM pores by being selective for a particular ion and being activated or opened by a specific physical or chemical stimulus. The pore-lining residues of the homotetrameric (N-terminal ectodomain) A/M2 channel have been studied with cysteine scanning mutagenesis and coordination with transition elements, and it was found that Ala_{30}, Gly_{34}, His_{37}, and Trp_{41} are pore-lining residues (Pinto *et al.*, 1997; Gandhi *et al.*, 1999; Shuck *et al.*, 2000). Amantadine-resistant "escape" mutants are found to occur at the location of two of these pore lining residues (Ala_{30} and Gly_{34}), and the apparent Ki for externally applied amantadine is lower for the wild-type channel when the pH of the bathing solution is high than when the pH is low, suggesting that externally applied amantadine fits into the pore of the channel. We tested the ability of amantadine to inhibit the channel when applied from the *inside* of oocytes expressing the A/M2 channel by injecting amantadine, together with a fluorescent tracer to measure the volume injected into the cytoplasm. We found that for intracellular concentrations of amantadine as high as 1 mM (many times higher than the concentration needed to inhibit when applied externally), there was no significant inhibition (Figure 8.1).

The A/M2 protein is very highly selective for protons (Chizhmakov *et al.*, 1996; Shimbo *et al.*, 1996; Mould *et al.*, 2000). This high proton selectivity was blunted by replacement of TM His_{37} with Ala, Gly, or Glu (Pinto *et al.*, 1992; Wang *et al.*, 1995). In order to help elucidate the mechanism for the action of His_{37}, the kinetic isotope effect was measured in experiments in which water was replaced by deuterium oxide. The magnitude was found to be between 1.8 and 2.2, values that are much higher than the ratio of the viscosities of these liquids (1.3). This result is consistent with the possibility that protons either bind to the histidine and are released (Mould *et al.*, 2000) or form short-lived proton "wires" across the His_{37} barrier (Smondyrev and Voth, 2002). The single channel conductance of the channel is very low (Mould *et al.*, 2000; Lin and Schroeder, 2001), and it is thus not likely that long-lived proton "wires" form, as this mechanism would result in a high proton conductance.

When oocytes expressing the A/M2 protein are bathed in solutions of low pH, the cytoplasm of the expressing cells becomes acidified, and it takes many minutes for their internal pH to return to normal (Shimbo *et al.*, 1996; Mould *et al.*, 2000). In contrast, the internal pH of cells that are exposed to protonophores decreases rapidly when bathed in solutions of low pH and also recovers rapidly when the pH of the bathing medium is restored to neutral. Several lines of evidence point to Trp_{41} of the A/M2 protein acting as a "gate" that shuts the path for proton outflow when the pH of the bathing medium is high (Tang *et al.*, 2002).

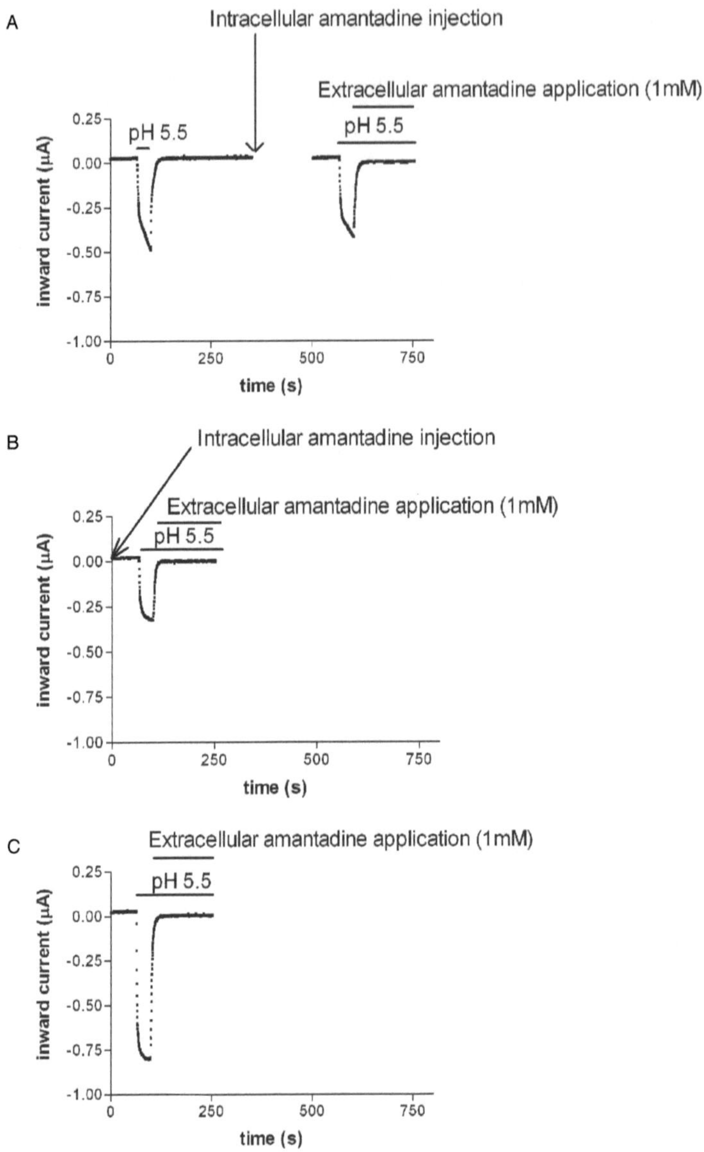

Figure 8.1. Intracellular injection of amantadine does not inhibit inward currents of oocytes expressing the wild-type A/M2 protein. (A) Extracellular application of amantadine inhibits inward currents elicited by oocytes expressing the wild-type A/M2 protein. An oocyte expressing A/M2 protein was bathed for 30 s in Barth's solution at low pH (5.5), resulting in inward current. This current was inhibited by extracellular application of amantadine (1 mM). (B) Oocytes expressing the wild-type A/M2 protein were injected with 50 nl of 20 mM amantadine, resulting in a final intracellular concentration of about 1 mM (assuming total oocyte volume of 1 μl). After injection, the oocytes were maintained in Barth's solution (pH 8.5) for 2.5 min and then placed in the recording chamber and

First, outflux of protons is speeded in mutants in which Trp_{41} is replaced by amino acids with smaller side chains. Second, intracellularly injected Cu^{2+} (which inhibits the wt A/M2 channel when applied extracellularly by coordination with His_{37}; see Gandhi *et al.*, 1999) is ineffective in inhibition of the wt channel but is able to inhibit the A/M2-W_{41}A mutant channel. Third, the efflux of protons from the A/M2-W_{41}C mutant channel can be influenced by intracellular injection of a cysteine-specific methanethiosulfonate reagent. However, one additional possibility for the reduced efflux is that the acidification of the cytoplasmic domain of the channel itself might reduce the conductance of the channel (desensitize or inactivate the channel). We tested this possibility by measuring the conductance of the A/M2 channel during the process of acidification and found no evidence for decreased conductance resulting from acidification that reduced the cytoplasmic pH of the expressing oocyte by as much as 1.5 pH units below the resting value of *c.*7.5 (Figure 8.2). Thus, Trp_{41} of the A/M2 channel seems to act as a "gate" that shuts the pathway for efflux of protons as long as the pH of the medium bathing the ectodomain of the channel is high. This would trend to "trap" protons in the interior of the virion while it is contained in the endosome (Tang *et al.*, 2002).

4. The BM2 Ion Channel of Influenza B Virus

The influenza B virus, like the influenza A virus, is endocytosed and uncoating occurs after fusion of the virion with the endosomal membrane (reviewed in Lamb and Krug, 2001). Uncoating of the virus requires a low pH step to allow dissociation of the RNPs from the matrix protein (Zhirnov, 1992). The influenza B virion was long thought to contain three integral membrane proteins, hemagglutinin, neuraminidase, and NB. A fourth small, virally encoded protein, BM2, was mistakenly thought to be found in the cytoplasm of expressing cells (Odagiri *et al.*, 1999). The primary amino acid sequence of the BM2 protein bears little resemblance to that of the A/M2 protein, but upon examination of the sequence, it was found to contain a predicted TM domain with a H–X–X–X–W motif. The BM2 protein was found to be an integral membrane protein with N-out C-in orientation, bearing a single TM domain and probably forms a homotetramer (Paterson *et al.*, 2003). The BM2 protein was studied in

---

Figure 8.1. Continued
exposed to Barth's solution at low pH (5.5), followed by extracellular application of amantadine (1 mM). The inward currents measured were in the range seen for the oocytes expressing wild-type A/M2 protein. Co-injecting 6-carboxyfluorescein with amantadine, and measuring the fluorescence from an excitation scan (400–515 nM) from these oocytes allowed calculation of the resulting concentrations. (C) Oocytes expressing the wild-type A/M2 protein were bathed for 30 s in Barth's solution at low pH (5.5), resulting in inward current. They were then washed with Barth's solution at high pH (8.5) for 4 min and injected with 50 nl of 20 mM amantadine, resulting in a final intracellular concentration of about 1 mM (measured by co-injection of 6-carboxyflourescein and assuming total oocyte volume of 1 μl). The oocytes were maintained in Barth's solution (pH 8.5) for 2.5 min and then returned to the recording chamber. They were then bathed in Barth's solution at low pH (5.5), followed by extracellular application of amantadine (1 mM). While intracellular injection of amantadine does not inhibit inward currents elicited by oocytes that express the wild-type protein ($N = 3$; $p > 0.05$), extracellular application of amantadine (1 mM) resulted in complete inhibition of the currents ($N = 3$; $p < 0.05$). In additional experiments, oocytes expressing the wild-type A/M2 protein were bathed for 30 s in Barth's solution at low pH (5.5), resulting in inward current. They were then injected with 50 nl of 20 mM amantadine, resulting in a final intracellular concentration of about 1 mM (measured by co-injection of 6-carboxyfluorescein and assuming total oocyte volume of 1μl). The oocytes were maintained in Barth's solution (pH 8.5) for 2.5 min and then returned to the recording chamber containing Barth's solution at low pH (5.5). Finally, amantadine (1 mM) was applied extracellularly. Intracellular amantadine injections did not inhibit inward currents elicited by oocytes expressing the wild-type A/M2 protein ($N = 3$; $p > 0.05$).

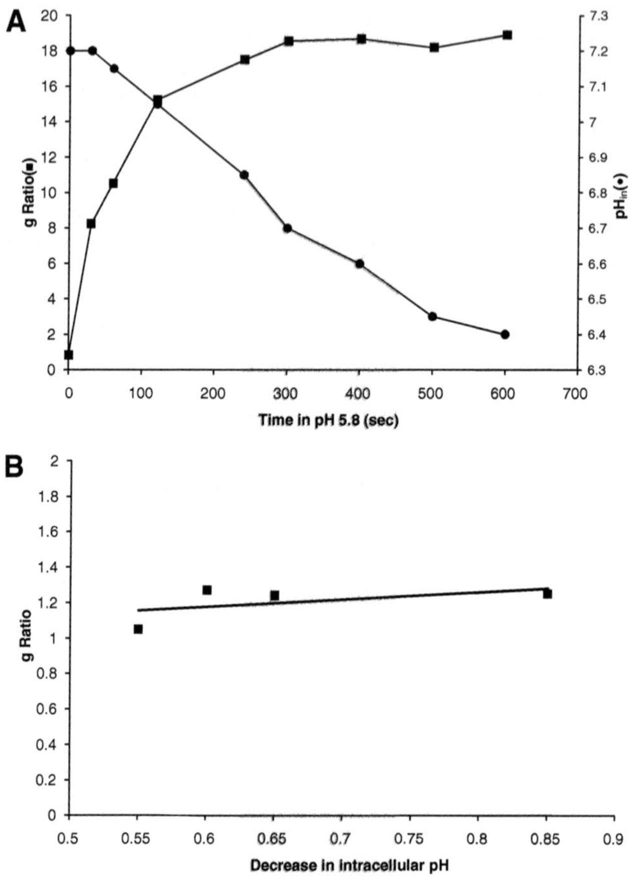

Figure 8.2. Acidification of the solution bathing the cytoplasmic domain of the wild-type A/M2 protein does not cause desensitization of the ion channel. (A) Time course of the acidification and membrane conductance of *Xenopus* oocytes that express the protein. Note that the conductance increases the first 200 s because acidification brings an increased concentration of the conducting ion to the cytoplasmic mouth of the channel. (B) Oocytes expressing the A/M2 protein were bathed continuously in low pH (5.8) solution. The ratio of the amantadine-sensitive conductance measured 600 s after the beginning of the low pH exposure to the conductance measured 120 s after the beginning of exposure is plotted against the decrease in intracellular pH that occurred between 120 and 600 s from the beginning of the low pH exposure. During this interval, the pH of the oocytes decreased substantially (see A) but the conductance did not decrease, showing that desensitization did not occur. Conductance was measured with two-electrode voltage clamp using the standard methods and internal pH was measured with a liquid ion exchanger electrode.

oocytes and mammalian cells and was found to be capable of mediating acidification of the cells when bathed in a solution of low pH (Mould *et al.*, 2003). The acidification activity was eliminated by mutation of the TM domain His_{19} to Cys. The dependence of current on external pH was found to be very similar to that of the A/M2 protein. One difference between the two proteins, however, lies in the inability of amantadine to inhibit the BM2 channel. This difference is not surprising in light of the great differences between the amino acid sequences of the two channels. The pore-lining residues of the A/M2 channel with which the hydrophobic domain of amantadine probably interacts are V_{27}, A_{30}, S_{31}, and G_{34}. In a helical wheel model of the TM domain of the BM2 channel, the analogous pore-lining residues are all serines, which are very unlikely to coordinate with the hydrophobic portion of the amantadine molecule.

The NB molecule, the "third" integral membrane protein of the influenza B virus, and once thought to be the needed ion channel protein of the virus, has been studied extensively in bilayers, where it induces cation currents (Sunstrom *et al.*, 1996; Fischer *et al.*, 2001) that can be reduced by high concentrations of amantadine (Fischer *et al.*, 2001). This molecule, however, does not provide for acidification of oocytes (Mould *et al.*, 2003) and its expression causes an upregulation of endogenous oocyte currents (Shimbo *et al.*, 1995). In light of these findings and our recent findings with the BM2 ion channel protein of influenza B virus, it is unlikely that the NB protein is an ion channel of influenza B virus that causes acidification during uncoating.

5. Conclusion

In order to conclude that a protein encoded by a virus serves as an ion channel in the life cycle of that virus, it is necessary to perform experiments in which the protein is perturbed and its function measured with *in vitro* systems as well as experiments that perturb the viral genome and measure the effect upon replication. Of great help is the ability to test the effect of an inhibitor of the ion channel upon both the function of the channel and the replication of the virus. No one single measurement is sufficient because each of the methods that are suitable for studies of virally encoded ion channels have shortcomings and are susceptible to artifacts. Comparison of the A/M2 and BM2 proteins shows that two virally encoded proteins are capable of achieving the same goal, virion acidification, in spite of having very different primary amino acid sequences but having a critical, five-residue-long motif in common within the TM domain of a homotetrameric protein. We thus believe that these two proteins are the first known members of the single-pass family of proton channels.

Acknowledgments

We thank Dr. Jorgen Mould for helpful discussions and Ms. Anne Andrews for technical help. Supported by NIAID R01AI-31882 (LHP) and R01AI-23173 (RAL).

References

Chizhmakov, I.V., Geraghty, F.M., Ogden, D.C., Hayhurst, A., Antoniou, M., and Hay, A.J. (1996). Selective proton permeability and pH regulation of the influenza virus M_2 channel expressed in mouse erythroleukaemia cells. *J. Physiol.* **494**, 329–336.

Ciampor, F., Thompson, C.A., Grambas, S., and Hay, A.J. (1992). Regulation of pH by the M2 protein of influenza A viruses. *Virus Res.* **22**, 247–258.

Fischer, W.B., Pitkeathly, M., and Sansom, M.S. (2001). Amantadine blocks channel activity of the transmembrane segment of the NB protein from influenza B. *Eur. Biophys. J.* **30**, 416–420.

Gandhi, C.S., Shuck, K., Lear, J.D., Dieckmann, G.R., DeGrado, W.F., Lamb, R.A. *et al.* (1999). Cu(II) inhibition of the proton translocation machinery of the influenza A virus M2 protein. *J. Biol. Chem.* **274**, 5474–5482.

Giffin, K., Rader, R.K., Marino, M.H., and Forgey, R.W. (1995). Novel assay for the influenza virus M_2 channel activity. *FEBS Lett.* **357**, 269–274.

Grambas, S., Bennett, M.S., and Hay, A.J. (1992). Influence of amantadine resistance mutations on the pH regulatory function of the M2 protein of influenza A viruses. *Virology* **191**, 541–549.

Grambas, S. and Hay, A.J. (1992). Maturation of influenza A virus hemagglutinin—estimates of the pH encountered during transport and its regulation by the M2 protein. *Virology* **190**, 11–18.

Hay, A.J. (1992). The action of adamantanamines against influenza A viruses: Inhibition of the M_2 ion channel protein. *Sem. Virol.* **3**, 21–30.

Hay, A.J., Wolstenholme, A.J., Skehel, J.J., and Smith, M.H. (1985). The molecular basis of the specific anti-influenza action of amantadine. *EMBO J.* **4**, 3021–3024.

Holsinger, L.J. and Lamb, R.A. (1991). Influenza virus M2 integral membrane protein is a homotetramer stabilized by formation of disulfide bonds. *Virology* **183**, 32–43.

Holsinger, L.J., Nichani, D., Pinto, L.H., and Lamb, R.A. (1994). Influenza A virus M2 ion channel protein: A structure-function analysis. *J. Virol.* **68**, 1551–1563.

Ito, T., Gorman, O.T., Kawaoka, Y., Bean, W.J., Jr., and Webster, R.G. (1991). Evolutionary analysis of the influenza A virus M gene with comparison of the M1 and M2 proteins. *J. Virol.* **65**, 5491–5498.

Lamb, R.A. and Choppin, P.W. (1981). Identification of a second protein (M_2) encoded by RNA segment 7 of influenza virus. *Virology* **112**, 729–737.

Lamb, R.A., Holsinger, L.J., and Pinto, L.H. (1994). The influenza A virus M_2 ion channel protein and its role in the influenza virus life cycle, In E. Wimmer (ed.), *Receptor-Mediated Virus Entry into Cells*, Harbor Press, Cold Spring Cold Spring Harbor, NY, pp. 303–321.

Lamb, R.A. and Krug, R.M. (2001). Orthomyxoviridae: The viruses and their replication. In D.M. Knipe and P.M. Howky (eds), *Fields Virology*, Lippincott, Williams & Wilkins, Philadelphia, pp. 1487–1531.

Lamb, R.A., Zebedee, S.L., and Richardson, C.D. (1985). "Influenza virus M_2 protein is an integral membrane protein expressed on the infected-cell surface." *Cell* **40**, 627–633.

Lin, T.I. and Schroeder, C. (2001). "Definitive assignment of proton selectivity and attoampere unitary current to the M2 ion channel protein of influenza A virus." *J. Virol.* **75**, 3647–3656.

Mould, J.A., Drury, J.E., Frings, S.M., Kaupp, U.B., Pekosz, A., Lamb, R.A. *et al.* (2000a). "Permeation and activation of the M2 ion channel of influenza A virus." *J. Biol. Chem.* **275**, 31038–31050.

Mould, J.A., Li, H.-C., Dudlak, C.S., Lear, J.D., Pekosz, A., Lamb, R.A. *et al.* (2000b). "Mechanism for proton conduction of the M2 ion channel of influenza A virus." *J. Biol. Chem.* **275**, 8592–8599.

Mould, J.A., Paterson, R.G., Takeda, M., Ohigashi, Y., Venkataraman, P., Lamb, R.A. *et al.* (2003). "Influenza B virus BM2 protein has ion channel activity that conducts protons across membranes." *Dev. Cell* **5**, 175–184.

Odagiri, T., Hong, J., and Ohara, Y. (1999). "The BM2 protein of influenza B virus is synthesized in the late phase of infection and incorporated into virions as a subviral component." *J. Gen. Virol.* **80**, 2573–2581.

Panayotov, P.P. and Schlesinger, R.W. (1992). Oligomeric organization and strain-specific proteolytic modification of the virion M2 protein of influenza A H1N1 viruses. *Virology* **186**, 352–355.

Paterson, R.G., Takeda, M., Ohigashi, Y., Pinto, L.H., and Lamb, R.A. (2003). "Influenza B virus BM2 protein is an oligomeric integral membrane protein expressed at the cell surface." *Virology* **306**, 7–17.

Pinto, L.H., Dieckmann, G.R., Gandhi, C.S., Shaughnessy, M.A., Papworth, C.G., Braman, J. *et al.* (1997). A functionally defined model for the M2 proton channel of influenza A virus suggests a mechanism for its ion-selectivity. *Proc. Natl. Acad. Sci. USA* **94**, 11301–11306.

Pinto, L.H., Holsinger, L.J., and Lamb, R.A. (1992). "Influenza virus M_2 protein has ion channel activity. *Cell* **69**, 517–528.

Plugge, B., Gazzarrini, S., Nelson, M., Cerana, R., Van Etten, J.L., Derst, C. *et al.* (2000). A potassium channel protein encoded by chlorella virus PBCV-1. *Science* **287**, 1641–1644.

Sakaguchi, T., Leser, G.P., and Lamb, R.A. (1996). "The ion channel activity of the influenza virus M2 protein affects transport through the Golgi apparatus." *J. Cell Biol.* **133**, 733–747.

Sakaguchi, T., Tu, Q., Pinto, L.H., and Lamb, R.A. (1997). The active oligomeric state of the minimalistic influenza virus M2 ion channel is a tetramer. *Proc. Natl. Acad. Sci. USA*. **94**, 5000–5005.

Schubert, U., Ferrer-Montiel, A.V., Oblatt-Montal, M., Henklein, P., Strebel, K., and Montal, M. (1996). "Identification of an ion channel activity of the Vpu transmembrane domain and its involvement in the regulation of virus release from HIV-1-infected cells." *FEBS Lett.* **398**, 12–18.

Shimbo, K., Brassard, D.L., Lamb, R.A., and Pinto, L.H. (1995). Viral and cellular small integral membrane proteins can modify ion channels endogenous to *Xenopus* oocytes. *Biophys. J.* **69**, 1335–1346.

Shimbo, K., Brassard, D.L., Lamb, R.A., and Pinto, L.H. (1996). Ion selectivity and activation of the M2 ion channel of influenza virus. *Biophys. J.* **70**, 1336–1346.

Shuck, K., Lamb, R.A., and Pinto, L.H. (2000). Analysis of the pore structure of the influenza A virus M2 ion channel by the substituted-cysteine accessibility method. *J. Virol.* **74**, 7755–7761.

Smondyrev, A.M. and Voth, G.A. (2002). Molecular dynamics simulation of proton transport through the influenza A virus M2 channel. *Biophys. J.* **83**, 1987–1996.

Sugrue, R.J., Bahadur, G., Zambon, M.C., Hall-Smith, M., Douglas, A.R., and Hay, A.J. (1990). Specific structural alteration of the influenza haemagglutinin by amantadine. *EMBO J.* **9**, 3469–3476.

Sugrue, R.J. and Hay, A.J. (1991). "Structural characteristics of the M2 protein of influenza A viruses: Evidence that it forms a tetrameric channel." *Virology* **180**, 617–624.

Sunstrom, N.A., Premkumar, L.S., Premkumar, A., Ewart, G., Cox, G.B., and Gage, P.W. (1996). Ion channels formed by NB, an influenza B virus protein. *J. Membr. Biol.* **150**, 127–132.

Takeda, M., Pekosz, A., Shuck, K., Pinto, L.H., and Lamb, R.A. (2002). Influenza a virus M2 ion channel activity is essential for efficient replication in tissue culture. *J. Virol.* **76**, 1391–1399.

Tang, Y., Zaitseva, F., Lamb, R.A., and Pinto, L.H. (2002). The gate of the influenza virus M2 proton channel is formed by a single tryptophan residue. *J. Biol. Chem.* **277**, 39880–39886.

Tobler, K., Kelly, M.L., Pinto, L.H., and Lamb, R.A. (1999). Effect of cytoplasmic tail truncations on the activity of the M(2) ion channel of influenza A virus. *J. Virol.* **73**, 9695–9701.

Tosteson, M.T., Pinto, L.H., Holsinger, L.J., and Lamb, R.A. (1994). Reconstitution of the influenza virus M2 ion channel in lipid bilayers. *J. Membr. Biol.* **142**, 117–126.

Wang, C., Lamb, R.A., and Pinto, L.H. (1994). Direct measurement of the influenza A virus M2 protein ion channel in mammalian cells. *Virology* **205**, 133–140.

Wang, C., Lamb, R.A., and Pinto, L.H. (1995). Activation of the M_2 ion channel of influenza virus: A role for the transmembrane domain histidine residue. *Biophys. J.* **69**, 1363–1371.

Wang, C., Takeuchi, K., Pinto, L.H., and Lamb, R.A. (1993). The ion channel activity of the influenza A virus M_2 protein: Characterization of the amantadine block. *J. Virol.* **67**, 5585–5594.

Watanabe, T., Watanabe, S., Ito, H., Kida, H., and Kawaoka, Y. (2001). Influenza A virus can undergo multiple cycles of replication without M2 ion channel activity. *J. Virol.* **75**, 5656–5662.

Zebedee, S.L., Richardson, C.D., and Lamb, R.A. (1985). Characterization of the influenza virus M2 integral membrane protein and expression at the infected-cell surface from cloned cDNA. *J. Virol.* **56**, 502–511.

Zhirnov, O.P. (1992). Isolation of matrix protein M1 from influenza viruses by acid-dependent extraction with nonionic detergent. *Virology* **186**, 324–330.

[illegible]

9

Influenza A Virus M2 Protein: Proton Selectivity of the Ion Channel, Cytotoxicity, and a Hypothesis on Peripheral Raft Association and Virus Budding

Cornelia Schroeder and Tse-I Lin

The influenza A virus M2 protein, the prototype viral ion channel, mediates passage through low-pH compartments during viral entry and maturation. Its proton channel activity is essential for virus uncoating and in certain cases for the maturation of viral hemagglutinin (HA). A fluorimetric assay of ion translocation by membrane-reconstituted M2 disclosed the nature of the conducted ions, protons, and allowed the determination of an average unitary current in the attoampere range. Upon hyperexpression in heterologous systems, M2 is cytotoxic in correlation to pH gradients at the cytoplasmic membrane. An M2 mutant with relaxed cation selectivity proved significantly more cytotoxic and was exploited as a conditional-lethal transgene. M2 has an additional function, not inhibited by channel blockers, as a cofactor in virus budding where it interacts with M1 to determine virus morphology—spherical or filamentous. The M2 protein has recently been shown to bind cholesterol, but cholesterol appeared nonessential for ion channel activity. The M2 endodomain contains a cholesterol-binding motif, which also occurs in HIV gp41. We propose that M2 is targeted to the raft periphery enabling it to co-locate with HA and NA during apical transport. In a new model of influenza virus morphogenesis M2 may cluster or merge separate rafts into viral envelope and act as a fission (pinching-off) factor during virus budding.

Cornelia Schroeder • Abteilung Virologie, Institut für Mikrobiologie und Hygiene, Universitätskliniken Homburg/Saar, Germany. **Tse-I Lin** • Tibotec BVDV, Gen. De. Wittelaan 11B-3, B-2800 Mechelen, Belgium.

Viral Membrane Proteins: Structure, Function, and Drug Design, edited by Wolfgang Fischer.
Kluwer Academic / Plenum Publishers, New York, 2005.

1. Determination of Ion Selectivity and Unitary Conductance

1.1. Background

Early electrophysiological recordings of M2 cation conductance revealed relatively weak cation channel activity (Pinto *et al.*, 1992). Significant background current measured in the presence of a selective inhibitor, amantadine, had to be subtracted. Despite technical advances (Chizhmakov *et al.*, 1996; Ogden *et al.*, 1999; Mould *et al.*, 2000) M2 single channels are not resolved by electrophysiological techniques. We adapted a fluorimetric method developed to monitor proton translocation by membrane-reconstituted bacterial rhodopsin (Dencher *et al.*, 1986). The hydrophilic fluorescent pH indicator pyranine is incorporated into the lumen of liposomes reconstituted with the ion channel; the dual wavelength fluorescence ratio is proportional to internal pH. This method delivered the proof that M2 translocated protons (Schroeder *et al.*, 1994a). A fluorimeter with greater temporal resolution enabled initial rate recordings and quantitative analyses of channel conductance (Lin and Schroeder, 2001). The method has also been used in a comparative inhibitor study of amantadine derivatives and polyamines (Lin *et al.*, 1997).

1.2. Method

1.2.1. Expression, Isolation, and Quantification of the M2 Protein

M2 protein is expressed from recombinant baculovirus in *Trichoplusia ni* (*T. ni*) insect cells. Yields are optimal (1–2 mg M2 per l) at a multiplicity of infection of two plaque-forming units per cell in the presence of amantadine. M2 expressed in this system carries the normal modifications, palmitoylation and phosphorylation (Schroeder *et al.*, 1994a). A detergent-extract of the total membranes is prepared and purified by immunoaffinity FPLC (Lin and Schroeder, 2001).

1.2.2. Reconstitution of M2 into Liposomes

Buffers contain either sodium or potassium ions: KPS (12 mM K_2HPO_4, 50 mM K_2SO_4, pH 7.4) or NaPS (12 mM Na_2HPO_4, 50 mM Na_2SO_4, pH 7.4). Complex liposomes are composed of L-α-dimyristoylphosphatidylcholine (DMPC), brain sphingomyelin, phosphatidylethanolamine, phosphatidylserine, phosphatidylinositol, gangliosides, and cholesterol (molar ratio 10 : 3 : 3 : 1 : 0.5 : 0.32 : 14). Simple liposomes contain DMPC/ phosphatidylserine (PS) (85:15). An ionophore, for example, 0.2 mol% valinomycin is included in the lipid mixture or added during incubation. The lipid film is taken up in 50 μl 400 mM 1-octyl-β-D-glucopyranoside (OG), followed immediately by 400 μl buffer and 50 μg M2 in 50 μl of the same buffer containing 40 mM OG at 37°C. Liposomes are formed at 4°C in dialysis cassettes (Slide-a-lyzer, Pierce) during step-wise dialysis against three changes of 3 vol. buffer, followed by three changes of 10 vol. and finally two changes of 5 l for 12 hr in the presence of Amberlite XAD-2. All buffers except the last contain 0.04% sodium azide. The fluorescent pH indicator pyranine (2 mM, Molecular Probes) is added during the first two steps of dialysis when it can still penetrate the detergent-containing membrane. The integrity of liposome-inserted M2 is checked by PAGE and Western blots with antibodies to both terminal peptides. Control liposomes are prepared in parallel without M2. The size of liposomes is determined by photon correlation spectroscopy. The buffer capacity of the liposome lumen is calculated as described by Dencher *et al.* (1986) from the decay kinetics of a pH gradient.

1.2.3. Proton Translocation Assay

Reagents are equilibrated at 18°C. Control or M2 vesicles (5–10 μl) are injected with a syringe into 2 ml incubation buffer, NaPS or KPS or NMDGH (*N*-methyl-D-glucamine-Hepes). Ionophores (monensin, 5 nM, or valinomycin, 50 nM) are added where required. Samples are stirred continuously. Pyranine emission at 510 nm at two excitation wavelengths (410 and 460 nm) is recorded at 1 s intervals. Three to five recordings are averaged. Pyranine fluorescence ratios are calibrated with standard buffers in increments of 0.1 to 0.2 pH units. For inhibitor studies vesicles can be pre-incubated in the presence of the test compound under incubation conditions and proton translocation triggered by addition of ionophore. Alternatively, inhibitors are added after triggering proton translocation (Lin and Schroeder, 2001).

1.3. Proton Selectivity

Since the permeabilities of channel proteins differ in the inward and outward direction, the orientation of a membrane-reconstituted protein to the liposome membrane is important. Under the conditions described, M2 inserts randomly (Lin and Schroeder, 2001), meaning that only half of the liposomal M2 is engaged in proton translocation, into or out of the vesicles, depending on concentration gradients.

M2 is active at 0.04–10 μM H^+ (pH 5 to 7.4). It is reasonable to monitor Na^+ and K^+ ions with comparable sensitivity as pH, but fluorescent probes for Na^+ and K^+ are too insensitive with a threshold of about 1 mM (Haugland, 1999). Therefore, the highly sensitive proton translocation assay was adapted to monitor metal ions: M2 proteoliposomes are prepared in a single-cation buffer, NaPS or KPS. To see whether M2 conducts a specific cation the liposomes are introduced into a buffer made up with a different cation or devoid of metal ions (NMDGH). In closed systems an ion flux must be coupled to a compensating counterflux, hence in the case of M2 coupled ion fluxes can be monitored via internal pH. Cation fluxes into and out of the vesicles are feasible since M2 is membrane-inserted in both orientations.

The principle is illustrated in Figure 9.1. Figure 9.1A depicts an M2 vesicle immersed in incubation buffer. Ionic conditions and/or pH differ on either side of the membrane. Figure 9.1B and C shows vesicles prepared in NaPS introduced into KPS of the same pH 7.4. Introduction of these M2 vesicles into NaPS did not result in pH change, but exposure to KPS also caused no pH change (Figure 9.2). M2 did not allow an influx of K^+or an efflux of Na^+ ions, either of which could have been compensated by proton counterflow. Only when ionophores specific for sodium (e.g., monensin, Figure 9.1B) or potassium ions (valinomycin, Figure 9.1C) were added, an immediate pH increase or decrease ensued. Proton flux was in reverse to metal ion flux carried by the ionophore, causing internal pH to rise or fall (Figure 9.2). Inverse fluxes were observed in M2 vesicles prepared in KPS (not shown). An external buffer without metal cations (Figure 9.1D) also did not permit metal ions to permeate M2 unless an ionophore was added. At pH 5.7 where M2 is more active (Pinto *et al.*, 1992; Chizhmakov *et al.*, 1996) the ion selectivity remained as stringent (cf. Lin and Schroeder, 2001). Thus, M2 proved essentially impermeable to K^+ and Na^+.

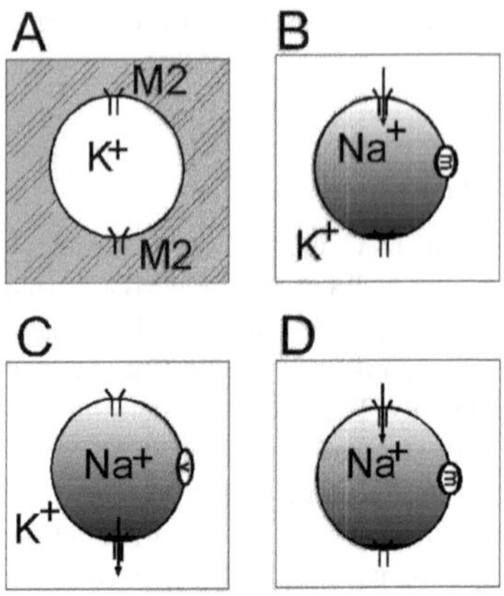

Figure 9.1. Setup of ion selectivity studies. A M2 vesicles prepared in neutral-pH buffer with a single metal ion, $K^+$is introduced into an incubation buffer with a different composition. M2 tetramers are present in both orientations with respect to the membrane. B M2 vesicles prepared in a Na^+ buffer are introduced into a K^+ buffer of the same pH. Proton influx (arrow) is elicited by adding a Na^+ ionophor (m=monensin) which allows Na^+ efflux. C A K^+ ionophor (v=valinomycin) elicits proton efflux. D Vesicles are introduced into a metal ion-free buffer pH 5.5. Addition of monensin allows Na^+ efflux and elicits proton influx. Adapted from Lin and Schroeder (2001) (with permission from the *J. Virol.*).

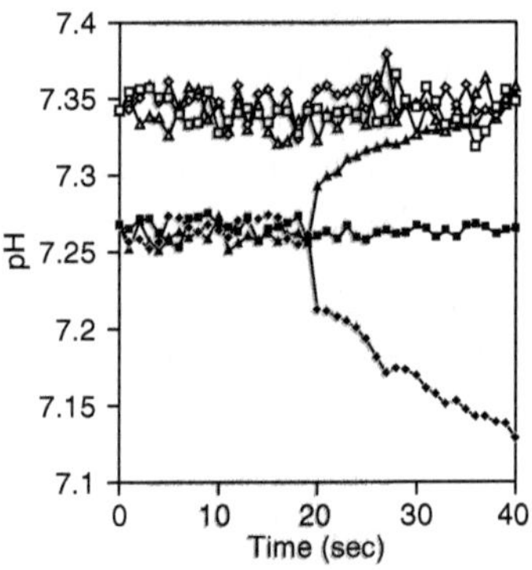

Figure 9.2. Demonstration of proton selectivity. M2 (solid symbols) or control (open symbols) vesicles containing Na^+ buffer were introduced into Na^+ or K^+ buffers; initial $pH_{out}=7.4$. Fluorimetrically monitored internal pH is plotted against time. Ionophores were added at 18 sec. Squares: Na^+buffer plus monensin; triangles: K^+ buffer plus valinomycin; diamonds: K^+ buffer plus monensin. Adapted from Lin and Schroeder (2001) (with permission from the *J. Virol.*).

An estimate of the proton selectivity of M2 (strain Weybridge) is at least 3×10^6 with respect to sodium and potassium ions. This is the ratio of the highest metal ion concentration (120 mM) at which ion flux was undetectable, and the lowest proton concentration (40 nM; pH7.4) where internal pH change was recorded (Lin and Schroeder, 2001). Similar estimates are 1.7×10^6 from whole-cell recordings on Weybridge M2-expressing MEL cells (Chizhmakov *et al.*, 1996; Ogden *et al.*, 1999) and 1.5 and 1.8×10^6 determined by patch-clamping of CV-1 cells and *Xenopus* oocytes expressing the M2 protein of influenza A/Udorn/72 (Mould *et al.*, 2000).

The arbitrary membrane insertion of M2 proved fortuitous for demonstrating its proton selectivity. However, the method would become more versatile if vectorial membrane insertion of the protein of interest were feasible, for example to study less selective channels, or functional coupling of two different channels.

1.4. Average Single-Channel Parameters and Virion Acidification During Uncoating

Given the low activity of this type of ion channel the presence of other ion channels in the preparation has to be excluded, since even minute contaminations would influence recordings. This is straightforward applying the selective M2 inhibitor rimantadine (reviewed by Hay, 1992). Following 5 min pre-incubation with 1 μM rimantadine, proton translocation is completely blocked confirming the identity of the channel and the absence of interfering activities.

Proton fluxes were calculated from the internal pH on the basis of the volume and buffer capacity of the liposome. Two types of liposome differing in composition and size were analyzed (Table 9.1). Cholesterol-containing vesicles of complex lipid composition had a significantly, 10-fold, larger volume than simple DMPC/PS vesicles. The buffer capacity, which depends mainly on the phospholipid head groups (Dencher *et al.*, 1986) was similar in both systems. Complex M2 vesicles contained 500, the simple vesicles 100 M2 tetramers.

Table 9.1. Average Single-Channel Parameters of Liposome-Reconstituted Influenza M2 Protein

	DMPC/PS vesicles[a]		Complex vesicles[a]	
Diameter (nm)	115 ± 38		256 ± 97	
Lipid content (2 × S/S_{PL})[b]	1.18×10^5		5.87×10^5	
Total liposomes (1 mg)[c]	6.64×10^{12}		1.33×10^{12}	
Liposome volume (nm^3)	7.89×10^5		8.75×10^6	
Total liposome volume (1 mg)	5.24 μl		11.7 μl	
M2 tetramers per liposome	100		500	
pH_{out}	7.4	5.7	7.4	5.7
Δø[mV]	150	−94	150	−94
ΔpH_{in} (initial pH change per s)	−0.1134 ±0.0362	−0.2664 ±0.0175	−0.0514 ±0.0210	−0.1701 ±0.0616
Initial proton translocation rate dH^+/dt [pmol s^{-1}]	40.5 ±12.9	95 ±6	42.9 ±17.5	142 ±51
Proton translocation rate per M2 tetramer [H^+ s^{-1}]	7.3 ±2.3	17.2 ±1.1	7.7 ±3.2	25.7 ±9.3
Unitary current at 18°C [aA][d]	1.2	2.7	1.2	4.1
Unitary current at 37°C [aA][d]	5	40	n.d[e]	n.d[e]
Unitary conductance at 18°C [aS][d]	8	29	8	44
Unitary conductance at 37°C [aS][d]	33	440	n.d[e]	n.d[e]

Notes:
[a]Liposome compositions: see chapter 1.2.2.
[b]Average lipid surface area S_{PL} = 0.7 nm^2.
[c]1 mg liposomes contains 7.83×10^{17} lipid molecules and 50 μg M2 (1.1 nmol = 6.65×10^{14} tetramers).
[d]aA = attoampere; aS = attosiemens.
[e]Not done. Adapted from Lin and Schroeder (2001) (with permission from the *J. Virol*).

Average single-channel currents were determined from the initial (1 sec) proton translocation rate at two pH_{out}, pH 7.4 where proton translocation is driven by the potassium ion concentration gradient and the channel is in its ground state, and pH 5.7, which activates the channel (Pinto *et al.*, 1992; Chizhmakov *et al.*, 1996). The average proton translocation rate for both types of vesicles is about 7 protons per second per tetramer. At a pH_{out} of 5.7, the flux increased to 17 protons per second in DMPC/PS, and 26 in complex vesicles (Lin and Schroeder, 2001). The lipid composition therefore had no significant influence on single-channel conductance (Table 9.1).

These proton currents represent 1.2–4.1 aA, four orders of magnitude below the noise level (<10 fA) of whole-cell patch clamp recordings defined as an upper boundary to M2 unitary currents (Ogden *et al.*, 1999). Based on a temperature dependence study an extrapolation to 37°C predicts a maximum of ≈40 aA for the low-pH activated state of the channel (Lin and Schroeder, 2001). Inactive M2 protein in the preparation will inevitably cause the activity to be underestimated.

Single-channel conductance is defined as the quotient of the current and the transmembrane (TM) potential. Assays were performed at 150 mV (K^+ gradient) and −94 mV (proton gradient); for comparison, the resting potential of cells is around −70 mV (reviewed in Lodish *et al.*, 1996). M2 average unitary conductance was between 8 and 44 aS (Table 9.1). pH activation enhanced the conductance 3.5- to 5-fold, which extrapolates to >10-fold (0.4 fS) at physiological temperature. Previous whole-cell recordings on M2-expressing MEL cells had not resolved single-channel conductance below 0.1 pS (Chizhmakov *et al.*, 1996; Ogden *et al.*, 1999). The unitary parameters of the M2 protein may be the lowest reported for any ion channel. M2 proton translocation rates of 7 to 26 per second (Table 9.1) resemble figures for transporters and pumps, orders of magnitude below rates of sodium or potassium channels, 10^7–10^8 per second (reviewed in Lodish *et al.*, 1996). However, the single-channel current of M2 appears especially minute because it is limited by the low physiological proton concentration, but its unitary proton permeability p turned out to be within the range of other, non-selective proton-conducting channels (Lin and Schroeder, 2001).

The generic function of the M2 ion channel in the influenza virus infectious cycle is the initiation of virus uncoating. Is its proton translocation rate sufficient to acidify the virus interior within a timescale of minutes? The initial pH decrease in DMPC/PS vesicles is 0.26 pH units per second (Table 9.1). Extrapolated to 37°C (≈15-fold increase) the virus interior may acidify in a minute or less. The virion is approximately the size of a DMPC/PS vesicle containing 75% to 90% less M2 protein (Zebedee and Lamb, 1988), yielding an acidification rate of 0.4–1 pH unit per sec. A coupling mechanism supporting a counterflow of cations or an influx of anions into the virion has not been discovered. Due to the site and mechanism of virus budding (see below) the balance of $K^+$and Na^+ in the virion interior should reflect concentrations in the cytoplasm (high K^+, low Na^+), the reverse of the endosomal K^+–Na^+ balance. A minute K^+ efflux from, or anion influx into the virion is conceivable, mediated by another protein or peptide as an ionophore, channel, or transporter. Possibilities worth testing are found in the literature. (1) The influenza B NB protein is an integral membrane protein of the virus envelope, which may have anion channel activity (for reviews see Lamb and Pinto, 1997; Fischer and Sansom, 2002). It would be interesting to see whether it couples with BM2, the functional equivalent of influenza A M2 (Mould *et al.*, 2003), (2) An amide hydrogen exchange study on TM peptides of influenza A HA suggests that several residues have access to water and the TM domain may form a pore (Tatulian and Tamm, 2000). Given the vast, about 50-fold excess of HA trimers over M2 tetramers in the virion even a very low level of potassium efflux through hypothetical HA pores would balance proton influx through M2. A hypothetical alternative not requiring coupling is that the pinching-off of budding virus establishes a concentration disequilibrium within the virion which is rectified by proton influx.

2. Cytotoxicity of Heterologous M2 Expression

Low single-channel conductance and stringent proton selectivity of the M2 protein cause minimal perturbation of ionic conditions during virus replication. However, M2 hyper-expression results in significant cytotoxicity to certain heterologous cell systems, for example, insect cells (Schroeder *et al.*, 1994a) and yeast (Kurtz *et al.*, 1995). M2 has also been shown to be cytotoxic in *E. coli* (Guinea and Carrasco, 1994), without however stringently establishing the relation of toxicity to ion channel activity by utilizing an M2-specific ion channel blocker. This was now done by cloning the M2 gene into the plasmid pTRC99A under a tight inducible promoter and selecting clones of M2-expressing *Escherichia coli* by their total inability to plate in the absence of 25 μM rimantadine (cp. Chapter 3; Schroeder *et al.*, 2004).

Heterologous cell systems in which the expression of wild-type M2 is toxic have in common a pH gradient at the cytoplasmic membrane. In *E. coli* and *Saccharomyces cerevisiae*, it is the proton electrochemical gradient (Kurtz *et al.*, 1995); in the case of insect cells the gradient is imposed by cultivation in pH 6 media. M2 is more active at pH < 7 and correspondingly more cytotoxic.

An M2 mutant with relaxed ion selectivity, H37A, was far more cytotoxic also to vertebrate cells, apparently as an unspecific cation channel (Smith *et al.*, 2002). Such mutants where the proton sensor, His37, has been replaced by Gly, Ala, or Glu retain amantadine sensitivity (Wang *et al.*, 1995; Smith *et al.*, 2002), indicating that His37 is not essential for amantadine binding and inhibition (cf. Figure 9.3). Thus, the mode of action proposed by Salom *et al.* (2000) whereby amantadine competes with protons for binding to His37 cannot be the exclusive mechanism of inhibition. Brian Thomas introduced M2 as a conditional-lethal transgene. *In vivo* expression of a construct with M2 H37A under the control of the T-cell specific $p56^{Lck}$ proximal promoter resulted in total ablation of T-cell development. *In vitro* development could be rescued with amantadine (Smith *et al.*, 2002). The technique's scalpel-like precision makes it an attractive tool for dissecting developmental gene expression, even in the era of RNAi.

3. The M2 Protein Associates with Cholesterol

Detergent-resistant cholesterol- and sphingolipid-rich membrane microdomains (DRM or rafts) are implicated in the budding of ortho- and paramyxo-, retro-, and other viruses (Sanderson *et al.*, 1995; Scheiffele *et al.*, 1999; Manié *et al.*, 2000; Zhang *et al.*, 2000; Pickl *et al.*, 2001; Suumalainen, 2002; for reviews see Nayak and Barman, 2002; Briggs *et al.*, 2003). In polarized host cells the three influenza A virus envelope proteins HA, neuraminidase (NA) and M2 are apically expressed and co-localize in the *trans*-Golgi (Zebedee *et al.*, 1985; Hughey *et al.*, 1992; Kundu *et al.*, 1996). The glycoproteins HA and NA are targeted to rafts (Skibbens *et al.*, 1989; Kurzchalia *et al.*, 1992; Kundu *et al.*, 1996; Scheiffele

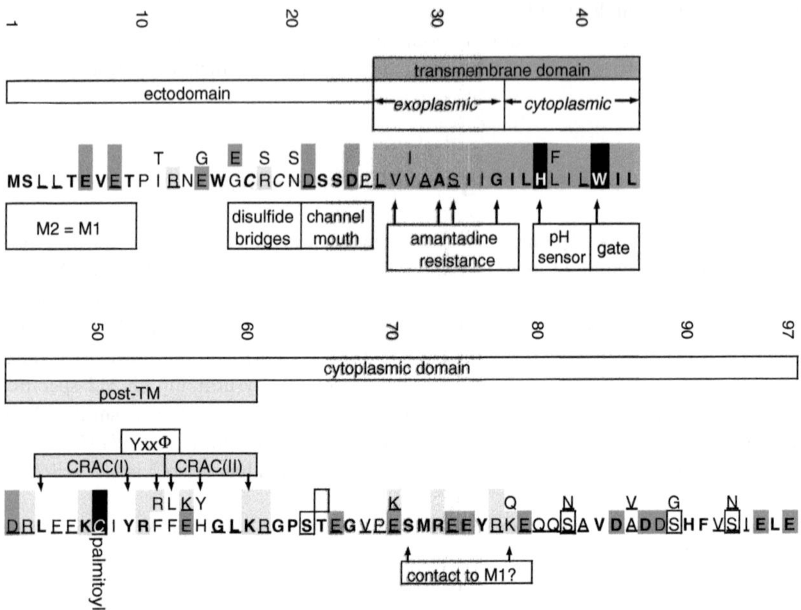

Figure 9.3. Functional domain organization and motifs of the M2 primary sequence. Top: residue number, domain structure and motifs. Center: M2 primary sequences, A/Udorn/72 (lower line) and A/Germany/29 (aka Weybridge) (upper line). The first nine residues of M1 and M2 are identical (M2 = M1). Descriptory code for residues: bold—conserved in sequenced M2s; underlined—mostly conserved; acidic—stippled background; basic—pale grey background; cysteine—italics; phosphorylated—framed (Udorn/72 M2: Holsinger *et al.*, 1995; Weybridge M2: Sugrue *et al.*, 1990a). CRAC motifs are indicated for Weybridge M2. Bottom: residue functions. Sites of substitutions determining amantadine resistance (cf. review by Hay, 1992), channel gate (Tang *et al.*, 2002), pH sensor (Wang *et al.*, 1995), disulfide bridges (Holsinger *et al.*, 1995), palmitoylation (Sugrue *et al.*, 1990a; Veit *et al.*, 1991), contact to M1: sites of mutations determining resistance to M2 N-terminus-specific IgG (Zebedee and Lamb, 1989).

et al., 1997; Harder *et al.*, 1998) where they assemble with viral matrix-RNP complexes (Scheiffele *et al.*, 1999; Zhang *et al.*, 2000; Barman *et al.*, 2001). In contrast, only about 10% of the M2 protein partition into DRM (Zhang *et al.*, 2000), correlating with the under-representation of M2 in the virus envelope (Zebedee and Lamb, 1988).

Cholesterol is a constitutive raft lipid essential for trafficking HA from the *trans*-Golgi network (TGN) to the apical surface (Keller and Simons, 1998). Cholesterol can also function as a protein-associated raft-targeting signal, binding very tightly but noncovalently to caveolin (Murata *et al.*, 1995) or covalently to hedgehog (Hh) proteins (Porter *et al.*, 1996). Several viral and host membrane proteins bind cholesterol, Sendai virus F protein (Asano and Asano, 1988), HIV gp41 (Vincent *et al.*, 2002), the mitochondrial peripheral benzodiazepine receptor (Li and Papadopoulos, 1998) and many others, most but not all of which are raft proteins. Caveolin and Shh are palmitoylated (Monier *et al.*, 1996; Pepinsky *et al.*, 1998), and so are influenza M2 (Sugrue *et al.*, 1990a; Veit *et al.*, 1991) and HA (Schmidt, 1982) while NA carries no lipid modifications.

Cleverley *et al.* (1997) reported that the cytotoxicity of the M2 protein to insect cells was abrogated by cholesterol depletion. Nevertheless, we found that M2 expression was toxic to intrinsically cholesterol-free *E. coli* (cp. chapter 2) and this toxicity was suppressed by rimantadine. We purified *E. coli* M2 carrying an N-terminal his-tag by Ni-NTA chromatography. Like authentic M2 and the insect cell-generated M2 protein *E. coli*-expressed M2 formed tetramers (Schroeder *et al.*, 2004) as required for ion channel activity (Sakaguchi *et al.*, 1997). Therefore, the toxicity exerted by M2 in *E. coli* was most likely caused by its ion channel activity, for which cholesterol is not essential.

Cholesterol was found associated to M2 protein isolated from influenza virus-infected cells and from insect cells (Schroeder *et al.*, 2004). Influenza virus M2 was purified from primary chick embryo cells pre-labeled with ^{3}H-cholesterol and neutral lipids extracted from the purified protein were analyzed by thin-layer chromatography. Only 50% of the label could be extracted. Of this about 70% migrated with cholesterol. Insect cell M2 protein stained with the cholesterol-binding fluorescent dye filipin (using the method of Smejkal and Hoff, 1994). The cholesterol content of M2 purified from its indigenous virus–host system or from insect cells was estimated at 0.5 or 0.9 molecules per monomer (Table 9.2).

Proton translocation rates of M2 reconstituted into cholesterol-free and cholesterol-rich, raft-like liposomes did not differ significantly (Table 9.1). Considering that conditions in those experiments were not absolutely cholesterol-free—*T. ni* M2 protein contained approximately equimolar cholesterol—we attempted functional reconstitution of cholesterol-free M2 isolated from *E. coli.* Cholesterol was not essential for membrane incorporation, ion channel function or the rimantadine sensitivity of *E. coli*-expressed M2, however, the specific activity of *E. coli* M2 was at least 5–times lower than that of *T. ni* M2 (Schroeder *et al.*, 2004). More subtle effects of cholesterol on M2 stability and activity can therefore not be ruled out.

Liposomes are widely used experimental models of lateral phase separation (Ahmed *et al.*, 1997) and the Triton X-100 (TX-100) insolubility of membrane microdomains and their associated proteins (Schroeder *et al.*, 1994b, 1998; Brown and London, 1998). Raft association is routinely analyzed by extraction with 1% TX-100 or CHAPS at 4°C (Fiedler *et al.*, 1993). We investigated the partition of insect cell- and *E. coli*-expressed M2 into synthetic raft membranes by detergent extraction and flotation. The complex lipid mixture was the same as used in the functional reconstitution of M2 (see chapter 1.2.2). 5–10% of the liposomal M2 protein partitioned into the detergent-resistant membranes, confirming the low degree of raft association of M2 protein expressed from influenza virus or constructs in MDCK cells (Zhang *et al.*, 2000). There was no significant difference in the behavior of *T. ni* and

Table 9.2. Cholesterol Content of M2 Expressed in Different Host Cells

Host cell	M2 expressed from	Cholesterol molecules per M2 tetramer
CEC[a]	Influenza virus	3.6[b]
T. ni	Baculovirus CFM2	2.2[c]
E. coli	p*TRC99A* Nx8′	0.1[c]

Notes:
[a]Primary chick embryo cells.
[b]Determined by labeling with ^{3}H-cholesterol.
[c]By densitometry of Coomassie and filipin-stained M2 bands in native gels (*E. coli* M2 serves as a cholesterol-free control for filipin staining).

E. coli-expressed M2, except that the latter was more prone to precipitation (Schroeder *et al.*, 2004).

4. M2 as a Peripheral Raft Protein and a Model of its Role in Virus Budding

The M2 protein has an independent, presumably structural role in virus morphogenesis (antiviral ion channel blockers have no effect, Zebedee and Lamb, 1989) and has therefore been termed a multifunctional protein (Hughey *et al.*, 1995). Antibodies to the M2 ectodomain inhibit virus release (Zebedee and Lamb, 1988; Hughey *et al.*, 1995) concomitantly tilting the balance of filamentous vs. spherical particles toward the latter (Hughey *et al.*, 1995; Roberts *et al.*, 1998). From mutations selected in the presence of such antibodies it is inferred that interactions between the M2 and the M1 protein influence the morphology of budding virus. The cytoskeleton (Roberts and Compans, 1998) is another significant factor in the generation of filamentous particles.

We explored the hypothesis that the cholesterol affinity of M2 is related to its enigmatic function in virus budding. Given that cholesterol is required for the apical transport of HA (Keller and Simons, 1998) and for efficient influenza virus budding (Scheiffele *et al.*, 1999) the cholesterol affinity of M2 may be the unknown (cf. Nayak and Barman, 2002) mechanism co-localizing M2 with viral glycoproteins in the TGN, in transport vesicles and, ultimately, ensuring low-level incorporation into budding virus. That M2 and HA are in fact co-transported in the same vesicles is implicated by the fact that M2 protects acid-labile HA from denaturation in the *trans*-Golgi (Sugrue *et al.*, 1990b; Ciampor *et al.*, 1992 a, b; Ohuchi *et al.*, 1994; Takeuchi *et al.*, 1994; Sakaguchi *et al.*, 1996).

Zhang *et al.* (2000) proposed that the exclusion of M2 from the virion be determined by its exclusion from rafts. But its essential role in virus uncoating (Kato and Eggers, 1969; Takeda *et al.*, 2002) and its influence on virus morphology argue in favor of a mechanism specifically incorporating (rather than excluding) the M2 protein in the virion: cholesterol affinity may guide its localization and function in virus assembly and budding.

Efficient budding of fully infectious influenza virus results from interactions of the matrix (M1) and the three envelope proteins (Latham and Galarza, 2001) and their association to membrane rafts (Scheiffele *et al.*, 1999). M1 is the major driving force of morphogenesis, since expression of M1 alone produces virus-like particles (Gomez-Puertas *et al.*, 2000) and budding is completely abolished in its absence (Lohmeyer *et al.*, 1979). Abrogation of HA and NA cytoplasmic sequences and of HA palmitoylation interferes with but does not prevent raft association and budding (Jin *et al.*, 1996, 1997; Zhang *et al.*, 2000). Palmitoylation at the HA TM and cytoplasmic sequence makes raft targeting more efficient (Melkonian *et al.*, 1999). The significance of M2 palmitoylation for virus budding is not known.

4.1. Interfacial Hydrophobicity and Potential Cholesterol and Raft Binding Motifs of the M2 Post-TM Region

The M2 sequence was scanned for potential cholesterol-binding hydrophobic areas. White and Wimley (1999) developed a scale for the affinity of peptide sequences to the water–membrane interface. Similar to the HIVgp41 pre-TM sequence (Suárez *et al.*, 2000) the M2 post-TM amino acid sequence has a high interfacial hydrophobicity extending >10 residues beyond the TM domain delineated by a classic Kyte and Doolittle (1982) hydrophobicity plot (Figure 9.4). Apart from the TM segment the post-TM sequence is the only other hydrophobic part of the M2 protein and thus a candidate cholesterol-binding site (Figure 9.3).

Vincent *et al.* (2002) pointed out that the cholesterol-binding pre-TM region of gp41 encompasses a cholesterol recognition/interaction amino acid consensus-L/V-$X_{(1-5)}$-Y-$X_{(1-5)}$-R/K- (the "CRAC" motif) first identified in the peripheral-type benzodiazepine receptor, a protein involved in mitochondrial cholesterol transport (Li and Papadopoulos, 1998; Li *et al.*, 2001). The M2 post-TM sequence exhibits one or two CRAC motifs immediately downstream of the TM domain (Figure 9.3). Of interest, the endodomain of the influenza B M2 protein (Mould *et al.*, 2003; Paterson *et al.*, 2003) also includes CRAC domains, albeit further downstream of the predicted TM segment. In HIV gp41 the motif is located adjacent but *upstream* of the TM domain, facing the extracellular space and the HIV exterior (Vincent *et al.*, 2002). In most sequenced influenza A M2 proteins, a CRAC motif encompasses the palmitoylation site Cys50, and in many it is followed by a second CRAC motif. In a majority of human influenza A strains the second motif is missing and the first motif appears to extend further downstream. In the strain A/Udorn/72 the consensus is violated (Figure 9.3). Whether these sequence variations have any impact on cholesterol binding remains to be seen. Similarly, the HIV gp41 CRAC consensus is violated in certain SIV strains (Vincent *et al.*, 2002).

The M2 post-TM sequence has been modeled as an alpha-helical extension of the TM domain by Kochendorfer *et al.* (1999). They predict that each amphiphilic helix of the post-TM makes a tight 90° turn and binds hydrophobically to the inner leaflet. The post-TM sequence also contains a conserved endocytic internalization motif YxxΦ (Collawn *et al.*, 1990) at amino acids 52–55, bridging the M2 CRAC tandem (Figure 9.3). YxxΦ frequently marks tight turns in the three-dimensional structures of internalized proteins (Collawn *et al.*, 1990). The two turns may confer the structural flexibility required for a role of the post-TM region in membrane fission (see below). Saldanha *et al.* (2002) generated a three-dimensional model of the M2 cytoplasmic domain, based on the three-dimensional

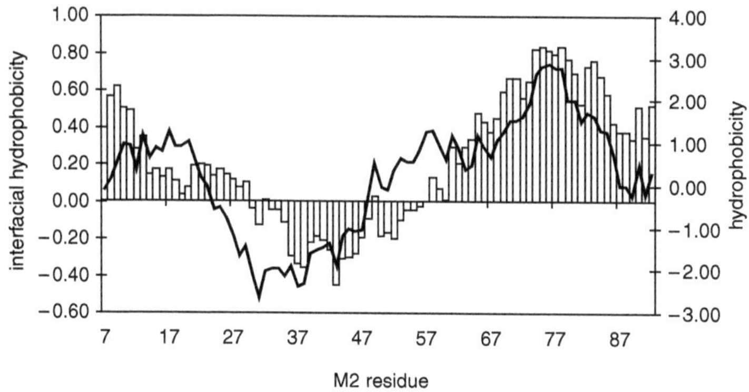

Figure 9.4. Interfacial hydrophobicity of the M2 protein. Columns: interfacial hydrophobicity (White and Wimley, 1999) of the influenza A/PR8/34 M2 sequence. Line: hydrophobicity according to Kyte and Doolittle (1982). Both were calculated for an 11-residue window.

structure of the cytoplasmic domain of the HIV Vpu protein. The cytoplasmic domains of M2 and Vpu exhibit significant sequence homology. Saldanha's model also predicts an alpha-helical secondary structure of the post-TM region with a membrane-proximal turn, and a kink at Lys60 (cf. Figure 9.3).

Kochendorfer *et al.* (1999) and Saldanha *et al.* (2002) pointed out the potential of the basic residues of the M2 post-TM region to coordinate with negatively charged phospholipid headgroups. Another pertinent observation is the striking homology of the M2 Weybridge post-TM sequence to the XIP region of Na/Ca exchangers: **DR**rl**LF**Y**K**YV*YK****R****Y*RA**G** (accession number AAD26362). The M2 post-TM and the XIP region are aligned with respect to the preceding TM segment and the internalization motif (italics), except for two additional residues in XIP (minor case). A CRAC motif is obvious, residues identical in M2 post-TM (cp. Figure 9.3) are in bold print, functionally similar residues, either hydrophobic or basic, are underscored. Moreover, the XIP domain exhibits specific affinity to phosphatidylinositol 4,5-bisphosphate (He *et al.*, 2000), which is enriched in the cytoplasmic leaflet of raft membranes (Liu *et al.*, 1998).

The gp41 pre-TM region is required for HIV fusion (Muñoz-Barroso *et al.*, 1999; Salzwedel *et al.*, 1999). It associates with liposome raft domains and forms pores in a cholesterol- and sphingomyelin-dependent manner (Sáez-Cirión *et al.*, 2002). In analogy, the M2 post-TM region may be involved in cholesterol binding and membrane restructuring, however, not during virus entry but morphogenesis.

4.2. M2 as a Peripheral Raft Protein

The 19-residue M2 TM helix (Lamb and Krug, 1996) is significantly shorter than the 26- to 29-residue TM segments of the raft proteins HA and NA. M2 is therefore unlikely to reside in the raft proper where membrane thickness increases with cholesterol content, the length of lipid acyl chains (Coxey *et al.*, 1993) and progression through the Golgi complex toward the plasma membrane (PM) (Ren *et al.*, 1997). The short TM domain of the M2 protein on the one hand and the post-TM sequence with the palmitate moiety on the other may confer dual affinity to raft and bulk membrane, thus a preference for the raft–non-raft interface. The short TM segment may be firmly rooted in non-raft membrane, while the palmitate moieties of the post-TM sequence may extend into rafts (Figure 9.5). The concept of peripheral raft binding is not new. Acetylcholinesterase was proposed to be targeted to the raft edge by its saturated acyl chain modification, which prefers liquid-ordered membrane (raft) whereas its polyunsaturated acyl chain preferentially inserts into liquid-disordered (non-raft) membrane (Schroeder *et al.*, 1998). For obvious steric reasons the individual palmitate chains of an M2 tetramer cannot all insert into the same raft (Figure 9.5B), unless the TM domains are surrounded by a narrow shell of non-raft membrane completely enclosed in raft membrane. M2 tetramers might thus bridge separate rafts, cluster or merge them. Dual membrane domain affinity may even enable M2 to shepherd membrane microdomains into larger arrays and to stabilize patches of non-raft membrane within raft macrodomains.

4.3. M2 as a Factor in the Morphogenesis and Pinching-Off of Virus Particles

Returning to the mutational evidence of physical interactions between the M1 and M2 proteins, antibodies to the M2 *ecto*domain inhibit virus release and shift the equilibrium

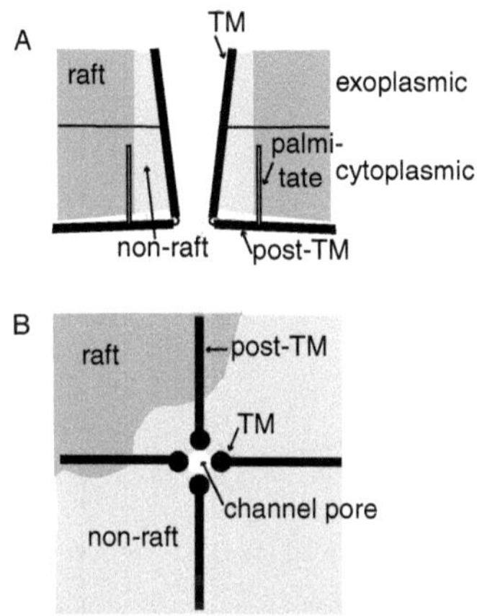

Figure 9.5. M2 as a peripheral Raft protein. (A) Schematic cross-section of the M2 tetramer. The TM helix is tilted (cf. review by Fischer and Sansom, 2002). Post-TM helices are associated with separate rafts or embedded in a small non-raft patch within surrounding raft membrane. (B) View from the cytoplasmic face of the membrane. In this example the post-TM regions of the upper and the left monomer are raft-associated.

toward the spherical morphology, but Fab fragments are not effective, suggesting the importance of cross-linking adjacent M2 molecules (Hughey *et al.*, 1995). Resistant mutants and laboratory strains (e.g., PR8, WSN) exhibit substitutions in M1 or M2 (Zebedee and Lamb, 1989). The fact that some of these map to the M2 *endo*domain implies that the cross-linking of M2 *ecto*domains interferes with interactions on the other side of the membrane between the M2 endodomain and M1 during virus budding. That the M2 endodomain binds more strongly to the matrix than the much shorter cytoplasmic domains of HA and NA has been observed following detergent (C12E8) extraction of purified influenza virus. HA and NA are quantitatively extracted within 10 min whereas M2 remains associated with the M1/RNP pellet (Bron *et al.*, 1993).

At the onset of budding the raft cluster maturing into viral envelope has a diameter of the order of 200 nm. The M2 protein is likely to concentrate at the edges (Figure 9.6). Initially M2 may also serve as a focus of attachment of M1/RNP and become trapped in surrounding raft membrane, as well as at the fault-lines between merging rafts (cp. Figure 9.5). As the periphery contracts during extrusion more and more M2 will be expelled into non-raft membrane. Before pinching-off nascent virus particles are connected to the PM by a tubular membrane surrounding the budding pore (Garoff *et al.*, 1998). M2 should concentrate here and in adjacent non-raft membrane (Figure 9.6), as seen in published electron micrographs of influenza virus budding. Immunogold-labeled antibodies to the M2 N-terminus cluster at the neck of virus buds (Hughey *et al.*, 1995; Lamb and Krug, 1996). The fact that antibodies to the M2 ectodomain favor the budding of spherical particles (Hughey *et al.*, 1995; Roberts *et al.*, 1998) may be rationalized by their linking the M2 tetramers into a ring, constricting the budding pore, which accelerates pinching-off. Likewise, dynamin and other proteins forming rings around membranous tubes promote membrane fission (for a review see Huttner and Zimmerberg, 2001).

An M2 molecule at the neck of the bud potentially interacts with M1, with raft and non-raft membrane. If the endodomain is restrained within the matrix, M2 is forced into

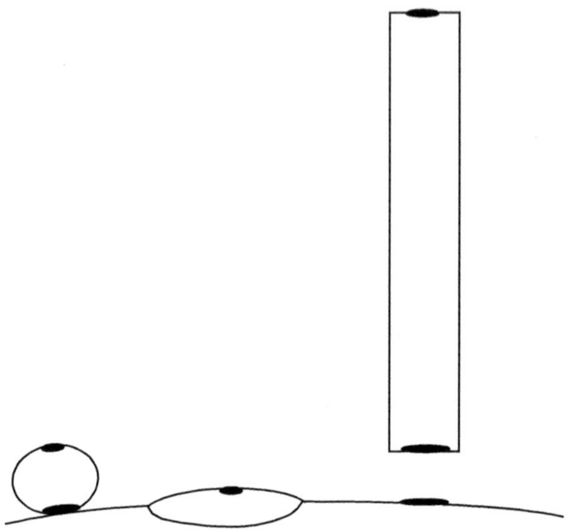

Figure 9.6. Budding and fission of Influenza Virus particles. A budding spherical particle before (left) and a filamentous particle (right) after pinching-off. M2 is shown clustered (black) at the budding pore and at opposite poles of virus particles. A nascent bud (center) with M2 at the raft (envelope)/non-raft membrane periphery.

an orientation perpendicular to the envelope. At the same time dynamic transitions of the palmitate group and/or the proposed post-TM raft-binding region as well as the TM domain may initiate membrane restructuring and fission. Post-TM segments and palmitate moieties can be envisaged dissociating from non-raft membrane and folding back into the envelope. Rearrangement of M2 clusters at the narrowing lipid domain boundary may seal both the viral and the plasma membrane. The proposed budding mechanism may produce virions with distinct polar caps of M2 clusters. The cartoon (Figure 9.6) depicts spherical and filamentous buds just before fission. M2 is shown at one or both ends of the bud, the latter version assuming that the long M2 endodomain formed the initial focus associating with the matrix. This model is presented in greater detail in Schroeder *et al.* (2004).

Acknowledgment

We are grateful to Larry Pinto for unpublished BM2 sequence information.

References

Ahmed, S.N., Brown, D.A. and London, E. (1997). On the origin of sphingolipid/cholesterol-rich detergent-insoluble cell membranes: Physiological concentrations of cholesterol and sphingolipid induce formation of a detergent-insoluble, liquid-ordered lipid phase in model membranes. *Biochemistry* **36**, 10944–10953.

Asano, K., and Asano, A. (1988). Binding of cholesterol and inhibitory peptide derivatives with the fusogenic hydrophobic sequence of F-glycoprotein of HVJ (Sendai Virus) Possible implication in the fusion reaction. *Biochemistry* **27**, 1321–1329.

Barman, S., Ali, A. Hui, E.K., Adhikary, L. and Nayak, D.P. (2001). Transport of viral proteins to the apical membranes and interaction of matrix protein with glycoproteins in the assembly of influenza viruses. *Virus Res.* **77**, 61–69.

Briggs, J.A.G., Wilk, T. and Fuller, S.D. (2003). Do lipid rafts mediate virus assembly and pseudotyping? *J. Gen. Virol.* **84**, 757–768.

Bron, R., Kendal, A.P., Klenk, H.D. and Wilschut, J. (1993). Role of the M2 protein in influenza virus membrane fusion: Effects of amantadine and monensin on fusion kinetics. *Virology* **195**, 808–811.

Brown, D.A. and London, E. (1998). Functions of lipid rafts in biological membranes. *Annu. Rev. Cell. Dev. Biol.* **14**, 111–136.

Chizhmakov, I.V., Geraghty, F.M., Ogden, D.C., Hayhurst, A., Antoniou, M. and Hay, A.J. (1996). Selective proton permeability and pH regulation of the influenza virus M2 channel expressed in mouse erythroleukemia cells. *J. Physiol.* **494**, 329–336.

Ciampor, F., Bayley, P.M., Nermut, M.V., Hirst, E.M.A., Sugrue, R.J. and Hay, A.J. (1992a). Evidence that the amantadine-induced, M2-mediated conversion of influenza A virus haemagglutinin to the low pH conformation occurs in an acidic trans Golgi compartment. *Virology* **188**, 14–24.

Ciampor, F., Thompson, C.A. and Hay, A.J. (1992b). Regulation of pH by the M2 protein of influenza A viruses. *Virus Res.* **22**, 247–258.

Cleverley, D.Z., Geller, H.M. and Lenard, J. (1997). Characterization of cholesterol-free insect cells infectible by baculoviruses: Effects of cholesterol on VSV fusion and infectivity and on cytotoxicity induced by influenza M2 protein. *Exp. Cell Res.* **233**, 288–296.

Collawn, J.F., Stangel, M., Kuhn, L.A., Esekogwu, V., Jing, S., Trowbridge, I.S. *et al.* (1990). Transferrin receptor internalization sequence YXRF implicates a tight turn as the structural recognition motif for endocytosis. *Cell* **63**, 1061–1072.

Coxey, R.A., Pentchev, P.G., Campbell, G. and Blanchette-Mackie, E.J. (1993). Differential accumulation of cholesterol in Golgi compartments of normal and Niemann-Pick type C fibroblasts incubated with LDL: A cytochemical freeze-fracture study. *J. Lipid Res.* **34**, 1165–1176.

Dencher, N.A., Burghaus, P.A. and Grzesiek, S. (1986). Determination of the net proton-hydroxide ion permeability across vesicular lipid bilayers and membrane proteins by optical probes. *Meth. Enzymol.* **127**, 746–760.

Fiedler, K., Kobayashi, T., Kurzchalia, T.V. and Simons, K. (1993). Glycosphingolipid-enriched, detergent-insoluble complexes in protein sorting in epithelial cells. *Biochemistry* **32**, 6365–6373.

Fischer, W.B. and Sansom, M.S.P. (2002). Viral ion channels: Structure and function. *Biochim. Biophys. Acta* **1561**, 27–45.

Garoff, H., Hewson, R. and Opstelten, D.E. (1998). Virus maturation by budding. *Microbiol. Molec. Biol. Revs.* **62**, 1171–1190.

Gómez-Puertas, P., Albo, C., Pérez-Pastrana, E., Vivo, A. and Portela, A. (2000). Influenza virus matrix protein is the major driving force in virus budding. *J. Virol.* **74**, 11538–11547.

Guinea, R., and Carrasco, L. (1994). Influenza virus M2 protein modifies membrane permeability in E coli cells. *FEBS Lett.* **343**, 242–246.

Harder, T., Scheiffele, P., Verkade, P. and Simons, K. (1998). Lipid domain structure of the plasma membrane revealed by patching of membrane components. *J. Cell Biol.* **141**, 929–942.

Haugland, R.P. (1999). Fluorescent Na^+ and K^+ indicators (Chapter 24). In *Handbook of Fluorescent Probes and Research Chemicals*, 7th edn. (CD ROM). Molecular Probes, Inc., Eugene, pp. 572–577.

Hay, A.J. (1992). The action of adamantanamines against influenza A viruses: Inhibition of the M2 ion channel protein. *Seminars Virol.* **3**, 21–30.

He, Z., S. Feng, Tong, Q., Hilgemann, D.W. and Philipson, K.D. (2000). Interaction of PIP(2) with the XIP region of the cardiac Na/Ca exchanger. *Am. J. Physiol. Cell. Physiol.* **278**, C661–C666.

Holsinger, L.J., Shaughnessy, M.A., Micko, A., Pinto, L.H. and Lamb, R.A. (1995). Analysis of posttranslational modifications of the influenza virus M2 protein. *J. Virol.* **69**, 1219–1225.

Hughey, P.G., Compans, R.W., Zebedee, S.L., and Lamb, R.A. (1992). Expression of the influenza A virus M2 protein is restricted to apical surfaces of polarized epithelial cells. *J. Virol.* **66**, 5542–5552.

Hughey, P.G., Roberts, P.C., Holsinger, L.J., Zebedee, S.L., Lamb, R.A., and Compans, R.W. (1995). Effects of antibody to the influenza A virus M2 protein on M2 surface expression and virus assembly. *Virology* **212**, 411–421.

Huttner, W.B. and Zimmerberg, J. (2001). Implications of lipid microdomains for membrane curvature, budding and fission. *Curr. Opin. Cell Biol.* **13**, 478–484.

Jin, H., Leser, G.P., Zhang, J. and Lamb, R.A. (1997). Influenza hemagglutinin and neuraminidase cytoplasmic tails control particle shape. *EMBO J.* **16**, 1236–1247.

Jin, H., K. Subbarao, Bagal, S., Leser, G.P., Murphy, B.R. and Lamb, R.A. (1996). Palmitylation of the influenza virus hemagglutinin (H3) is not essential for virus assembly or infectivity. *J. Virol.* **70**, 1406–1414.

Kato, N., and Eggers, H.J. (1969). Inhibition of uncoating of fowl plague virus by 1-adamantanamine hydrochloride. *Virology* **37**, 632–641.

Keller, P., and Simons, K. (1998). Cholesterol is required for surface transport of influenza virus hemagglutinin. *J. Cell Biol.* **140**, 1357–1367.

Kochendorfer, G.G., Salom, D., Lear, J.D., Wilk-Orescen, R., Kent, S.B.H., and DeGrado, W.F. (1999). Total synthesis of the integral membrane protein influenza A virus M2: Role of its C-terminal domain in tetramer assembly. *Biochemistry* **38**, 11905–11913.

Kundu, A., Avalos, R.T., Sanderson, C.M., and Nayak, D.P. (1996). Transmembrane domain of influenza virus neuraminidase, a type II protein, possesses an apical sorting signal in polarized MDCK cells. *J. Virol.* **70**, 6508–6515.

Kurtz, S., Luo, G., Hahnenberger, K.M., Brooks, C., Gecha, O., Ingalls, K. *et al.* (1995). Growth impairment resulting from expression of influenza virus M2 protein in Saccharomyces cerevisiae: Identification of a novel inhibitor of influenza virus. *Antimicrob. Agents Chemother.* **39**, 2204–2209.

Kurzchalia, T.V., Dupree, P., Parton, R.G., Kellner, R., Virta, H., Lehnert, M. *et al.* (1992). VIP21 a 21-kD membrane protein is an integral component of trans-Golgi-network-derived transport vesicles. *J. Cell Biol.* **118**, 1003–1014.

Kyte, J. and Doolittle, R.F. (1982). A simple method for displaying the hydropathic character of a protein. *J. Mol. Biol.* **157**, 105–132.

Lamb, R.A. and Krug, R.M. (1996). Orthomyxoviridae: The viruses and their replication. In B.N. Fields, D.M. Knipe, P.M. Howley, R.M. Chanock, J.L. Melnick, T.P. Monath, *et al.* (eds.), *Fields Virology*, 3rd ed. Lippincott-Raven Publishers, Philadelphia, pp. 1353–1395.

Lamb, R.A. and Pinto, L.A. (1997). Do vpu and vpr of human immunodeficiency virus type 1 and NB of influenza B virus have ion channel activities in the viral life cycles? *Virology* **229**, 1–11.

Latham, T. and Galarza, J.M. (2001). Formation of wild-type and chimeric influenza virus-like particles following simultaneous expression of only four structural proteins. *J. Virol.* **75**, 6154–6165.

Li, H. and Papadopoulos, V. (1998). Peripheral-type benzodiazepine receptor function in cholesterol transport. Identification of a putative cholesterol recognition/interaction amino acid sequence and consensus pattern. *Endocrinology* **139**, 4991–4997.

Li, H., Z. Yao, Degenhardt, B., Teper, G. and Papadopoulos, V. (2001). Cholesterol binding at the cholesterol recognition/interaction amino acid consensus (CRAC) of the peripheral-type benyodiazepine receptor and inhibition of steroidogenesis by an HIV TAT-CRAC peptide. *Proc. Natl. Acad. Sci. USA* **98**, 1267–1272.

Lin, T., and Schroeder, C. (2001). Definitive assignment of proton selectivity and attoampere unitary current to the M2 ion channel protein of influenza A virus. *J. Virol.* **75**, 3647–3656.

Lin, T., Heider, H. and Schroeder, C. (1997). Different modes of inhibition by adamantane amine derivatives and natural polyamines of the functionally reconstituted influenza virus M2 proton channel protein. *J. Gen. Virol.* **78**, 767–774.

Liu, Y., Casey, L. and Pike, L.J. (1998). Compartmentalization of phosphatidylinositol 4,5-bisphosphate in low-density membrane domains in the absence of caveolin. *Biochem. Biophys. Res. Commun.* **245**, 684–690.

Lodish, H., Beerk, H., Matsudaira, P., Baltimore, D., Zipurski, L. and Darnell, J. (1996). *Molecular Cell Biology 3.0*, CD-ROM. W.H. Freeman and Company, New York.

Lohmeyer, J., Talens, L.T. and Klenk H.D. (1979). Biosynthesis of influenza virus envelope in abortive infection. *J. Gen. Virol.* **42**, 73–88.

Manié, S.N., Debreyne, S., Vincent, S. and Gerlier, D. (2000). Measles virus structural components are enriched into lipid raft microdomains: A potential cellular location for virus assembly. *J. Virol.* **74**, 305–311.

Melkonian, K.A., Ostermeyer, A.G., Chen, J.Z., Roth, M.G. and Brown, D.A. (1999). Role of lipid modifications in targeting proteins to detergent-resistant membrane rafts. *J. Biol. Chem.* **274**, 3910–3917.

Monier, S., D.J. Dietzen, W.R. Hastings, D.M. Lublin, and T.V. Kurzchalia (1996). Oligomerization of VIP21-caveolin in vitro is stabilized by long chain fatty acylation or cholesterol. *FEBS Lett.* **388**, 143–149.

Mould, J.A., Drury, J.E., Frings, S.M., Kaupp, U.B., Pekosz, E., Lamb, R.A. *et al.* (2000). Permeation and activation of the M2 ion channel of influenza A virus. *J. Biol. Chem.* **275**, 31038–31050.

Mould, J.A., Paterson, R.G., Takeda, M., Ohigashi, Y., Venkataraman, P., Lamb, R.A., *et al.* (2003). Influenza B virus BM2 protein has ion channel activity that conducts protons across membranes. *Dev. Cell* **5**, 175–184.

Muñoz-Barroso, I., Salzwedel, K., Hunter, E., and Blumenthal, R. (1999). Role of the membrane-proximal domain in the initial stages of human immunodeficiency virus type I envelope glycoprotein-mediated membrane fusion. *J. Virol.* **73**, 6089–6092.

Murata, M., Peränen, J., Schreiner, R., Wieland, F., Kurzchalia, T.V., and Simons, K. (1995). VIP21/caveolin is a cholesterol-binding protein. *Proc. Natl. Acad. Sci. USA* **92**, 10339–10343.

Nayak, D.P. and Barman, S. (2002). Role of lipid rafts in virus assembly and budding. *Advances in Virus Res.* **58**, 1–28.

Ogden, D., I.V. Chizhmakov, F.M. Geraghty, and Hay, A.J. (1999). Virus ion channels. *Meth. Enzymol.* **294**, 490–506.

Ohuchi, M., A. Cramer, M. Vey, R. Ohuchi, M. Garten, and H.D. Klenk, (1994). Rescue of vector-expressed fowl plague virus hemagglutinin in biologically active form by acidotropic agents and coexpressed M2 protein. *J. Virol.* **68**, 920–926.

Paterson, R.G., Takeda, M., Ohigashi, Y., Pinto, L.H. and Lamb, R.A. (2003). Influenza B virus BM2 protein is an oligomeric integral membrane protein expressed at the cell surface. *Virology* **306**, 7–17.

Pepinsky, R.B., Zheng, C., Wen, D., Rayhorn, P., Baker, D.P., Williams, K.P. *et al.* (1998). Identification of a palmitic acid-modified form of human sonic hedgehog. *J. Biol. Chem.* **273**, 14037–14045.

Pickl, W.F., Pimentel-Muiños, F.X., and Seed, B. (2001). Lipid rafts and pseudotyping. *J. Virol.* **75**, 7175–7183.

Pinto, L.H., L.J. Holsinger, and R.A. Lamb, (1992). Influenza virus M2 protein has ion channel activity. *Cell* **69**, 517–528.

Porter, J.A., Young, K.E., and Beachy, P.A. (1996). Cholesterol modification of hedgehog signaling domains in animal development. *Science* **274**, 255–259.

Ren, J., S. Lew, Wang, Z., and London, E. (1997). Transmembrane orientation of hydrophobic alpha-helices is regulated both by the relationship of helix length to bilayer thickness and by cholesterol concentration. *Biochemistry* **36**, 10213–10220.

Roberts, P.C. and Compans, R.C. (1998). Host cell dependence of viral morphology. *Proc. Natl. Acad. Sci. USA* **95**, 5746–5751.

Roberts, P.C., Lamb, R.A., and Compans, R.W. (1998). The M1 and M2 proteins of influenza virus are important determinants in filamentous particle formation. *Virology* **240**, 127–137.

Sáez-Ciri6n, A., Nir, S., Lorizate, M., Agirre, A., Cruz, A., Pérez-Gil, J. *et al.* (2002). Sphingomyelin and cholesterol promote HIV gp41 pretransmembrane sequence surface aggregation and membrane restructuring. *J. Biol. Chem.* **277**, 21776–21785.

Sakaguchi, T., Leser, G.P., and Lamb, R.A. (1996). The ion channel activity of the influenza virus M2 protein affects transport through the Golgi apparatus. *J. Cell Biol.* **133**, 733–747.

Sakaguchi, T., Tu, Q., Pinto, L.H., and Lamb, R.A. (1997). The active oligomeric state of the minimalistic influenza virus M2 ion channel is a tetramer. *Proc. Natl. Acad. Sci. USA* **94**, 5000–5005.

Saldanha, J.W., Czabotar, P.E., Hay, A.J., and Taylor, W.R. (2002). A model for the cytoplasmic domain of the influenza A virus M2 channel by analogy to the HIV-1 vpu protein. *Protein Peptide Lett.* **9**, 495–502.

Salom, D., Hill, B.R., Lear, J.D., and DeGrado, W.F. (2000). pH-dependent tetramerization and amantadine binding of the transmembrane helix of M2 from the influenza A virus. *Biochemistry* **39**, 14160–14170.

Salzwedel, K., West, J.T. and Hunter, E. (1999). A conserved tryptophae-rich motif in the membrane-proximal region of the human immunodeficiency virus type 1 gp41 ectodomain is important for Env-mediated fusion and virus infectivity. *J. Virol.* **73**, 2469–2480.

Sanderson, C.M., Avalos, R., Kundu, A., and Nayak, D.P. (1995). Interaction of Sendai viral F, HN and M proteins with host cytoskeletal and lipid components in Sendai virus-infected BHK cells. *Virology* **209**, 701–707.

Scheiffele, P., Rietveld, A., Wilk, T., and Simons, K. (1999). Influenza viruses select ordered lipid domains during budding from the plasma membrane. *J. Biol. Chem.* **274**, 2038–2044.

Scheiffele, P., Roth, M.G., and Simons, K. (1997). Interaction of influenza virus haemagglutinin with sphingolipid-cholesterol membrane domains via its transmembrane domain. *EMBO J.* **16**, 5501–5508.

Schmidt, M.F.G., (1982). Acylation of viral spike glycoproteins: A feature of enveloped RNA viruses. *Virology* **116**, 327–338.

Schroeder, C., Ford, C.F., Wharton, S.A., and Hay, A.J. (1994a). Functional reconstitution in lipid vesicles of influenza virus M2 protein expressed by baculovirus: evidence for proton transfer activity. *J. Gen. Virol.* **75**, 3477–3484.

Schroeder, C., Heider, H., Möncke-Buchner, E., and Lin, T. (2004). The influenza virus ion channel and maturation cofactor M2 is a cholesterol-binding protein. *Eur. Biophys. J.* DOI: 10.1007/s00249-004-0424-1.

Schroeder, R., London, E. and Brown, D. (1994b). Interactions between saturated acyl chains confer detergent resistance on lipids and glycosylphosphatidylinositol (GPI)-anchored proteins: GPI-anchored proteins in liposomes and cells show similar behaviour. *Proc. Natl. Acad. Sci. USA* **91**, 12130–12134.

Schroeder, R.J., Ahmed, S.N., Zhu, Y., London, E. and Brown, D.A. (1998). Cholesterol and sphingolipid enhance the Triton X-100 insolubility of glycosylphosphatidylinositol-anchored proteins by promoting the formation of detergent-insoluble ordered membrane domains. *J. Biol. Chem.* **273**, 1150–1157.

Skibbens, J.E., Roth, M.G., and Matlin, K.S. (1989). Differential extractability of influenza virus hemagglutinin during intracellular transport in polarized epithelial cells and nonpolar fibroblasts. *J. Cell Biol.* **108**, 821–832.

Smejkal, G.B., and Hoff, H.F. (1994). Filipin staining of lipoproteins in polyacrylamide gels: Sensitivity and photobleaching of the fluorophore and its use in a double staining method. *Electrophoresis* **15**, 922–925.

Smith, C.A., Graham, C.M., Mathers, K., Skinner, A., Hay, A.J., Schroeder, C. *et al.* (2002). Conditional ablation of T-cell development by a novel viral ion channel transgene. *Immunology* **105**, 306–313.
Suárez, T., Gallaher, W.R., Agirre, A., Goñi, F.M., and Nieva, J.L. (2000). Membrane interface-interacting sequences within the ectodomain of the human immunodeficiency virus type 1 envelope glycoprotein: Putative role during viral fusion. *J. Virol.* **74**, 8038–8047.
Sugrue, R.J. and Hay, A.J. (1991). Structural characteristics of the M2 protein of influenza A viruses: Evidence that it forms a tetrameric channel. *Virology* **180**, 617–624.
Sugrue, R.J., Bahadur, G., Zambon, M.C., Hall-Smith, M., Douglas, A.R., and Hay, A.J. (1990b). Specific structural alteration of the influenza haemagglutinin by amantadine. *EMBO J.* **9**, 3469–3476.
Sugrue, R.J., Belshe, R.B., and Hay, A.J. (1990a). Palmitoylation of the influenza A virus M2 protein. *Virology* **179**, 51–56.
Suumalainen, M. (2002). Lipid rafts and assembly of enveloped viruses. *Traffic* **3**, 705–709.
Takeda, M., Pekosz, A., Shuck, K., Pinto, L.H., and Lamb, R.A. (2002). Influenza A virus M2 ion channel activity is essential for efficient replication in tissue culture. *J. Virol.* **76**, 1391–1399.
Takeuchi, K. and Lamb, R.A. (1994). Influenza virus M2 protein ion channel activity stabilizes the native form of fowl plague virus hemagglutinin during intracellular transport. *J. Virol.* **68**, 911–919.
Tang, Y., Zaitseva, F., Lamb, R.A., and Pinto, L.H. (2002). The gate of the influenza virus M2 protein channel is formed by a single tryptophan residue. *J. Biol. Chem.* **277,** 39880–39886.
Tatulian, S.A. and Tammm, L.K. (2000). Secondary structure, orientation, oligomerization, and lipid interactions of the transmembrane domain of influenza hemagglutinin. *Biochemistry* **39**, 496–507.
Veit, M., Klenk, H.D., Rott, R., and Kendal, A. (1991). The M2 protein of influenza A virus is acylated. *J. Gen. Virol.* **72**, 1461–1465.
Vincent, N., Genin, C., and Malvoisin, E. (2002). Identification of a conserved domain of the HIV-1 transmembrane protein gp41 which interacts with cholesteryl groups. *Biochem. Biophys. Acta* **1567**, 157–164.
Wang, C., Lamb, R.A., and Pinto, L.H. (1995). Activation of the M2 ion channel of influenza virus, a role for the transmembrane domain histidine residue. *Biophys. J.* **69**, 1363–1371.
White, S.H. and Wimley, W.C. (1999). Membrane protein folding and stability: physical principles *Annu. Rev. Biophys. Biomol. Struct.* **28**, 319–365.
Zebedee, S.L. and Lamb, R.A. (1988). Influenza A virus M2 protein: Monoclonal antibody restriction of virus growth and detection of M2 in virions. *J. Virol.* **62**, 2762–2772.
Zebedee, S.L. and Lamb, R.A. (1989). Growth restriction of influenza A virus by M2 protein antibody is genetically linked to the M1 protein. *Proc. Natl. Acad. Sci. USA* **86,** 1061–1065.
Zebedee, S.L., Richardson, C.D., and Lamb, R.A. (1985). Characterization of the influenza virus M2 integral membrane protein and expression at the infected-cell surface from cloned cDNA. *J. Virol.* **56**, 502–511.
Zhang, J., Pekosz, A., and Lamb, R.A. (2000). Influenza virus assembly and lipid raft microdomains: A role for the cytoplasmic tails of the spike glycoproteins. *J. Virol.* **74**, 4634–4644.

10

Computer Simulations of Proton Transport Through the M2 Channel of the Influenza A Virus

Yujie Wu and Gregory A. Voth

1. Introduction

The M2 channel of the influenza A virus is formed by a viral integral membrane protein. This channel is highly proton-selective and low pH-gated. It plays important roles in the viral life cycle, which have been illustrated by many experiments as will be further elaborated upon in the next section, making it of great interest to drug design and medicine. These characteristics of the M2 channel have attracted a significant amount of research effort in the past couple of decades. Today, experimental studies have covered various aspects, including, but not limited to, virology, cellular biology, electrophysiology, molecular genetics, biochemistry, and structural biology. Despite these efforts, some of the most important properties of this important channel remain unclear. One key example is the detailed mechanism for proton selectivity and gating. Furthermore, information concerning the explicit dynamics of proton transport (PT) inside the channel is largely unavailable from experiments.

Recent computer simulation studies from our group have shed some light on these issues. From the structural and dynamical properties of the protein and pore water to the dynamics of an excess proton inside the channel, these studies have presented us with a more detailed picture of the M2 system at the atomic level and helped to improve our understanding. This chapter summarizes our recent computer simulation results on the M2 channel with an emphasis on the PT process through the channel and the molecular mechanisms underlying its selectivity and gating properties.

In the following section, an overview of current experimental results is presented. We then review the computer simulations of explicit PT through the M2 channel and discuss its implication for the selectivity and gating mechanisms in the following section. Finally, we present and discuss our latest computational results concerning the possible closed and open structures of the channel and the proton conductance mechanism.

Yujie Wu and Gregory A. Voth • Department of Chemistry and Henry Eyring Center for Theoretical Chemistry, University of Utah, 315 S. 1400 E. Rm 2020, Salt Lake City, Utah.

Viral Membrane Proteins: Structure, Function, and Drug Design, edited by Wolfgang Fischer.
Kluwer Academic / Plenum Publishers, New York, 2005.

2. Overview of Experimental Results for the M2 Channel

2.1. The Roles of the M2 Channel in the Viral Life Cycle

The M2 channel is formed by the viral M2 protein that can be found in the membrane of both the viral particle and the infected cells. When the pH at the N-terminal/extracellular side of the channel goes lower than 5.8 or so, the channel opens and selectively transports protons across the membrance from its N-terminal side to its C-terminal side (Pinto *et al.*, 1992; Chizhmakov *et al.*, 1996).

This function of the M2 channel provides a pH-regulating mechanism that has been found crucial in the viral life cycle. The dissociation of the viral matrix protein from the ribonucleoprotein, which is part of the viral uncoating process, requires a low pH environment produced through the M2 channel (Bukrinskaya *et al.*, 1982; Sugrue *et al.*, 1990; Martin and Helenius, 1991; Wang *et al.*, 1993). Some influenza virus subtypes also need the M2 channel in the viral maturation process, where it elevates the intravesicular pH of the *trans*-Golgi network, preventing the viral protein haemagglutinin from incorrect folding in an otherwise low pH environment (Ciampor *et al.*, 1992; Grambas and Hay, 1992; Grambas *et al.*, 1992; Takeuchi and Lamb, 1994). Experiments have shown that both parts of the viral life cycle can be interrupted by blocking the M2 channel with the antiflu drug amantadine (1-aminoadamantine hydrochloride, AMT) (Ciampor *et al.*, 1992; Grambas and Hay, 1992; Grambas *et al.*, 1992; Sugrue *et al.*, 1990).

2.2. The Architecture of the M2 Channel

The gene for the M2 peptide has been cloned and sequenced, revealing a primary structure containing 97 amino acids (Lamb and Choppin, 1981; Lamb *et al.*, 1981; Hull *et al.*, 1988). The peptide can fold into three structural domains: a 24-residue N-terminal/extracellular domain, a 19-residue transmembrane (TM) domain, and a 54-residue C-terminal/cytoplasmic domain (Lamb *et al.*, 1985). The whole M2 protein is a parallel homotetramer of the M2 peptide (Holsinger and Lamb, 1991; Sugrue and Hay, 1991; Pinto *et al.*, 1977; Sakaguchi *et al.*, 1997; Bauer *et al.*, 1999; Kochendoerfer *et al.*, 1999; Salom *et al.*, 2000; Tian *et al.*, 2002). The tetramer is held together mainly by noncovalent interactions. Intersubunit disulfide links, existing in the N-terminal domain (Holsinger and Lamb, 1991; Sugrue and Hay, 1991; Castrucci *et al.*, 1997; Kochendoerfer *et al.*, 1999), further stabilizes the tetrameric structure. The TM domain can fold into an α-helix and is the main channel-formation structure (Holsinger *et al.*, 1994; Pinto *et al.*, 1997). In the form of tetramer, the helices are able to align an aqueous pore through which ions can be transported across the membrane. A Cys-scanning mutagenesis study by Pinto *et al.* (1997) suggested that the overall structure of the tetramer is a left-handed coiled-coil or helix bundle; and the tilt angle of the α-helices with respect to the membrane normal has recently been determinied for the M2 protein in liposomes, revealing a value of $25° \pm 3$ (Tian *et al.*, 2002).

Interestingly, a synthetic 25-residue peptide (M2-TMP) with the amino acid sequence corresponding to the residues 22–46 (encompassing the segment for the TM domain, residues 25–43) of the M2 peptide has been found to be able to form proton-selective channels in lipid bilayers (Duff and Ashley, 1992). The M2-TMP channel is also AMT-sensitive and has similar ion selectivity and transport efficiency to the M2 channel (Duff and Ashley, 1992; Salom *et al.*, 2000). Furthermore, many evidences have indicated that they are also very similar in structure. For example, it was found to form tetramers in lipid micelles (Kochendoerfer *et al.*, 1999; Salom *et al.*, 2000). Circular dichroism and solid-state NMR data have shown that its secondary structure is predominantly α-helix (Duff *et al.*, 1992; Song *et al.*, 2000; Wang *et al.*, 2001). The helix tilt angle with respect to the membrane normal and the rotational angle around the helix axis are roughly 30–40° and $-50°$ respectively, as determined by both site-directed infrared dichroism spectra (Kukol *et al.*, 1999; Torres *et al.*, 2000; Torres and Arkin, 2002) and solid-state NMR (Kovacs and Cross, 1997; Kovacs *et al.*, 2000; Song *et al.*, 2000; Wang *et al.*, 2001). The tetramer of M2-TMP was also confirmed to be left-handed so that the hydrophilic residues can be oriented toward the pore lumen (Kovacs and Cross, 1997). Therefore, the M2-TMP channel has been used as a simplified model for the M2 channel in many studies.

2.3. Ion Conductance Mechanisms

The M2 channel is highly proton-selective—at least 10^6-fold more conductive than other cations (Chizhmakov *et al.*, 2003; Mould *et al.*, 2000); and it is low-pH gated—undergoing a 50-fold conductance increase from pH 8.2 to pH 4.5 (Wang *et al.*, 1995; Mould *et al.*, 2000). Though the detailed structure responsible for these functions remains unclear, several lines of evidence have suggested that a highly conservative residue His37 plays a crucial role. This residue is the only one that has a pKa around 6 in the TM domain. Mutagenesias studies have shown that replacing it with Ala, Glu, or Gly can result in a large increase in proton conductance and loss of pH-induced gating behavior (Pinto *et al.*, 1992; Wang *et al.*, 1995). Replacing it with Glu can also reduce its selectivity (Wang *et al.*, 1995), while mutating it to Cys completely abolishes its channel function (Shuck *et al.*, 2000). The pH titration (Wang *et al.*, 1995) and Cu^{2+} inhibition experiments (Gandhi *et al.*, 1999) also suggest that the His37 residue is implicated in the selective filter.

Two conductance mechanisms have been proposed. One is called the gating mechanism (Sansom *et al.*, 1997), which suggests that each of the His37 side chains may acquire an additional proton to be positively charged when the pH goes lower than the pKa. Then due to the electrosatic repulsion, the side chains sway from each other, thus opening the otherwise occuluded pore to let the pore water penetrate through to form a continuous proton-conductive water wire (proton wire). The other mechanism (known as the shuttle mechanism) suggests that the His37 residues are directly involved in a proton relay such that they accept the excess proton at one side while release a proton at the other side of the imidazole ring (Pinto *et al.*, 1997). In contrast to the gating mechanism, the bi-protonated intermediate is presumed to be relatively short-lived and tends to release either the ε- or δ-hydrogen back to the pore water to become neutral again (proton shuttling). To transport the next proton, the initial state might be regenerated through tautomerization or flipping of the imidazole ring.

The existence of bi-protonated histidines in the open state of the M2-TMP channel has been reported by a UV Resonance Raman (UVRR) spectroscopy study (Okada *et al.*, 2001). This may lend support to the gating mechanism. Futhermore, their data suggested that the histidine residues would be fully bi-protonated since the addition of Sodium Dodecyl Sulfate (SDS), which was expected to disrupt the bundle structure and expose the subunits to the low pH buffer, did not yield a change in the corresponding Raman intensity. However, no evidence was provided to show the subunits were really exposed to the buffer upon adding SDS.

Several recent experimental studies also indicated that Trp41 is another important residue involved in the gating mechanism. This includes the UVRR experimental study mentioned above (Okada *et al.*, 2001), which suggested that the Trp41 residue may have cation–π interaction with the bi-protonated histidine residues in the open state. A mutagenesis–electrophysiological study found that mutating Trp41 to Phe, Cys, or Ala results in "leaky" channels that can transport protons outward (i.e., from the C-end to the N-end), suggesting that Trp41 may be the actual channel gate (Tang *et al.*, 2002). Cross and coworkers determined that the distance between the N_δ-His37 and C_γ-Trp41 should be less than 3.9 Å for the closed channel (Nishimura *et al.*, 2002), implying the involvement of Trp41, and furthermore they proposed a closed structure with the (**t-160, t-105**)* conformation for His37 and Trp41, in which the Trp41 residues form the actual channel gate.

3. Molecular Dynamics Simulations of Proton Transport in the M2 Channel

The gating mechanism was proposed through a restrained molecular dynamics (MD) simulation (Sansom *et al.*, 1997), where the four charged histidines were observed to move away to open the pore. However, a later MD simulation without restraints illustrated that a fully biprotonated state destabilizes the overall channel structure while a doubly biprotonated state leads to a deformed, yet closed, structure (Schweighofer and Pohorille, 2000). The latter result suggests that a fully biprotonated channel is unlikely and the shuttle mechanism seems more favorable.

The above simulations mainly focused on the structure and/or dynamics of the protein and pore water molecules, and the excess proton was not included explicitly. No doubt, explicit PT simulation in the M2 channel may yield more information regarding the mechanism and the PT dynamics.

3.1. Explicit Proton Transport Simulation and Properties of the Excess Proton in the M2 Channel

Explicit PT simulations in the M2 channel were successfully performed by our group (Smondyrev and Voth, 2002) with the aid of a Multistate Empirical-Valence Bond (MS-EVB) model for PT in aqueous systems (Schmitt and Voth, 1998, 1999a,b, 2000; Cuma *et al.*, 2000, 2001; Day *et al.*, 2002). The biomolecular system for these simulations included an excess proton, the TM domain, and a fully solvated dimyristoylphosphatidyl choline (DMPC) bilayer. The system was simulated under the condition of constant temperature and volume, and no restraints were used. With a presumption that one stable biprotonated His37 might somewhat open the channel for protons, the TM domain was constructed to contain one His^+ and three neutral His's. These histidine residues were treated as chemically stable in the MD simulations—they can neither accept nor donate a proton. Seven roughly 1 nsec MD simulation trajectories were produced from different starting configurations, in all of which the excess proton was placed inside the channel near the N-end.

*The notation— (**t-160, t-105**)—means that the conformations of His37 and Trp41 are the **t-160** and **t-105** rotamers, respectively. The nomenclature for rotamers used here follows the Penultimate Rotamer Library (Lovell *et al.*, 2000). In the text, when the monoprotonation state of his histidine is taken into account, the symbol δ or ϵ is added—for example, (**t-160, t-105, δ**)—to indicate the histidine is δ- or ϵ-monoprotonated, respectively.

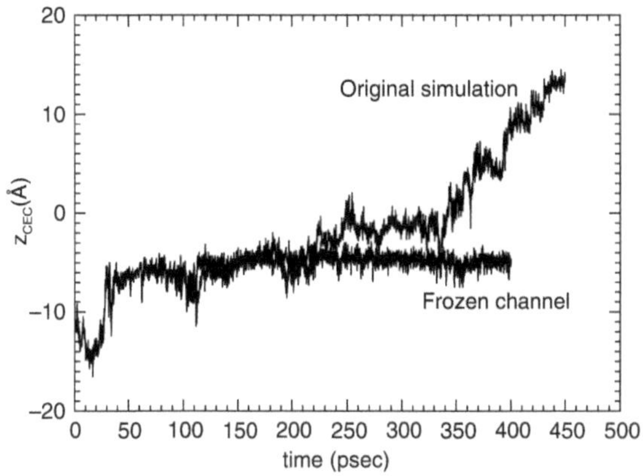

Figure 10.1. The z coordinate of the protonic center of excess charge (CEC) as a function of time for two simulations, in one of which the channel was frozen. Notice how the CEC was effectively immobilized in the frozen channel.

Detailed pictures on the dynamics and solvation structure of the excess proton in the M2 channel were obtained from these simulations. Here the most important results are reviewed. It was found that the excess proton in the channel still favors an Eigen-like solvation structure most of the time. This property is similar to that of the bulk water. However, the overall diffusion coefficient of the excess proton was reduced by the channel by up to a factor of three; in some cases, it was even immobilized for long periods. It was confirmed that the excess proton in the channel is transferred via the so-called Grotthuss *structural diffusion* mechanism as in bulk water, where the excess proton's solvation structure rather than the hydronium is propagated in space (Tuckerman *et al.*, 1995; Day *et al.*, 2000).

It was also found that the dynamics of the protein has dramatic influence on the excess proton's motion in the channel: A frozen channel was observed in the simulations to effectively immobilize the otherwise transferring excess proton (Figure 10.1). The same phenomenon has also been observed in explicit PT simulations of the gramicidin A channel (unpublished data). Though this result may not be too surprising, it is not fully understood. A recent simulation study on the excess proton in a leucine–serine synthetic channel (LS2) has reported the significance of the channel's local microenvironment on the excess proton's hydration properties (Wu and Voth, 2003b). Further studies on this important topic are currently underway in our group.

3.2. Implications for the Proton Conductance Mechanism

Though detailed side chains' structure and dynamics of the amino acid residues are not the target of the Smondyrev–Voth work, the simulation results do present a different view of the conductance mechanism in terms of PT. The MD simulation trajectories have reached roughly 1 nsec. Within this timescale, the excess proton was observed to pass through the channel at different times in three of the seven trajectories. A simulation snapshot when the excess proton was passing through the gating region by hopping through a transient water wire is given in Figure 10.2. Concerted widening of the pore radius was observed in the three trajectories when the proton was passing through the gating region. This may be ascribed to

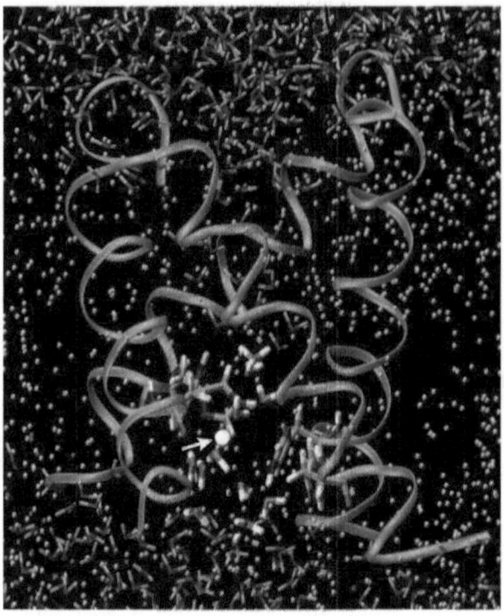

Figure 10.2. An MD simulation snapshot showing the excess proton (white ball indicated by arrow) passing the gating region by hopping through a transient water wire. The backbone of the channel is displayed by coils. The water molecules are shown as the small angled sticks. The dots in between the two water layers are the DMPC lipids.

the conformational change of the side chains of the His37 residues. Although the applied TM electric field (100–200 mV) may complicate these conclusions, this result lends direct support to the gating mechanism. However, the picture here also seems rather different from that described previously. The MD simulations suggest that one (or maybe two) biprotonated histidine may be sufficient for opening the channel for protons while still keeping it closed for other ions. Moreover, it seems that the presence of one positive charge near the constrictive region formed by the His37 residues does not to lead to a barrier too high for protons to pass through.

4. Possible Closed and Open Conformations

Comparing the M2 channel model used in our previous work (Smondyrev and Voth, 2002) with the available NMR backbone structure determined by Cross *et al.* (2001) showed good agreement in the helix tilt angles (Wu and Voth, 2003a). However, the rotational angles deviate substantially. Though this result is already quite good for computer simulations without previous knowledge of experimental structural data, a better starting structure for further MD simulations is certainly desirable. Furthermore, it is important to include accurate conformations of the key residues involved in the proposed conductance mechanism.

4.1. A Possible Conformation for the Closed M2 Channel

As mentioned in Section 2.3, Cross and coworkers have recently proposed a conformation of His37 and Trp41—**(t-160, t-105)**—for the closed M2 channel (Nishimura *et al.*, 2002). Nevertheless, based on the same structural information and our computational results, an alternative conformation—(**t60, t90, δ**)—has recently been proposed (Wu and Voth, 2003a). This result was obtained via a thorough scan over the conformational space followed by energetic and functional assessment. A representative structure of this conformation is shown in Figure 10.3a while Figure 10.3b shows the pore radius as a function of the channel axis. From these figures, it can be seen that a constrictive region is formed by both His37 and Trp41.

It was found that this conformation is consistent with nearly all relevant published experimental results. For example, the mutagenesis, results by Pinto and colleagues (Tang *et al.*, 2002) can be interpreted with this conformation: The δ-hydrogen of His37 is pointed toward the indole ring of Trp41, implying a hydrogen–π interaction that can stabilize each

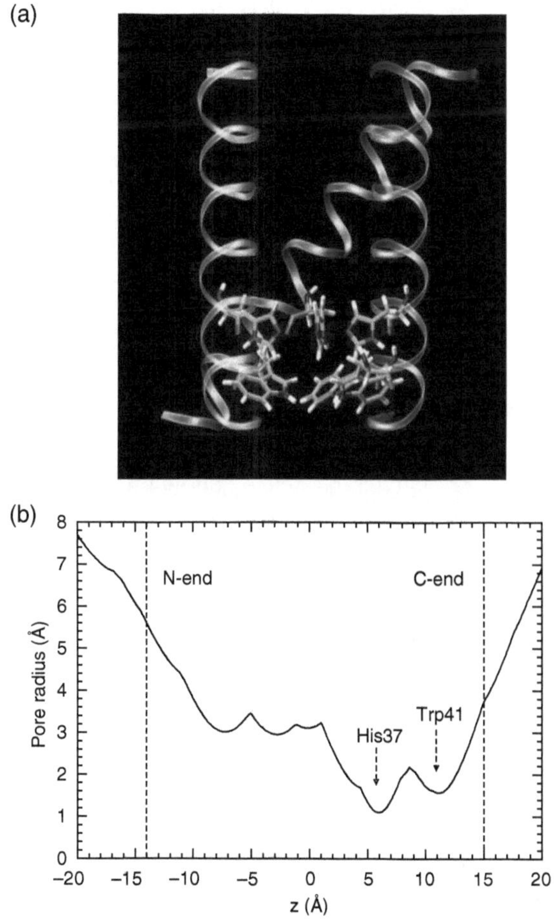

Figure 10.3. (a) A typical structure of the (**t60, t90, δ**) conformation of His37 and Trp41 for the closed channel and (b) the corresponding pore radius profile. The backbone of the channel is displayed by coils. For the sake of clarity, one helix and one pair of His37 and Trp41 are not shown.

other's conformation; changing the tryptophan to Phe, Cys, or Ala would abolish this interaction, destabilizing the conformation of His37 and resulting in a leaky channel. In contrast, if the Trp41 residues are mutated to Tyr, the conformation of His37 can be maintained because the Tyr residues can provide hydroxyl groups to form hydrogen bonds with the δ-hydrogens of the His37 residues. This explains why mutating Trp41 to Tyr does not lead to a leaky channel.

In this proposed closed conformation, the four ε-nitrogens of the His37 residues are not protonated and are pointed to one another (Figure 10.4). The distance between two

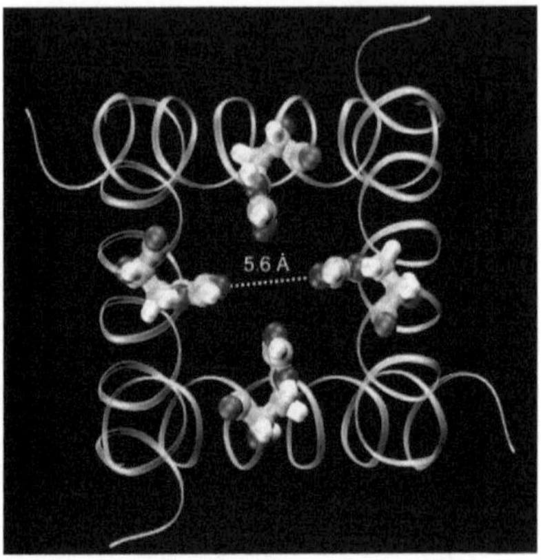

Figure 10.4. The view of the same structure in Figure 10.3 from the C-end. Notice the close distance between the two diagonal ε-nitrogens.

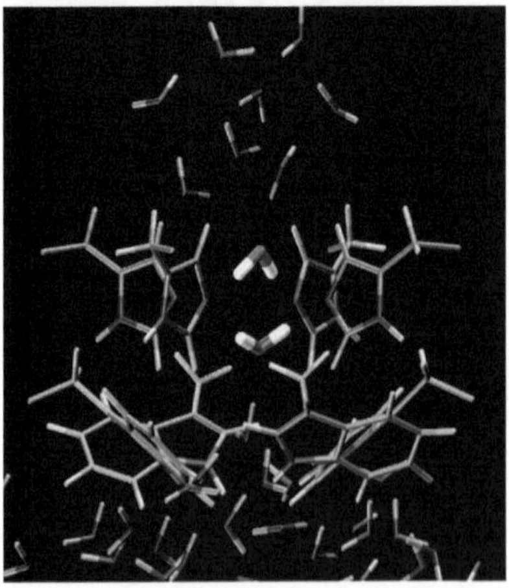

Figure 10.5. A snapshot from a MD simulation in which the backbone and heavy atoms of His37 and Trp41 were position-restrained and the His37 and Trp41 residues took the (**t60, t90, δ**) conformation. Notice the two pore waters near the His37 residues have opposite orientation.

ε-nitrogens in the diagonal is within the range of 4.8–6 Å. This structure can be a good chelating site for Cu^{2+}, explaining why the M2 channel can be inhibited by Cu^{2+} ions (Gandhi *et al.*, 1999).

The constrictive region formed by His37 and Trp41 is narrow enough to prevent cations effectively larger than an excess proton from passing through the channel. To seek an answer to how it can stop proton transport, an MD simulation has been carried out. The positions of the backbone atoms and the heavy atoms of His37 and Trp41 were restrained in the simulation to stabilize the conformation, while the other atoms were free to move. An interesting pore water structure in the constrictive region was seen in the simulation (Figure 10.5). The pore waters near His37 have opposite orientations. Similar pore water structure has been found in an MD simulation of aquaporin and has been used to explain why that channel does not transport protons (Tajkhorshid *et al.*, 2002). This conformation also leads to an explanation for the channel's gating behavior: The opposite orientations of the two water molecules interrupt the proton wire through which the excess proton can hop; and recovery of the proton wire must involve at least one His37 residue being protonated. The simulation also shows that the conformation of Trp41 prevents His37 from being exposed to the bulk water at the C-end, explaining why protons cannot be transported outward in a wild type M2 channel.

4.2. A Possible Conformation for the Open M2 Channel

Efficient proton shuttling by His37 requires its δ-nitrogen to be pointed toward the N-end of the channel so that both δ- and ε-nitrogens can simultaneously form hydrogen bonds with pore water. Furthermore, the δ-nitrogen should be deprotonated so that it can accept a proton coming from the N-end of the channel. For these reasons, the orientation of His37 in our proposed closed structure does not necessarily imply the shuttle mechanism. It was therefore necessary to find an open structure into which the closed structure can evolve upon protonation. We found a small rotation of the histidine's χ_2 angle from 60° to 0° and slight adjustment of Trp41's dihedral angles can lead to a conformation with the pore wide enough for water to penetrate through (Figure 10.6) (Wu and Voth, 2003a). Our density functional theory calculations showed a pair of histidine and tryptophan in this conformation has much lower energy than that obtained by rotating the histidine's χ_2 angle from 60° to 180°, which can yield a conformation having the same pore radius.

The close contact between His37 and Trp41 in this open structure makes it possible to have a cation–π interaction between the two residues. This is in accord with the UVRR study (Okada *et al.*, 2001). To examine how this structure may open the channel and at the same time maintain the proton-selectivity, we performed an MD simulation in which the positions of the backbone and the heavy atoms of the His37 and Trp41 residues were restrained and all His37 residues were biprotonated. Figure 10.7 shows the pore water structure of a snapshot after equilibration. The figure illustrates that the pore radius at the constrictive region is large enough to allow a water molecule to fit in while small enough to prevent ions larger than an excess proton from passing through efficiently. It can also be clearly seen that the pore water molecules are highly ordered and their orientations forbid proton transport from the N-end to the histidine residues. This can be ascribed to the four positively charged histidine residues. From this result, it may be concluded that the speculative open conformation can open the channel without loss of proton-selectivity and that it is not very likely for the open channel to have fully biprotonated histidines unless considerable conformational change of the backbone occur during protonation.

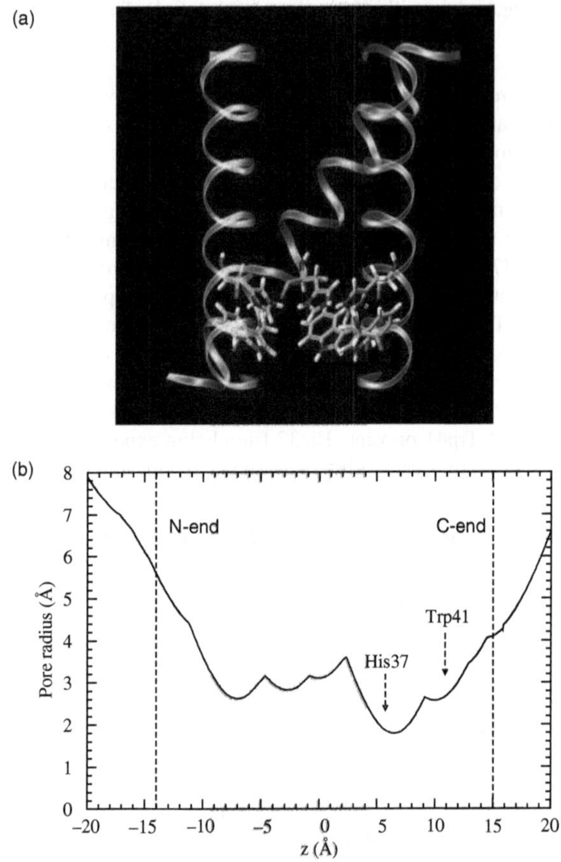

Figure 10.6. (a) The conformation of His37 and Trp41 for the speculative open channel and (b) the corresponding pore radius profile. The backbone of the channel is displayed by coils. For the sake of clarity, one helix and one pair of His37 and Trp41 are not shown.

4.3. Further MD Simulations with the Closed Structure

In order to examine the stability of the closed conformation, we have recently performed two MD simulations with the Amber 99 force field. One of them has position-restraints only on the backbone atoms, while the other has no restraint. The conformational stability can be evaluated through the calculated root mean squared deviations (RMSD) as functions of time for the side chain atoms of His37 and Trp41 (Figure 10.8). Interestingly, the conformation is more stable in the restrained simulation, while it is not in the free simulation. Visual examination of the final structure also confirmed that the conformation of His37 and Trp41 was maintained very well in the restrained simulation, while one pair of His37 and Trp41 in the free one lost their original conformation. The high stability in the restrained simulation suggests that dihedral potentials of Amber 99 force field for the two amino acids

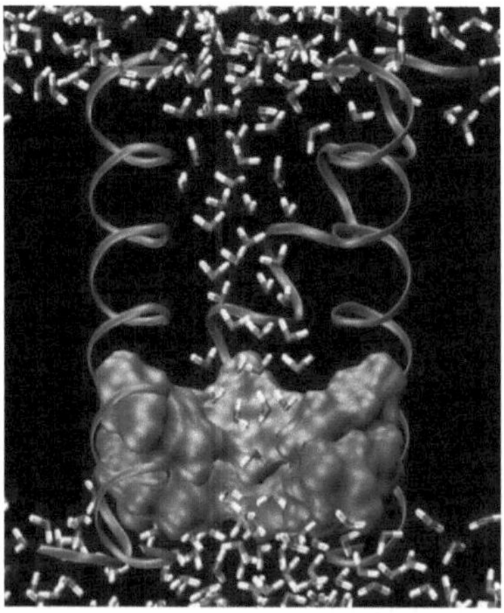

Figure 10.7. A snapshot from a MD simulation of the possible open structure. The backbone and heavy atoms of His37 and Trp41 were position-restrained in the simulation. The His37 residues (in gray color) and the Trp41 residues (dotted area) are displayed as their molecular surface. Notice that the pore is open enough for a single water molecule to fit into the constrictive region and that the orientation of the pore water molecules does not allow proton to hop through from the N-end to the histidine residues.

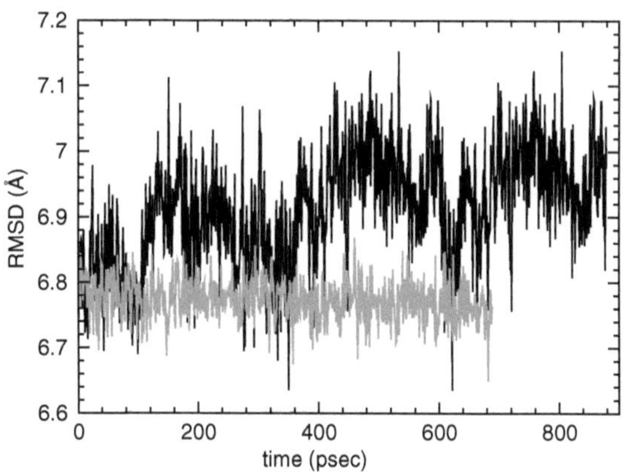

Figure 10.8. The root mean squared deviation (RMSD) of the side chains of the His37 and Trp41 residues as a function of the time for two simulations: The one with position-restraint on the backbone atoms (lower one) and the other without restraint (upper one).

favor this conformation. However, the energy barrier for maintaining this conformation is not high; they can be overcome easily by, for example, entropy effects as the restraints on the backbone are removed, which is reflected by the observed relative instability in the free simulation.

One explanation for this result can be that our proposed closed structure may be incorrect. Some preliminary solid-state NMR data on the AMT/M2-TMP complex in liposome suggest the ε-nitrogen of His37 is protonated at the closed state (Hu and Cross, personal communication), thus in contradistinction to our proposed structure. Another explanation is that the lack of polarizability of the force field does not treat correctly the hydrogen–π interaction between the δ-hydrogen of His37 and the indole ring of Trp41. Hydrogen–π interaction has been recognized as a significant contributor to molecular structural stability (Babu, 2003; and references cited therein). The second explanation can be examined through re-parameterization for the particular conformation, which is currently underway in our group.

5. A Revised Gating Mechanism and Future Work

The MS-EVB simulation results and our recent structural and simulation data have led us to propose a revised gating mechanism (Wu and Voth, 2003a), described briefly as follows: The closed channel adopts the (**t60, t90 δ**) conformation. When the pH goes down, one (or two) histidine residue becomes bi-protonated and undergoes a small conformational change by rotating its χ_2 angle from 60° to 0°. The resultant conformation provides a widened pore so that pore waters penetrate, forming a proton-conductive water wire through the gate region. Protons can then diffuse through channel by hopping through the pore water molecules via the Grotthuss shuttle mechanism.

The proposed mechanism will be subject to examinations by both future MD simulations and experiments. Our most recent development of an MS-EVB for histidine will allow us to examine the role of the key residue His37 in the conductance mechanism in more detail. Furthermore, we have recently been able to simulate multiple excess protons in the M2 channel with the completion of an MS-EVB simulation framework for multiple chemical reactions in system. These future simulation studies will certainly provide valuable microscopic details and raise new questions and concepts concerning the proton conductance mechanism in the M2 channel.

Acknowledgments

This work was supported by the National Institutes of Health (GM53148). Computational support from the Center for High Peformance Computing of the University of Utah is gratefully acknowledged. We also thank Jun Hu and Dr. Timothy A. Cross for offering their preliminary data and helpful discussion.

References

Babu, M.M. (2003). NCI: A server to identify non-canonical interactions in protein structures. *Nucl. Acids. Res.* **31**, 3345–3348.

Bauer, C.M., Pinto, L.H., Cross, T.A., and Lamb, R.A. (1999). The influenza virus M2 ion channel protein: Probing the structure of the transmembrane domain in intact cells by using engineered disulfide cross-linking. *Virology* **254**, 196–209.

Bukrinskaya, A.G., Vorkunova, N.K., Kornilayeva, G.V., Narmanbetova, R.A., and Vorkunova, G.K. (1982). Influenza virus uncoating in infected cells and effect of rimantadine. *J. Gen. Virol.* **60**, 49–59.

Castrucci, M.R., Hughes, M., Calzoletti, L., Donatelli, I., Wells, K., Takada, A. *et al.* (1997). The cysteine residues of the M2 protein are not required for influenza A virus replication. *Virology* **238**, 128–134.

Chizhmakov, I.V., Geraghty, F.M., Ogden, D.C., Hayhurst, A., Antoniou, M., and Hay, A.J. (1996). Selective proton permeability and pH regulation of the influenza virus M2 channel expressed in mouse erythroleukaemia cells. *J. Physiol.* **494 (Pt 2)**, 329–336.

Chizhmakov, I.V., Ogden, D.C., Geraghty, F.M., Hayhurst, A., Skinner, A., Betakova, T. *et al.* (2003). Differences in conductance of M2 proton channels of two influenza viruses at low and high pH. *J. Physiol.* **546**, 427–438.

Ciampor, F., Thompson, C.A., Grambas, S. and Hay, A.J. (1992). Regulation of pH by the M2 protein of influenza A viruses. *Virus. Res.* **22**, 247–258.

Cuma, M., Schmitt, U.W., and Voth, G.A. (2000). A multi-state empirical valence bond model for acid-base chemistry in aqueous solution. *Chem. Phys.* **258**, 187–199.

Cuma, M., Schmitt, U.W., and Voth, G.A. (2001). A multi-state empirical valence bond model for weak acid dissociation in aqueous solution. *J. Phys. Chem. A* **105**, 2814–2823.

Day, T.J.F., Schmitt, U.W., and Voth, G.A. (2000). The mechanism of hydrated proton transport in water. *J. Am. Chem. Soc.* **122**, 12027–12028.

Day, T.J.F., Soudackov, A.V., Schmitt, U.W., and Voth, G.A. (2002). A second generation multistate empirical valence bond model for proton transport in aqueous systems. *J. Chem. Phys.* **117**, 5839–5849.

Duff, K.C. and Ashley, R.H. (1992). The transmembrane domain of influenza A M2 protein forms amantadine-sensitive proton channels in planar lipid bilayers. *Virology* **190**, 485–489.

Duff, K.C., Kelly, S.M., Price, N.C., and Bradshaw, J.P. (1992). The secondary structure of influenza A M2 transmembrane domain. A circular dichroism study. *FEBS Lett.* **311**, 256–258.

Gandhi, C.S., Shuck, K., Lear, J.D., Dieckmann, G.R., DeGrado, W.F., Lamb, R.A. *et al.* (1999). Cu^{++} inhibition of the proton translocation machinery of the influenza A virus M2 protein. *J. Biol. Chem.* **274**, 5474–5482.

Grambas, S. and Hay, A.J. (1992). Maturation of influenza A virus hemagglutinin—estimates of the pH encountered during transport and its regulation by the M2 protein. *Virology* **190**, 11–18.

Grambas, S., Bennett, M.S., and Hay, A.J. (1992). Influence of amantadine resistance mutations on the pH regulatory function of the M2 protein of influenza A viruses. *Virology* **191**, 541–549.

Holsinger, L.J. and Lamb, R.A. (1991). Influenza virus M2 integral membrane protein is a homotetramer stabilized by formation of disulfide bonds. *Virology* **183**, 32–43.

Holsinger, L.J., Nichani, D., Pinto, L.H., and Lamb, R.A. (1994). Influenza A virus M2 ion channel protein: A structure-function analysis. *J. Virol.* **68**, 1551–1563.

Hull, J.D., Gilmore, R., and Lamb, R.A. (1988). Integration of a small integral membrane protein, M2, of influenza virus into the endoplasmic reticulum: Analysis of the internal signal-anchor domain of a protein with an ectoplasmic NH_2 terminus. *J. Cell. Biol.* **106**, 1489–1498.

Kochendoerfer, G.G., Salom, D., Lear, J.D., Wilk-Orescan, R., Kent, S.B., and DeGrado, W.F. (1999). Total chemical synthesis of the integral membrane protein influenza A virus M2: Role of its C-terminal domain in tetramer assembly. *Biochemistry* **38**, 11905–11913.

Kovacs, F.A. and Cross, T.A. (1997). transmembrane four-helix bundle of influenza A M2 protein channel: Structural implications from helix tilt and orientation. *Biophys. J.* **73**, 2511–2517.

Kovacs, F.A., Denny, J.K., Song, Z., Quine, J.R., and Cross, T.A. (2000). Helix tilt of the M2 transmembrane peptide from influenza A virus: An intrinsic property. *J. Mol. Biol.* **295**, 117–125.

Kukol, A., Adams, P.D., Rice, L.M., Brunger, A.T., and Arkin, T.I. (1999). Experimentally based orientational refinement of membrane protein models: A structure for the influenza A M2 H^+ channel. *J. Mol. Biol.* **286**, 951–962.

Lamb, R.A. and Choppin, P.W. (1981). Identification of a second protein (M2) encoded by RNA segment 7 of influenza virus. *Virology* **112**, 729–737.

Lamb, R.A., Lai, C.J., and Choppin, P.W. (1981). Sequences of mRNAs derived from genome RNA segment 7 of influenza virus: Colinear and interrupted mRNAs code for overlapping proteins. *Proc. Natl. Acad. Sci. USA* **78**, 4170–4174.

Lamb, R.A., Zebedee, S.L., and Richardson, C.D. (1985). Influenza virus M2 protein is an integral membrane protein expressed on the infected-cell surface. *Cell* **40**, 627–633.

Lovell, S.C., Word, J.M., Richardson, J.S., and Richardson, D.C. (2000). The penultimate rotamer library. *Proteins* **40**, 389–408.

Martin, K. and Helenius, A. (1991). Transport of incoming influenza virus necleocapsids into the nucleus. *J. Virol.* **65**, 232–244.

Mould, J.A., Drury, J.E., Frings, S.M., Kaupp, U.B., Pekosz A., Lamb, R.A. *et al.* (2000). Permeation and activation of the M2 ion channel of influenza A virus. *J. Biol. Chem.* **275**, 31038–31050.

Nishimura, K., Kim, S., Zhang L., and Cross, T.A. (2002). The closed state of a H^+ channel helical bundle combining precise orientational and distance restraints from solid state NMR. *Biochemistry* **41**, 13170–13177.

Okada, A., Miura, T., and Takeuchi, H. (2001). Protonation of histidine and histidine-tryptophan interaction in the activation of the M2 ion channel from influenza A virus. *Biochemistry* **40**, 6053–6060.

Pinto, L.H., Dieckmann, G.R., Gandhi, C.S., Papworth, C.G., Braman, J., Shaughnessy, M.A. *et al.* (1997). A functionally defined model for the M2 proton channel of influenza A virus suggests a mechanism for its ion selectivity. *Proc. Natl. Acad. Sci. USA* **94**, 11301–11306.

Pinto, L.H., Holsinger, L.J., and Lamb, R.A. (1992). Influenza virus M2 protein has ion channel activity. *Cell* **69**, 517–528.

Sakaguchi, T., Tu, Q., Pinto, L.H., and Lamb, R.A. (1997). The active oligomeric state of the minimalistic influenza virus M2 ion channel is a tetramer. *Proc. Natl. Acad. Sci. USA* **94**, 5000–5005.

Salom, D., Hill, B.R., Lear, J.D., and DeGrado, W.F. (2000). pH-dependent tetramerization and amantadine binding of the transmembrane helix of M2 from the influenza A virus. *Biochemistry* **39**, 14160–14170.

Sansom, M.S., Kerr, I.D., Smith, G.R., and Son, H.S. (1997). The influenza A virus M2 channel: A molecular modeling and simulation study. *Virology* **233**, 163–173.

Schmitt, U.W. and Voth, G.A. (1998). Multistate empirical valence bond model for proton transport in water. *J. Phys. Chem. B* **102**, 5547–5551.

Schmitt, U.W. and Voth, G.A. (1999a). The computer simulation of proton transport in water. *J. Chem. Phys.* **111**, 9361–9381.

Schmitt, U.W. and Voth, G.A. (1999b). Quantum properties of the excess proton in liquid water. *Isr. J. Chem.* **39**, 483–492.

Schmitt, U.W. and G.A. Voth (2000). The isotope substitution effect on the hydrated proton. *Chem. Phys. Lett.* **329**, 36–41.

Schweighofer, K.J. and Pohorille, A. (2000). Computer simulation of ion channel gating: The M2 channel of influenza A virus in a lipid bilayer. *Biophys. J.* **78**, 150–163.

Shimbo, K., Brassard, D.L., Lamb, R.A., and Pinto, L.H. (1996). Ion selectivity and activation of the M2 ion channel of influenza virus. *Biophys. J.* **70**, 1335–1346.

Shuck, K., Lamb, R.A., and Pinto, L.H. (2000). Analysis of the pore structure of the influenza A virus M2 ion channel by the substituted-cysteine accessibility method. *J. Virol.* **74**, 7755–7761.

Smondyrev, A.M. and Voth, G.A. (2002). Molecular dynamics simulation of proton transport through the influenza A virus M2 channel. *Biophys. J.* **83**, 1987–1996.

Song, Z., Kovacs, F.A., Wang, J., Denny, J.K., Shekar S.C., Quine, J.R. *et al.* (2000). Transmembrane domain of M2 protein from influenza A virus studies by solid-state ^{15}N polarization inversion spin exchange at magic angle NMR. *Biophys. J.* **79**, 767–775.

Sugrue, R.J. and Hay, A.J. (1991). Structural characteristics of the M2 protein of influenza A viruses: Evidence that it forms a tetrameric channel. *Virology* **180**, 617–624.

Sugrue, R.J., Bahadur, G., Zambon, M.C., Hall-Smith, M., Douglas, A.R., and Hay, A.J. (1990). Specific structural alteration of the influenza haemagglutinin by amantadine. *EMBO J.* **9**, 3469–3476.

Tajkhorshid, E., Nollert, P., Jensen, M.O., Miercke, L.J., O'Connell, J., Stroud, R.M., and Schulten, K. (2002). Control of the selectivity of the aquaporin water channel family by global orientational tuning. *Science* **296**, 525–530.

Takeuchi, K. and Lamb, R.A. (1994). Influenza virus M2 protein ion channel activity stabilizes the native form of fowl plague virus hemagglutinin during intracellular transport. *J. Virol.* **68**, 911–919.

Tang, Y., Zaitseva, R., Lamb, R.A., and Pinto, L.H. (2002). The gate of the influenza virus M2 proton channel is formed by a single tryptophan residue. *J. Biol. Chem.* **277**, 39880–39886.

Tian, C., Tobler, K., Lamb, R.A., Pinto, L.H., and Cross, T.A. (2002). Expression and initial structural insights from solid-state NMR of the M2 proton channel from influenza A virus. *Biochemistry* **41**, 11294–11300.

Torres, J., Kukol, A., and Arkin, I.T. (2000). Use of a single glycine residue to determine the tilt and orientation of a transmembrane helix. A new structural label for infrared spectroscopy. *Biophys. J.* **79**, 3139–3143.

Torres, J. and Arkin, I.T. (2002). C-deuterated alanine: A new label to study membrane protein structure using site-specific infrared dichroism. *Biophys. J.* **82**, 1068–1075.

Tuckerman, M., Laasonen, K., Sprik, M., and Parrinello, M. (1995). Ab initio molecular dynamics simulation of the solvation and transport of H_3O^+ and OH^- ions in water. *J. Chem. Phys.* **103**, 150–161.

Wang, C., Lamb, R.A., and Pinto, L.H. (1995). Activation of the M2 ion channel of influenza virus: A role for the transmembrane domain histidine residue. *BioPhys. J.* **69**, 1363–1371.

Wang, C., Takeuchi, K., Pinto, L.H., and Lamb, R.A. (1993). Ion channel activity of influenza A virus M2 protein: Characterization of the amantadine block. *J. Virol.* **67**, 5585–5594.
Wang, J., Kim, S., Kovacs, F., and Cross, T.A. (2001). Structure of the transmembrane region the M2 protein H^+ channel. *Protein Sci.* **10**, 2241–2250.
Wu, Y. and Voth, G.A. (2003a). Computational studies of proton transport through the M2 channel. *FEBS Lett.* **552**, 23–27.
Wu, Y. and Voth, G.A. (2003b). A computer simulation study of the hydrated proton in a synthetic proton channel. *Biophys. J.* **85**, 864–875.

11

Structure and Function of Vpu from HIV-1

S.J. Opella, S.H. Park, S. Lee, D. Jones, A. Nevzorov, M. Mesleh, A. Mrse, F.M. Marassi, M. Oblatt-Montal, M. Montal, K. Strebel, and S. Bour

1. Introduction

Virus protein "u" (Vpu) contributes to the virulence of HIV-1 infections of humans by enhancing the production and release of progeny virus particles. Its biological activities are associated with the two distinct structural domains of the protein. Since the entire polypeptide consists of only 81 amino acid residues, each of the biological activities is associated with a relatively small and well-defined structural entity. This suggests that the three-dimensional structure of the protein will lead to a detailed understanding of its biological functions, and potentially to the identification of small molecules that act as drugs by interfering with its functions (Miller and Sarver, 1997) as has been done for other HIV-1 encoded proteins (Turner and Summers, 1999; Wlodawer, 2002). The many structure determinations of HIV protease alone and complexed with inhibitors led to the development of the highly effective drugs that are a mainstay of current therapy for AIDS (Erickson and Burt, 1996; Vondrasek *et al.*, 1997). Even though the protease is about 20% larger than Vpu, its structure was determined very soon after its discovery (Navia *et al.*, 1989; Wlodawer *et al.*, 1989), while the structure of Vpu is yet to be determined. The reasons that Vpu has not followed quickly in the path of protease have their roots in the most fundamental aspects of experimental structural biology and biochemistry. Vpu is a helical membrane protein, and it requires the presence of lipids and water to adopt its native functional structure. The lipids interfere with the formation of crystals for X-ray diffraction as well as the preparation of samples suitable for multidimensional solution NMR spectroscopy. In contrast, protease is a globular, soluble protein well suited for experimental structure determination by both X-ray crystallography and

S.J. Opella, S.H. Park, S. Lee, D. Jones, A. Nevzorov, M. Mesleh, and A. Mrse • Department of Chemistry and Biochemistry, University of California, San Diego, La Jolla, California. **F.M. Marassi** • The Burnham Institute, La Jolla, California. **M. Oblatt-Montal, M. Montal** • Section of Neurobiology Division of Biology, University of California, San Diego, La Jolla, California. **S. Bour** • Bioinformatics and Cyber Technology Center, Office of Technology and Information Systems, National Institute of Allergy and Infectious Diseases, National Institutes of Health, Bethesda, Maryland. **K. Strebel** • Viral Biochemistry Section, Laboratory of Molecular Microbiology, National Institute of Allergy and Infectious Diseases, National Institutes of Health, Bethesda, Maryland.

Viral Membrane Proteins: Structure, Function, and Drug Design, edited by Wolfgang Fischer.
Kluwer Academic / Plenum Publishers, New York, 2005.

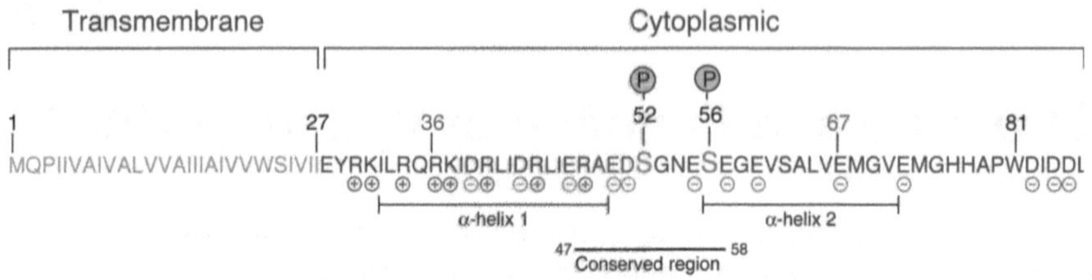

Figure 11.1. The primary structure of Vpu from HIV-1 (Bour and Strebel, 2003).

solution NMR spectroscopy (Ishima *et al.*, 2003). In addition, protease is an enzyme with a well-defined active site. Not only is Vpu a helical membrane protein, but also it has multiple biological functions that involve two different domains that act as channels or interact with other proteins, making it a much more complex system for structural biology.

In addition to the prototypical retroviral *gag, pol*, and *env* genes, HIV-1 encodes a number of so-called accessory genes that perform essential functions during the viral life cycle. The term accessory genes was coined following the finding that their inactivation resulted in little or no impairment of virus replication in continuous cell lines (Strebel *et al.*, 1988; Fan and Peden, 1992). However, subsequent studies have shown that accessory gene products can dramatically change the course and severity of the viral infection (Bour and Strebel, 2000, 2003) and thus have the potential to be novel targets for drugs (Miller and Sarver, 1997).

The *vpu* gene is found exclusively in HIV-1. The Rev-dependent bicistronic mRNA that encodes Vpu also contains the downstream Env ORF, which is translated by leaky scanning of the Vpu initiation codon (Schwartz *et al.*, 1990). In primary isolates, the *vpu* gene is not always functional due to the presence of mutated initiation codons or internal deletions (Korber *et al.*, 1997), suggesting a mechanism by which Vpu expression is regulated by the virus.

Vpu is an 81 amino acid type 1 integral membrane protein (Cohen *et al.*, 1988; Strebel *et al.*, 1988). As shown with its annotated primary sequence in Figure 11.1 and the schematic representations in the other figures, it has an N-terminal hydrophobic transmembrane helix and a C-terminal cytoplasmic domain that has two amphipathic in-plane helices. The highly conserved region spanning residues 47 to 58 between the two amphipathic helices in the cytoplasmic domain contains a pair of serine residues that are constitutively phosphorylated by casein kinase II (Schubert *et al.*, 1994) and is of particular interest. Based on structural similarities between Vpu and the influenza virus M2 ion-channel protein it was speculated that homo-oligomeric complexes of Vpu might possess pore-forming abilities (Maldarelli *et al.*, 1993). Thus, the three-dimensional structures of the polypeptides and their oligomeric arrangement in membranes are of interest for understanding the functions of Vpu (Cordes *et al.*, 2001; Fischer and Sansom, 2002; Fischer, 2003).

2. Structure Determination of Vpu

Initial attempts at characterizing the structures of polypeptides corresponding to portions of Vpu were limited by the intrinsic properties of hydrophobic helical membrane proteins, the use of synthetic peptides, solvent rather than lipid environments, and the experimental methods.

Only recently has the first atomic-resolution structure of a Vpu polypeptide become available (Park *et al.*, 2003), and that was the result of integrating the studies of Vpu into the development of new experimental methods for determining the structures of membrane proteins by NMR spectroscopy (Opella, 2002). In addition, substantial progress has been made toward the structure determination of the cytoplasmic domain as well as the full-length protein with these methods. The three-dimensional structures of full-length Vpu and of its structural and functional domains will be compared to the findings of previous investigations of the structure and dynamics of a variety of Vpu polypeptides, including by infrared dichroism (Kukol and Arkin, 1999), NMR spectroscopy (Wray *et al.*, 1995, 1999; Federau *et al.*, 1996; Willbold *et al.*, 1997; Henklein *et al.*, 2000; Coadou *et al.*, 2002, 2003), and molecular dynamics simulations (Grice *et al.*, 1997; Moore *et al.*, 1998; Torres *et al.*, 2001; Cordes *et al.*, 2002; Lopez *et al.*, 2002). Both biochemical experiments and computer simulations and modeling have provided insights into how the oligomerization of Vpu could lead to the formation of ion-conductive membrane pores, and these efforts will benefit from the availability of the atomic-resolution structures of the channel-forming domain and the full-length protein.

Vpu is a typical helical membrane protein. A major challenge in structural biology is the determination of the structures of membrane proteins. Despite substantial efforts, Vpu has not been crystallized and continues to remain outside the realm of X-ray crystallography. Thus, NMR is the only method that is applicable, since it does not require crystals, and is capable of determining the three-dimensional structures of proteins with atomic resolution. However, the conventional multidimensional solution NMR methods that work well with globular proteins, such as HIV protease, are problematic for membrane proteins, even relatively small ones like Vpu, because the slow reorientation rates that accompany solubilization in micelles result in broad resonance line widths, short relaxation times, and efficient spin-diffusion, all of which contribute to the difficulty in observing and assigning long-range NOEs as distance constraints for determining the three-dimensional structures of these proteins. Proteins reconstituted into the other well-characterized model membrane environment of lipid bilayers are more attractive from a biochemical perspective, but are even less well suited for solution NMR methods; since the polypeptides are effectively immobilized on the relevant NMR timescales by their interactions with the phospholipids they give spectra with broad, ill-behaved resonances.

However, it is important to recognize that NMR studies of membrane proteins are difficult only because of the motional properties of the samples and not the properties of the polypeptides. We are developing two complementary NMR approaches to structure determination to deal with this problem (Opella, 1997; Opella *et al.*, 2002), both of which are being applied in parallel to Vpu polypeptides (Marassi *et al.*, 1999; Ma *et al.*, 2002; Park *et al.*, 2003). The first utilizes completely aligned bilayers samples and solid-state NMR separated local field experiments (Waugh, 1976; Wu *et al.*, 1994; Nevzorov and Opella, 2003) that measure orientationally dependent dipolar coupling and chemical shift frequencies. The second utilizes weakly aligned micelle samples (Ma and Opella, 2000; Veglia and Opella, 2000; Chou *et al.*, 2002) and multidimensional solution NMR spectroscopy for the measurement of residual dipolar couplings and chemical shift anisotropies as orientational constraints (Bax *et al.*, 2001; Prestegard *et al.*, 2001). The use of orientational constraints is resulting in the convergence of solution NMR and solid-state NMR approaches to the structure determination of membrane proteins (Opella *et al.*, 2002; Mesleh *et al.*, 2003).

Many of the NMR methods have been developed with Vpu in mind, including the expression, sample preparation, and spectroscopic and functional studies of individual structural and functional domains. The association of the two principal activities of Vpu with different portions of the protein molecule is a strong influence on the plan for structure determination. The degradation of CD4 appears to be affected by the cytoplasmic domain, and the ion-channel activity resides in the transmembrane helix, which oligomerizes to form channels. The NMR studies of Vpu are illustrated with results from polypeptides corresponding to the transmembrane helix, the cytoplasmic domain, and full-length Vpu with both domains. All of the polypeptides are prepared by expression in bacteria, which enables uniform and selective isotopic labeling.

Proteins associated with phospholipid bilayers are immobile on the 10^4 Hz timescales of the dominant chemical shift and heteronuclear dipolar interactions. Therefore, solid-state NMR methods originally developed for crystals and powders are applicable to fully hydrated samples of protein-containing lipid bilayers as well as other biological supramolecular assemblies (Pake, 1948; Waugh, 1976; Opella *et al.*, 1987). Figure 11.2 compares the one-dimensional solid-state NMR spectra of uniformly ^{15}N labeled full-length Vpu(2–81) to that of Vpu(28–81), the polypeptide corresponding to the cytoplasmic domain. The frequency of each backbone amide nitrogen resonance is determined by the anisotropic chemical shift interaction because the dipolar couplings to the nearby hydrogens are decoupled, and this is sufficient to qualitatively determine the orientations of the three helical segments of Vpu in membrane bilayers. The resonances in the solid-state NMR spectrum of full-length Vpu are clearly segregated into two distinct bands at chemical shift frequencies associated with NH bonds in the transmembrane helix perpendicular to the membrane surface (200 ppm) and with NH bonds in both cytoplasmic helices parallel to the membrane surface (70 ppm). Moreover, these data show that the cytoplasmic domain adopts a unique conformation in the presence of phospholipids and binds strongly to the membrane surface in only one way. In phospholipid bilayers, the protein is fully immobilized in its native conformation, and there is no evidence in these spectra of a substantial number of residues undergoing local motions. Moreover, the fine structure of the resonances is very similar for the in-plane (70 ppm) regions in both the full-length and cytoplasmic domain polypeptides, which suggests that the in-plane helices have the same conformations and orientations with and without the presence of the N-terminal transmembrane helix. These data serve to verify the secondary structure and establish the

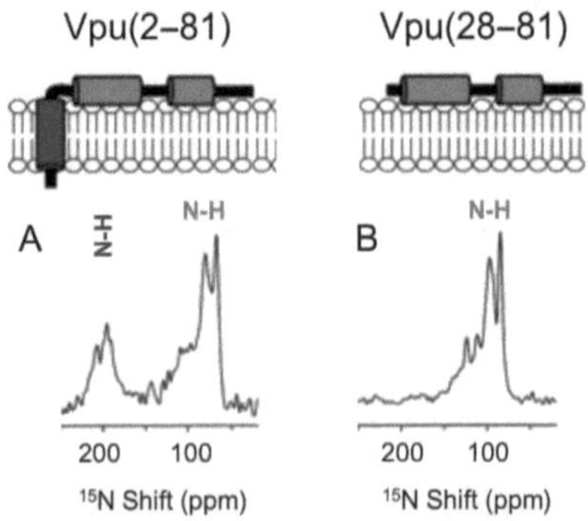

Figure 11.2. One-dimensional ^{15}N solid-state NMR spectra of uniformly ^{15}N-labeled polypeptides in completely aligned bilayers. A. Full-length Vpu (residues 2–81). B. Cytoplasmic domain of Vpu (residues 28–81) (Marassi *et al.*, 1999).

overall topology of Vpu in membrane bilayers. Resolution of individual resonances is essential for obtaining structural information with atomic resolution, and this requires multidimensional solid-state NMR experiments.

In the two-dimensional PISEMA (polarization inversion spin exchange at the magic angle) (Wu *et al.*, 1994) spectrum of uniformly ^{15}N labeled Vpu(2–30+) shown in Figure 11.3, each amide site contributes a single correlation resonance that is characterized by unique ^{1}H–^{15}N dipolar coupling and ^{15}N chemical shift frequencies. The unmistakable "wheel-like" pattern of resonances in the two-dimensional PISEMA spectrum of Vpu(2–30+) in a completely aligned bilayers sample is typical of those observed for transmembrane helices in uniformly ^{15}N labeled helical membrane proteins. These PISA (polarity index slant angle) wheel patterns correspond to helical wheel projections and are direct indices of secondary structure and topology (Marassi and Opella, 2000; Wang *et al.*, 2000). The magnitudes of the ^{1}H–^{15}N dipolar coupling frequencies can be measured directly from the spectra, and provide input for direct characterization of the helices and comparisons among the domains. The combination of the chemical shift and dipolar coupling frequencies enable the determination of the three-dimensional structure of the protein with atomic resolution (Park *et al.*, 2003).

Figure 11.4 compares the solution NMR spectra of the same polypeptides in micelles. Most of the correlation resonances of the polypeptide corresponding to the cytoplasmic domain overlap with the corresponding resonances in the spectrum of the full-length protein. This suggests that the residues that constitute the cytoplasmic domain of Vpu have very similar local environments and properties whether or not the transmembrane domain is attached at the N-terminus. Similarly, the resonances from the transmembrane helical residues are unaffected by the presence or absence of the cytoplasmic domain (Marassi *et al.*, 1999). Thus, both the solid-state NMR data in Figure 11.2 and the solution NMR data in Figure 11.4 indicate that Vpu is a modular protein with two essentially independent structural domains. In separate experiments, it was found that there is a continuous stretch of 17 amino acid residues in the middle of the polypeptide that are resistant to hydrogen exchange (Park *et al.*, 2003). By this criterion, the transmembrane helix of Vpu consists of residues 8–25 in micelles in agreement with the finding of the resonances from the same residues in the appropriate region of the PISEMA spectrum in Figure 11.3 of the polypeptide in lipid bilayers.

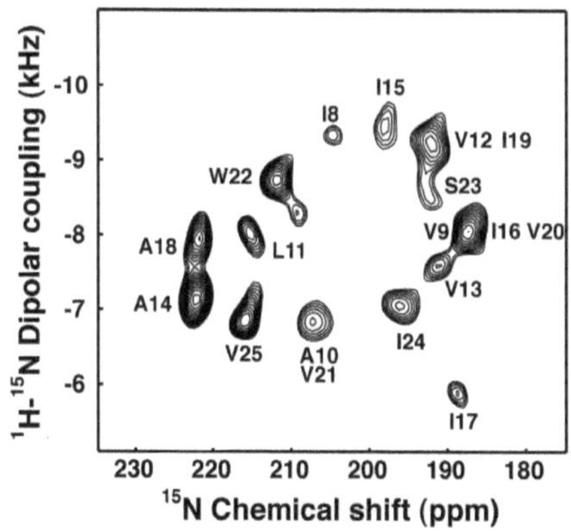

Figure 11.3. Two-dimensional PISEMA spectrum of uniformly ^{15}N-labeled transmembrane domain of Vpu (residues 2–30+) (Park *et al.*, 2003).

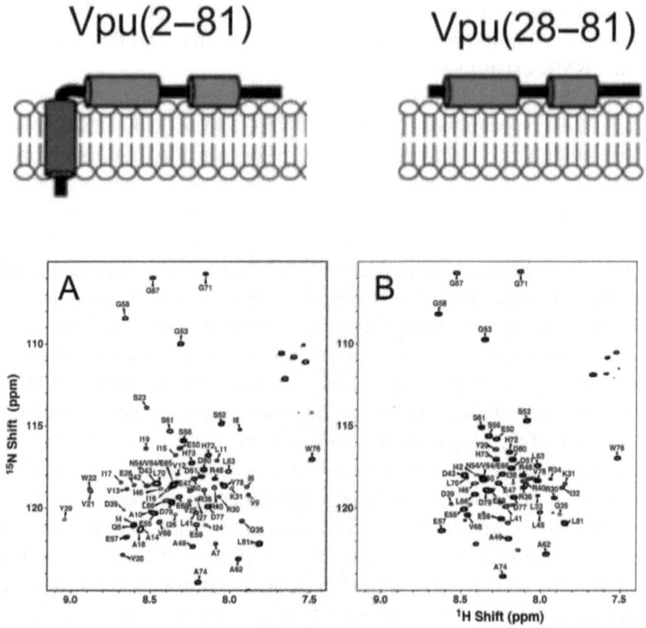

Figure 11.4. Two-dimensional HSQC solution NMR spectra of uniformly ^{15}N-labeled polypeptides in micelles. A. Full-length Vpu (residues 2–81). B. Cytoplasmic domain of Vpu (residues 28–81). (Reproduced with permission from Marassi *et al.* (1999). *Proc. Natl. Acad. Sci. USA* **96**, © 1999.)

The polypeptide in micelles was weakly aligned by soaking it into a polycrylamide gel. Then, by comparisons between measurements on the protein in the stressed gel and in isotropic micelle solutions, it is possible to measure residual dipolar couplings. The unaveraged dipolar couplings measured in solid-state NMR experiments and the residual dipolar couplings measured in solution NMR experiments are completely analogous representations of the anisotropic dipole–dipole interactions first described by Pake (1948) and brought to high resolution by Waugh (1976) with separated local field spectroscopy. One-dimensional Dipolar Waves are an extension of two-dimensional PISA wheels, both of which map protein structure from data obtained on weakly aligned and completely aligned samples. Dipolar Waves describe the periodic variation of the magnitudes of dipolar couplings as function of residue number in aligned samples (Mesleh *et al.*, 2002, 2003; Mesleh and Opella, 2003). We utilize the fit of the ^{1}H–^{15}N dipolar coupling from the backbone amide sites to sinusoids of periodicity 3.6 to characterize the length, curvature, orientation, and rotation of α-helices. They also serve as a sensitive index of the regularity and ideality of the helices in proteins. Dipolar waves from solid-state NMR data give absolute measurements of helix orientations because the polypeptides are immobile and the samples have a known alignment in the magnetic field. In contrast, solution NMR data give relative orientations of helices in a common molecular frame (Mesleh *et al.*, 2003).

Experimental dipolar couplings measured from the PISEMA spectrum of a uniformly ^{15}N labeled sample of Vpu(2–30+) are plotted as a function of residue number for the transmembrane helix of Vpu in Figure 11.5. The residues in the helix are identified on the basis of scoring with a 6-residue sliding window that measures the quality of fit to a sinusoid with

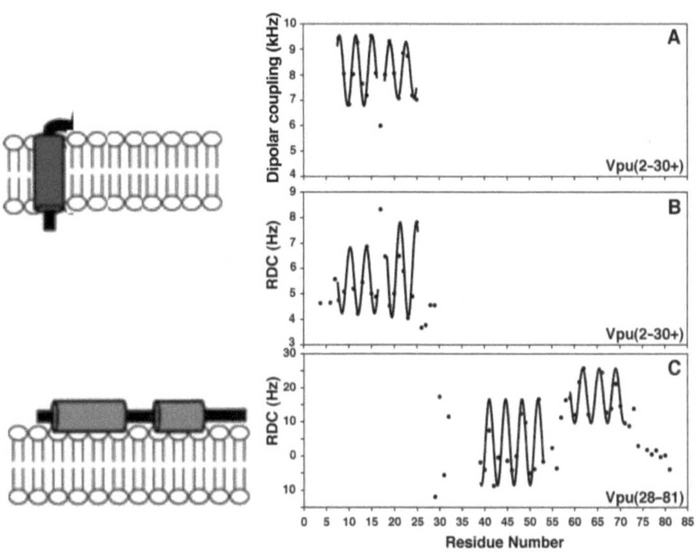

Figure 11.5. Dipolar Waves for N–H couplings in aligned samples of uniformly ^{15}N-labeled polypeptides. A. and B. are from the transmembrane domain of Vpu (2–30+). C. is from the cytoplasmic domain of Vpu (residues 28–81). A. represents unaveraged dipolar couplings from a completely aligned bilayers sample. B. and C. represent residual dipolar couplings from weakly aligned micelle samples (Park *et al.*, 2003).

a periodicity of 3.6 residues. The transmembrane helix of Vpu has two distinct components because there is a kink at isoleucine 17, and this is observed in data from both micelle and bilayer samples. This suggests that all of the detailed structural features of the helix, that is, the constituent residues, polarity, tilt, and kink, are properties of the polypeptide itself and are not induced by specific intermolecular interactions with lipids.

The measured value of the dipolar coupling of Ile17 deviates markedly from the sinusoidal functions that fit well to the neighboring sites. Higher values of the scoring function and changes in phase are observed near Ile17 relative to those observed for other residues in the transmembrane helical region. These data clearly indicate that there is a deviation from ideality of the helix near Ile17. The different amplitudes and average values of the sine waves for the residues show that there is a kink in the helix at residue 17, and that the two helical segments have slightly different orientations. The tilt angle of the helical segments in the lipid bilayers can only be defined from the solid-state NMR data where the alignment frame is established by the placement of the sample in the magnet. Residues 8–16 have a tilt angle of 12° and residues 18–26 have a tilt angle of 15° with a slightly different rotation angle. The two components of the transmembrane helix are represented by tubes in Figure 11.6.

3. Correlation of Structure and Function of Vpu

In addition to its destabilizing effect on CD4, Vpu mediates the efficient release of viral particles from HIV-1-infected cells (Strebel *et al.*, 1989; Klimkait *et al.*, 1990). These two

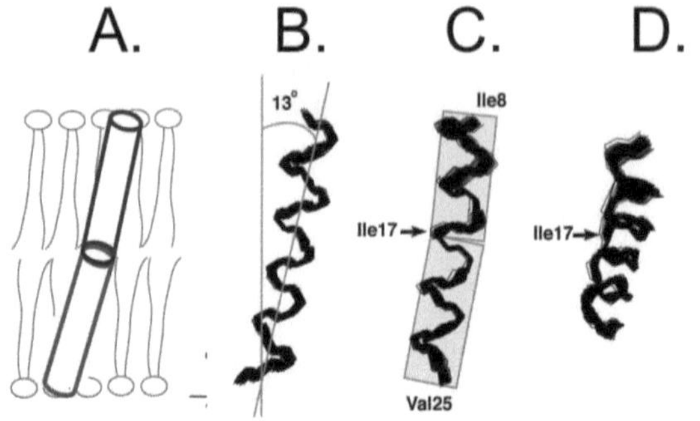

Figure 11.6. Structural representations of the transmembrane helix of Vpu. A. A tube representation of the alignment in lipid bilayers. B. Superposition of 100 calculated backbone three-dimensional structures. C. A 90° rotation of panel B to the vertical axis. D. A 30° tilt of panel C toward the reader (Park *et al.*, 2003).

biological activities of Vpu appear to be mechanistically distinct and involve different structural domains in Vpu. For example, the particle release enhancing activity of Vpu is independent of CD4 and does not require the envelope glycoprotein. Also, mutations of serine residues 52 and 56, which are crucial for CD4 degradation, only partially affect virus release (Schubert and Strebel, 1994). In addition, while the determinants for CD4 degradation are all contained in the cytoplasmic domain of Vpu, the transmembrane domain has been shown to play an essential role for the particle release activity (Schubert *et al.*, 1996a; Paul *et al.*, 1998). The presence of Vpu does not affect the synthesis, processing or stability of the viral structural proteins (Strebel *et al.*, 1989).

3.1. Vpu-Mediated Enhancement of Viral Particle Release

Two documented scenarios can account for the enhanced viral progeny production in the presence of Vpu. First, Vpu appears to improve targeting of the plasma membrane as the main site of viral assembly. Indeed, Vpu-defective viruses show a higher rate of intracellular budding and accumulation of immature viral particles in intracytoplasmic vesicles (Klimkait *et al.*, 1990). Second, Vpu may affect a late stage of viral budding at the plasma membrane. In the absence of Vpu, both single particles and chains of tethered virions remain attached to the plasma membrane and fail to be released (Klimkait *et al.*, 1990). This phenomenon is reminiscent of late domain defects in the p6 Gag protein (Gottlinger *et al.*, 1991) but the two events appear to be unrelated (Schwartz *et al.*, 1996). Indeed, while in the case of p6 Gag mutations, virions that accumulate at the cell surface have an immature appearance and non-condensed cores (Gottlinger *et al.*, 1991), Vpu-defective tethered virions appear mature (Klimkait *et al.*, 1990). These data suggest that Vpu is not directly involved in virion formation or maturation but purely in the late release step involving the separation of the virion and cellular membranes.

Vpu ion-channel activity was experimentally demonstrated in two independent studies by measuring current fluctuations across an artificial lipid bilayer containing either full-length recombinant Vpu protein or synthetic peptides corresponding to the transmembrane domain of Vpu (Ewart *et al.*, 1996). In addition, voltage clamp analysis on amphibian oocytes

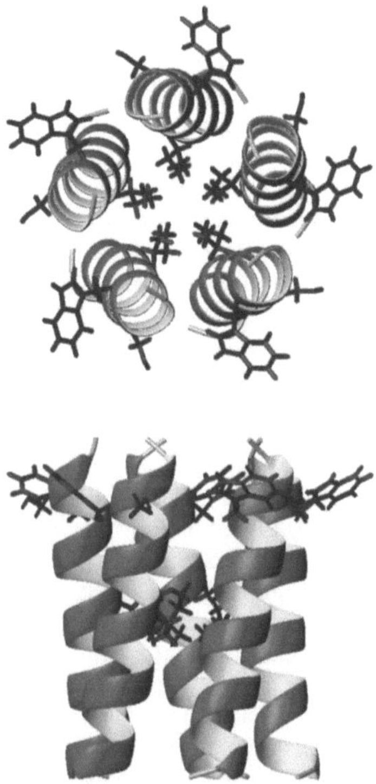

Figure 11.7. An optimized structure for the transmembrane domain of Vpu as a pentamer (Park *et al.*, 2003).

expressing full-length Vpu support the notion that Vpu forms ion-conductive pores (Schubert *et al.*, 1996b). The Vpu channel appears to be selective for monovalent cations such as sodium and potassium. There is an intriguing correlation between the ability of Vpu to form ion-conductive channels and its ability to enhance viral particle release *in vivo*. Indeed, a Vpu mutant bearing a transmembrane domain with a scrambled amino acid sequence lacked ion-channel activity and was unable to enhance virus particle release, yet retained full CD4 degradation activity (Schubert *et al.*, 1996b). Nevertheless, how an ion-channel activity of Vpu could lead to enhanced viral particle production is still unclear (Lamb and Pinto, 1997). It is conceivable that the channel activity of Vpu locally modifies the electric potential at the plasma membrane, leading to facilitated release of membrane budding structures. Alternatively, the action of the Vpu channel could induce cellular factors involved in the late stages of virus release or exclude cellular factors inhibitory to the viral budding process. Regardless of the exact mechanism of how the ion-channel activity affects the release of HIV particles, it is the best characterized function of the protein (Montal, 2003). Vpu-mediated enhancement of virus release is strictly dependent on the integrity of the transmembrane domain, and the ion-channel activity strongly correlates with the three-dimensional structure of the domain as a pentamer shown in Figure 11.7.

3.2. Vpu-Mediated Degradation of the CD4 Receptor

Receptor interference is a hallmark of retroviral infections that involves specific removal of the cellular receptor used for entry into the host. HIV-1 has been shown to effectively interfere with the transport, stability, and cell-surface localization of its specific receptor, CD4 (Bour *et al.*, 1995a). The gp160 envelope glycoprotein precursor (Env) and Vpu both significantly contribute to the viral effort to downregulate CD4. Gp160 is a major player in CD4 down-modulation that can, in most instances, quantitatively block the bulk of newly synthesized CD4 in the endoplasmic reticulum (ER) (Jabbar and Nayak, 1990; Crise *et al.*, 1990; Bour *et al.*, 1995a). However, the formation of CD4-gp160 complexes in the ER blocks the transport and maturation of not only CD4 but of the Env protein itself (Bour *et al.*, 1991). An important function of Vpu is to induce the degradation of CD4 molecules trapped in intracellular complexes with Env thus allowing gp160 to resume transport toward the cell surface (Willey *et al.*, 1992). Co-immunoprecipitation experiments showed that CD4 and Vpu physically interact in the ER and that this interaction is essential for targeting CD4 to the degradation pathway (Bour *et al.*, 1995b). Mutagenesis studies delineated a domain extending from residues 416 to 418 (EKKT) in the CD4 cytoplasmic domain required for degradation and Vpu binding (Lenburg and Landau, 1993; Vincent *et al.*, 1993; Bour *et al.*, 1995b). While two conserved serine residues at positions 52 and 56 in the cytoplasmic domain of Vpu are critically important for CD4 degradation (Schubert and Strebel, 1994; Paul and Jabbar, 1997), they are not required for CD4 binding since phosphorylation-defective mutants of Vpu retained the capacity to interact with CD4 (Bour *et al.*, 1995b). The role of the Vpu phosphoserine residues in the induction of CD4 degradation was elucidated when yeast two-hybrid assays as well as co-immunoprecipitation studies revealed an interaction of Vpu with the human beta Transducin-repeat Containing Protein (βTrCP) (Margottin *et al.*, 1998). Vpu variants mutated at serines 52 and 56 were unable to interact with βTrCP (Margottin *et al.*, 1998), providing a mechanistic explanation for the requirement for Vpu phosphorylation and strongly suggesting that βTrCP was directly involved in the degradation of CD4. Structurally, the Dipolar Waves in Figure 11.5 suggest that the two amphipathic in-plane helices of the cytoplasmic domain are not collinear and that the phosphorylation sites are well situated for interactions with other proteins. More details will be revealed when the three-dimensional structure of the phosphorylated and non-phosphorylated forms of the cytoplasmic domain become available for comparisons.

Structurally, βTrCP shows a modular organization. Similar to its *Xenopus laevis* homolog (Spevak *et al.*, 1993), human βTrCP contains seven C-terminal WD repeats that mediate interactions with Vpu in a phosphoserine-dependent fashion (Margottin *et al.*, 1998). In addition to the WD repeats, βTrCP contains an F-box domain that functions as a connector between proteins targeted for degradation and the ubiquitin-dependent proteolytic machinery (Figure 11.8) (Bai *et al.*, 1996). A number of proteasome degradation pathways involving βTrCP have recently been deciphered that resemble, at least in part, that of Vpu-mediated CD4 degradation. For example, ubiquitination and proteasome targeting of the NFκB inhibitor IκBα was shown to involve the TrCP-containing SkpI, Cullin, F-box protein (SCF^{TrCP}) E3 complex also involved in CD4 degradation (Yaron *et al.*, 1998). Interestingly, the recognition motif on all known cellular substrates of βTrCP consists of a pair of conserved phosphoserine residues similar to those present in Vpu (Margottin *et al.*, 1998). These serine residues are arranged in a consensus motif: DSpGYXSp; where Sp stands for phosphoserine, Y stands for a hydrophobic residue and X stands for any residue. Serine-phosphorylation plays the major

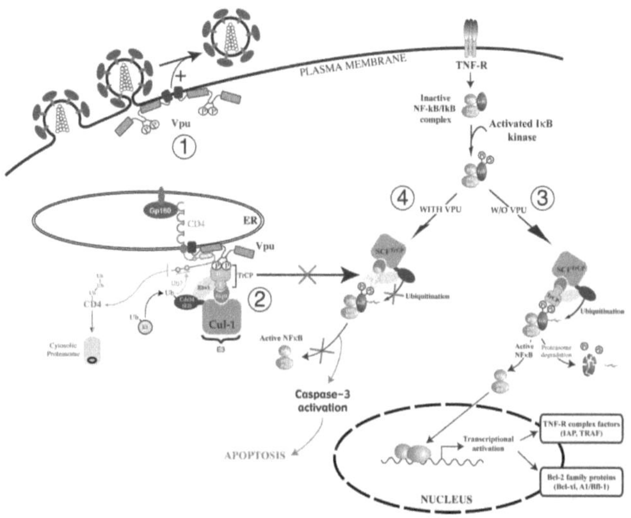

Figure 11.8. Model of the biological roles of Vpu.

regulatory role in the stability of SCF^{TrCP} target proteins. For example, activation of the IκB kinase complex (IKK) by external stimuli such as TNFα induces the serine-phosphorylation of IκBα followed by rapid TrCP-mediated proteasome degradation (Figure 11.8). Although the molecular mechanisms by which Vpu targets CD4 for degradation are now reasonably well defined, it remains unclear how the membrane-anchored CD4 is ultimately brought into contact with cytoplasmic proteasome complexes (Figure 11.8). There is only indirect evidence that CD4 ubiquitination precedes its degradation by Vpu (Fujita *et al.*, 1997; Schubert *et al.*, 1998). It is also not clear at present whether Vpu-induced degradation involves dislocation of CD4 from the ER membrane, as shown for other membrane-bound proteasome substrates such as MHC class I heavy chains (Wiertz *et al.*, 1996).

Vpu has one intriguing property that distinguishes it from all other known substrates of βTrCP: it is resistant to proteasome degradation. Indeed, while the SCF^{TrCP} usually degrades the serine-phosphorylated protein directly bound to the TrCP WD domains (i.e., Vpu), CD4, bound to the Vpu cytoplasmic domain, is degraded instead. Vpu therefore appears to have evolved a decoy mechanism by which Vpu domains that might be targeted for poly-ubiquitination are masked and those present in CD4 are presented instead to the Cdc34 ubiquitin ligase. This phenomenon has serious implications for the regulation and availability of the SCF^{TrCP} in cells that express Vpu. Indeed, due to the fact that Vpu is constitutively phosphorylated (Schubert *et al.*, 1994), binds βTrCP with high affinity (Margottin, 1998) and is not released from the complex by degradation (Bour *et al.*, 2001), Vpu expression in HIV-infected cells was shown to perturb the physiological function of the SCF^{TrCP} and prevent NF-κB activation through competitive trapping of βTrCP (Bour *et al.*, 2001). As depicted in Figure 11.8, this inhibition of NF-κB activation was further shown to induce apoptosis by inhibiting the NF-κB-dependent expression of anti-apoptotic genes such as *Bcl-2*, leading to enhanced intracellular levels of the apoptosis-promoting caspase-3 (Akari *et al.*, 2001).

4. Summary

Although the *vpu* gene is unique to HIV-1, the activity Vpu provides for enhanced viral particle release is not. Indeed, the envelope proteins of several HIV-2 isolates, including ROD10 and ST2, were shown to promote viral particle release in a manner indistinguishable from that of HIV-1 Vpu (Bour *et al.*, 1996; Ritter *et al.*, 1996). Both Vpu and the ROD10 Env are functionally interchangeable and each augment the release of HIV-1, HIV-2 and simian immunodeficiency virus (SIV) particles, suggesting a common mechanism of action for these two proteins (Gottlinger *et al.*, 1993; Bour and Strebel, 1996; Bour *et al.*, 1996; Bour *et al.*, 1999b). Site-directed mutagenesis revealed that the ability of the HIV-2 ROD Env protein to enhance viral particle release is regulated by a single amino acid substitution (position 598) in the ectodomain of the gp36 TM subunit (Bour *et al.*, 2003). Substituting the threonine at that position in the inactive ROD14 Env by the alanine found at the same position in the active ROD10 Env restored full particle release activity to the ROD14 Env in transfected HeLa cells (Bour *et al.*, 2003).

Unlike Vpu, the HIV-2 Env protein is unable to induce CD4 degradation (Bour and Strebel, 1996). The absence of a degradative activity in the ROD10 Env suggests that this additional function may have evolved in Vpu from the ancestral particle release activity in response to increased affinity between the HIV-1 Env and CD4 (Bour and Strebel, 1996; Willey *et al.*, 1992). Additional evidence in favor of this hypothesis comes from examining the sequence of SIVcpz isolates. The serine residues at positions 52 and 56 essential for interaction with TrCP are less conserved in SIVcpz than in the prototypical subtype C HIV-1 isolates (McCormick-Davis *et al.*, 2000b). The Vpu proteins from SIVcpz isolates are therefore unlikely to induce CD4 degradation. Since SIV isolates bearing a pseudo *vpu* gene are viewed as potential ancestors of HIV-1, it is tempting to speculate that the CD4 degradation ability of Vpu appeared late in the evolution of HIV-1.

Studies in pig-tailed macaque using SIV/HIV chimeric viruses (SHIV) have shown that mutation of the *VPU* initiation codon rapidly reverts to give rise to a functional *vpu* ORF (Stephens *et al.*, 1997). Such reversion occurs as early as 16 weeks post infection and correlates with a phase of profound loss of CD4-positive cells (McCormick-Davis *et al.*, 1998). Similar results were obtained in cynomolgus monkeys where the presence of Vpu was correlated with a vast increase in the plasma viral RNA levels 2 weeks post-infection (Li *et al.*, 1995). The increased viral fitness and pathogenicity conferred by Vpu is bimodal. First, Vpu increases viral loads in the plasma, thereby contributing to viral spread. Second, the higher frequency of *de novo* infections that results from these higher viral loads leads to increased rates of mutations in the *env* gene (Li *et al.*, 1995; Mackay *et al.*, 2002). This in turn leads to more rapid and efficient escape from neutralizing antibodies and accelerated disease progression (Li *et al.*, 1995). In animals infected with viruses where *vpu* deletions were large enough to prevent reversions, investigators observed long term nonprogressing infections characterized by a lack of circulating CD4+ T cells loss (Stephens *et al.*, 2002). Finally, studies in pig-tailed macaques showed that in the presence of large deletions in *vpu*, additional mutations in the *env* gene were acquired that partially compensated for the lack of Vpu (McCormick-Davis *et al.*, 2000a; Singh *et al.*, 2001). Although the mechanism by which Env would recapitulate the activity of Vpu in these animals is unclear, it is tempting to speculate that Env might have acquired a particle release activity similar to that displayed by some HIV-1 macrophage tropic isolates (Schubert *et al.*, 1999) and some HIV-2 isolates (Bour *et al.*, 1996; Ritter *et al.*, 1996).

As details of Vpu's action on CD4 degradation and particle release emerge, the question arises as to why such apparently unrelated activities have evolved within a single protein.

One possibility is that enhancement of particle release and CD4 degradation are two unrelated activities, each performing its own function in the viral life cycle. The main role of the CD4 degradation activity would thus be to liberate envelope protein precursors trapped in intracellular complexes with CD4 (Bour *et al.*, 1991; Willey *et al.*, 1992). As the rate of viral particle production augments, the action of Vpu would guarantee that enough mature envelope proteins are available for incorporation into virions. In addition, it is possible that CD4 degradation was selected as a supporting feature to the main particle release activity. Indeed, there is experimental evidence that the presence of CD4 at the cell surface actively interferes with the ability of Vpu to promote viral particle release (Bour *et al.*, 1999a). Another purpose of intracellular degradation of CD4 by Vpu would therefore be to prevent interference by cell surface CD4 of the Vpu particle release activity (Bour *et al.*, 1999a).

Much work remains to be done to fully characterize the particle release activity of Vpu. This includes a better characterization of the ion-channel activity and the identification of cellular partners of Vpu involved in promoting viral egress. We have recently paid special attention to the cellular specificity of Vpu action. Indeed, we have shown that certain cell types, such as the 293T embryonic kidney cell line, do not require Vpu for efficient HIV-1 particle release. While this may indicate that cell type-specific protein factors may be involved in the mechanism of Vpu action, it is also conceivable that membrane composition characteristics such as cholesterol concentration determine the need for Vpu in promoting viral release.

There is clear evidence that Vpu-mediated enhancement of viral particle production, downregulation of cell-surface CD4 and raising viral loads *in vivo* play key roles in the fitness and pathogenesis of HIV-1. It would therefore be beneficial to patients if drugs that target Vpu particle release activity were available. Providing the particle release activity of Vpu is indeed dependent on the presence of an ion-channel activity, it is conceivable that drugs similar to the Influenza M2 channel blocker amantidine could be developed to specifically target Vpu. The opportunities to develop drugs that target the biological functions of Vpu will be greatly increased as the structures of the Vpu and its structural and functional domains become available.

Acknowledgments

We thank C. Wu and C. Grant for assistance with the instrumentation and the experiments. This research was supported by grants RO1GM066978 and PO1GM64676 and the Biomedical Technology Resource for NMR Molecular Imaging of Proteins supported by grant P41EB002031 from the National Institutes of Health.

References

Akari, H., Bour, S., Kao, S., Adachi, A., and Strebel, K. (2001). The human immunodeficiency virus type 1 accessory protein Vpu induces apoptosis by suppressing the nuclear factor kappaB-dependent expression of antiapoptotic factors. *J. Exp. Med.* **194**, 1299–1311.

Bai, C., Sen, P., Hofmann, K., Ma, L., Goebl, M., Harper, J.W. *et al.* (1996). SKP1 connects cell cycle regulators to the ubiquitin proteolysis machinery through a novel motif, the F-box. *Cell* **86**, 263–274.

Bax, A., Kontaxis, G., and Tjandra, N. (2001) Dipolar couplings in macromolecular structure determination. *Meth. Enzymol.* **339**, 127–174.

Bour, S., Akari, H., Miyagi, E., and Strebel, K. (2003). Naturally occurring amino acid substitutions in the HIV-2 ROD envelope glycoprotein regulate its ability to augment viral particle release. *Virology* **309**, 85–98.

Bour, S., Boulerice, F., and Wainberg, M.A. (1991). Inhibition of gp160 and CD4 maturation in U937 cells after both defective and productive infections by human immunodeficiency virus type 1. *J. Virol.* **65**, 6387–6396.

Bour, S., Geleziunas, R., and Wainberg, M.A. (1995a). The human immunodeficiency virus type 1 (HIV-1) CD4 receptor and its central role in the promotion of HIV-1 infection. *Microbiol. Rev.* **59**, 63–93.

Bour, S., Perrin, C., Akari, H., and Strebel, K. (2001). The human immunodeficiency virus type 1 Vpu protein inhibits NF-kappa B activation by interfering with beta TrCP-mediated degradation of Ikappa B. *J. Biol. Chem.* **276**, 15920–15928.

Bour, S., Perrin, C., and Strebel, K. (1999a). Cell surface CD4 inhibits HIV-1 particle release by interfering with Vpu activity. *J. Biol. Chem.* **274**, 33800–33806.

Bour, S., Schubert, U., Peden, K., and Strebel, K. (1996). The envelope glycoprotein of human immunodeficiency virus type 2 enhances viral particle release: A Vpu-like factor? *J. Virol.* **70**, 820–829.

Bour, S., Schubert, U., and Strebel, K. (1995b). The human immunodeficiency virus type 1 Vpu protein specifically binds to the cytoplasmic domain of CD4: Implications for the mechanism of degradation. *J. Virol.* **69**, 1510–1520.

Bour, S. and Strebel, K. (1996). The human immunodeficiency virus (HIV) type 2 envelope protein is a functional complement to HIV type 1 Vpu that enhances particle release of heterologous retroviruses. *J. Virol.* **70**, 8285–8300.

Bour, S. and Strebel, K. (2000). HIV accessory proteins: Multifunctional components of a complex system. *Adv. Pharmacol.* **48**, 75–120.

Bour, S. and Strebel, K. (2003). The HIV-1 Vpu protein: A multifunctional enhancer of viral particle release. *Microbes Infect.* **5**, 1029–1039.

Bour, S. P., Aberham, C., Perrin, C., and Strebel, K. (1999b). Lack of effect of cytoplasmic tail truncations on human immunodeficiency virus type 2 ROD env particle release activity. *J. Virol.* **73**, 778–782.

Chou, J.J., Kaufman, J.D., Stahl, S.J., Wingfield, P.T., and Bax, A. (2002). Micelle-induced curvature in a water-insoluble HIV-1 Env peptide revealed by NMR dipolar coupling measurement in stretched polyacrylamide gel. *J. Amer. Chem. Soc.* **124**, 2450–2451.

Coadou, G., Evrard-Todeschi, N., Gharbi-Benarous, J., Benarous, R., and Girault, J.P. (2002). HIV-1 encoded virus protein U (Vpu) solution structure of the 41–62 hydrophilic region containing the phosphorylated sites Ser52 and Ser56. *Int. J. Biol. Macromol.* **30**, 23–40.

Coadou, G., Gharbi-Benarous, J., Megy, S., Bertho, G., Evrard-Todeschi, N., Seberal, E. *et al.* (2003). NMR studies of the phosphorylation motif of the HIV-1 protein Vpu bound to the F-box protein B-TrCP. *Biochemistry* **42**, 14741–14751.

Cohen, E.A., Terwilliger, E.F., Sodroski, J.G., and Haseltine, W.A. (1988). Identification of a protein encoded by the vpu gene of HIV-1. *Nature* **334**, 532–534.

Cordes, F.S., Kukol, A., Forrest, L.R., Arkin, I.T., Sansom, M.S.P., and Fischer, W.B. (2001). The structure of the HIV-1 Vpu ion channel: Modelling and simulation studies. *Biochim. Biophys. Acta* **1512**, 291–298.

Cordes, F.S., Tustian, A.D., Sansom, M.S.P., Watts, A., and Fischer, W.B. (2002). Bundles consisting of extended transmembrane segments of Vpu from HIV-1: Computer simulations and conductance measurements. *Biochemistry* **41**, 7359–7365.

Crise, B., Buonocore, L., and Rose, J. K. (1990). CD4 is retained in the endoplasmic reticulum by the human immunodeficiency virus type 1 glycoprotein precursor. *J. Virol.* **64**, 5585–5593.

Erickson, J.W. and Burt, S.K. (1996). Structural mechanisms of HIV drug resistance. *Annu. Rev. Pharmacol. Toxicol.* **36**, 545–571.

Ewart, G. D., Sutherland, T., Gage, P. W., and Cox, G. B. (1996). The Vpu protein of human immunodeficiency virus type 1 forms cationselective ion channels. *J. Virol.* **70**, 7108–7115.

Fan, L. and Peden, K. (1992). Cell-free transmission of Vif mutants of HIV-1. *Virology* **190**, 19–29.

Federau, T., Schubert, U., Flossdorf, J., Henklein, P., Schomburg, D., and Wray, V. (1996). Solution structure of the cytoplasmic domain of the human immunodeficiency virus type 1 encoded virus protein U (Vpu). *Int. J. Pept. Protein Res.* **47**, 297–310.

Fischer, W.B. (2003). Vpu from HIV-1 on an atomic scale: Experiments and computer simulations. *FEBS Lett.* **552**, 39–46.

Fischer, W.B. and Sansom, M.S.P. (2002). Viral ion channels: Structure and function. *Biochim. Biophys. Acta* **1561**, 27–45.

Fujita, K., Omura, S., and Silver, J. (1997). Rapid degradation of CD4 in cells expressing human immunodeficiency virus type 1 Env and Vpu is blocked by proteasome inhibitors [published erratum appears in *J. Gen. Virol.* 1997 Aug; **78**(Pt 8), 2129–2130. *J. Gen. Virol.* **78**, 619–625.

Gottlinger, H.G., Dorfman, T., Cohen, E.A., and Haseltine, W.A. (1993). Vpu protein of human immunodeficiency virus type 1 enhances the release of capsids produced by gag gene constructs of widely divergent retroviruses. *Proc. Natl. Acad. Sci. USA* **90**, 7381–7385.

Gottlinger, H.G., Dorfman, T., Sodroski, J.G., and Haseltine, W.A. (1991). Effect of mutations affecting the p6 gag protein on human immunodeficiency virus particle release. *Proc. Natl. Acad. Sci. USA* **88**, 3195–3199.

Grice, A.L., Kerr, I.D., and Sansom, M.S. (1997). Ion channels formed by HIV-1 Vpu: A modelling and simulation study. *FEBS Lett.* **405**, 299–304.

Henklein, P., Kinder, R., Schubert, U., and Bechinger, B. (2000). Membrane interactions and alignment of structures within the HIV-1 Vpu cytoplasmic domain: Effect of phosphorylation of serines 52 and 56. *FEBS Lett.* **482**, 220–224.

Ishima, R., Torchia, D.A., Lynch, S.M., Gronenborn, A.M., and Louis, J.M. (2003). Solution structure of the mature HIV-1 protease monomer. *J. Biol. Chem.* **278**, 43311–43319.

Jabbar, M.A. and Nayak, D.P. (1990). Intracellular interaction of human immunodeficiency virus type 1 (ARV-2) envelope glycoprotein gp160 with CD4 blocks the movement and maturation of CD4 to the plasma membrane. *J. Virol.* **64**, 6297–6304.

Klimkait, T., Strebel, K., Hoggan, M.D., Martin, M.A., and Orenstein, J.M. (1990). The human immunodeficiency virus type 1-specific protein vpu is required for efficient virus maturation and release. *J. Virol.* **64**, 621–629.

Korber, B., Foley, B. Leitner, T., McCutchan, F., Hahn, B., Mellors, J.W. *et al.* (1997). Human retroviruses and AIDS. *In Theoretical Biology and Biophysics*. Los Alamos National Laboratory, Los Alamos.

Kukol, A. and Arkin, I.T. (1999). Vpu transmembrane peptide structure obtained by site-specific Fourier transform infrared dichroism and global molecular dynamics searching. *Biophys. J.* **77**, 1594–1601.

Lamb, R.A. and Pinto, L.H. (1997). Do Vpu and Vpr of human immunodeficiency virus type 1 and NB of influenza B virus have ion channel activities in the viral life cycle? *Virology* **229**, 1–11.

Lenburg, M.E. and Landau, N.R. (1993). Vpu-induced degradation of CD4: Requirement for specific amino acid residues in the cytoplasmic domain of CD4. *J. Virol.* **67**, 7238–7245.

Li, J.T., Halloran, M., Lord, C.I., Watson, A., Ranchalis, J., Fung, M. *et al.* (1995). Persistent infection of macaques with simian-human immunodeficiency viruses. *J. Virol.* **69**, 7061–7067.

Lopez, C.F., Montal, M., Blasie, J.K., Klein, M.L., and Moore, P.B. (2002). Molecular dynamics investigation of membrane-bound bundles of the channel-forming transmembrane domain of viral protein U from the human immunodeficiency virus HIV-1. *Biophys. J.* **83**, 1259–1267.

Ma, C. and Opella, S.J. (2000). Lanthanide ions bind specifically to an added EF-hand and orient a membrane protein in micelles for solution NMR spectroscopy. *J. Magn. Reson.* **146**, 381–384.

Ma, C., Marassi, F.M., Jones, D.H., Straus, S.K., Bour, S., Strebel, K. *et al.* (2002). Expression, purification and activities of full-length and truncated versions of the integral membrane protein Vpu from HIV-1, *Protein Sci.* **11**, 556–557.

Mackay, G.A., Niu, Y., Liu, Z.Q., Mukherjee, S., Li, Z., Adany, I. *et al.* (2002). Presence of intact vpu and nef genes in nonpathogenic SHIV is essential for acquisition of pathogenicity of this virus by serial passage in macaques. *Virology* **295**, 133–146.

Maldarelli, F., Chen, M.Y., Willey, R.L., and Strebel, K. (1993). Human immunodeficiency virus type 1 Vpu protein is an oligomeric type I integral membrane protein. *J. Virol.* **67**, 5056–5061.

Marassi, F.M. and Opella, S.J. (2000). A solid-state NMR index of helical membrane protein structure and topology. *J. Magn. Reson.* **144**, 162–167.

Marassi, F.M., Ma, C., Gratkowski, H., Straus, S.K., Strebel, K., Oblatt-Montal, M. *et al.* (1999). Correlation of the structural and functional domains in the membrane protein Vpu from HIV-1. *Proc. Natl. Acad. Sci. USA* **96**, 14336–14341.

Margottin, F., Bour, S.P., Durand, H., Selig, L., Benichou, S., Richard, V. *et al.* (1998). A novel human WD protein, h-beta TrCp, that interacts with HIV-1 Vpu connects CD4 to the ER degradation pathway through an F-box motif. *Mol. Cell* **1**, 565–574.

McCormick-Davis, C., Dalton, S.B., Hout, D.R., Singh, D.K., Berman, N.E., Yong, C. *et al.* (2000a). A molecular clone of simian-human immunodeficiency virus (DeltavpuSHIV(KU-1bMC33)) with a truncated, non-membrane-bound vpu results in rapid CD4(+) T cell loss and neuro-AIDS in pig-tailed macaques. *Virology* **272**, 112–126.

McCormick-Davis, C., Dalton, S.B., Singh, D.K., and Stephens, E.B. (2000b). Comparison of Vpu sequences from diverse geographical isolates of HIV type 1 identifies the presence of highly variable domains, additional invariant amino acids, and a signature sequence motif common to subtype C isolates. *AIDS Res. Hum. Retroviruses* **16**, 1089–1095.

McCormick-Davis, C., Zhao, L.J., Mukherjee, S., Leung, K., Sheffer, D., Joag, S.V. *et al.* (1998). Chronology of genetic changes in the vpu, env, and Nef genes of chimeric simian-human immunodeficiency virus (strain HXB2) during acquisition of virulence for pig-tailed macaques. *Virology* **248**, 275–283.

Mesleh, M.F. and Opella, S.J. (2003). Dipolar waves as NMR maps of helices in proteins. *J. Magn. Reson.* **163**, 288–299.

Mesleh, M.F., Lee, S., Veglia, G., Thiriot, D.S., Marassi, M.M., and Opella, S.J. (2003). Dipolar waves map the structure and topology of helices in membrane proteins. *J. Amer. Chem. Soc.* **125**, 8928–8935.

Mesleh, M.F., Veglia, G. DeSilva, T.M., Marassi, F.M., and Opella, S.J. (2002). Dipolar waves as NMR maps of protein structure. *J. Amer. Chem. Soc.* **124**, 4206–4207.

Miller, R.H. and Sarver, N. (1997). HIV accessory proteins as therapeutic targets. *Nature Med.* **3**, 389–394.

Montal, M. (2003). Structure-function correlates of Vpu, a membrane protein of HIV-1. *FEBS Lett.* **552**, 47–53.

Moore, P. B., Zhong, Q., Husslein, T., and Klein, M. L. (1998). Simulation of the HIV-1 Vpu transmembrane domain as a pentameric bundle. *FEBS Lett.* **431**, 143–148.

Navia, M.A., Fitzgerald, P.M., McKenner, B.M. *et al.* (1989). Three-dimensional structure of asparatyl protease from human immunodeficiency virus HIV-1. *Nature* **337**, 615–620.

Nevzorov, A.A. and Opella, S.J. (2003). A "magic sandwich" pulse sequence with reduced offset dependence for high-resolution separated local field spectroscopy. *J. Magn. Reson.* **164**, 182–186.

Opella, S.J. (1997). NMR and membrane proteins. *Nat. Struct. Biol. NMR suppl.*, 845–848.

Opella, S.J., Stewart, P., and Valentine K. (1987). Protein structure by solid-state NMR spectroscopy. *Q. Rev. Biophys.* **19**, 7–49.

Opella, S.J., Nevzorov, A., Mesleh, M.F., and Marassi, F.M. (2002). Structure determination of membrane proteins by NMR spectroscopy. *Biochem. Cell Biol.* **80**, 597–604.

Pake, G. (1948). Nuclear resonance absorption in hydrated crystals: Fine structure of the proton line. *J. Chem. Phys.* **16**, 327–336.

Park, S.H., Mrse, A.A., Nevzorov, A.A., Mesleh, M.G., Oblatt-Montal, M., Montal, M., and Opella, S.J. (2003). Three-dimensional structure of the channel-forming trans-membrane domain of virus proteins "u" (Vpu) from HIV-1. *J. Mol. Biol.* **333**, 409–424.

Paul, M. and Jabbar, M.A. (1997). Phosphorylation of both phosphoacceptor sites in the HIV-1 Vpu cytoplasmic domain is essential for Vpu-mediated ER degradation of CD4. *Virology* **232**, 207–216.

Paul, M., Mazumder, S., Raja, N., and Jabbar, M.A. (1998). Mutational analysis of the human immunodeficiency virus type 1 Vpu transmembrane domain that promotes the enhanced release of virus-like particles from the plasma membrane of mammalian cells. *J. Virol.* **72**, 1270–1279.

Prestegard, J.H., Al-Hashimi, H.M., and Tolman, J.R. (2001). NMR structures of biomolecules using field oriented media and residual dipolar couplings. *Q. Rev. Biophys.* **33**, 371–424.

Ritter, G.D., Jr., Yamshchikov, G., Cohen, S.J., and Mulligan, M.J. (1996). Human immunodeficiency virus type 2 glycoprotein enhancement of particle budding: Role of the cytoplasmic domain. *J. Virol.* **70**, 2669–2673.

Schubert, U., Anton, L.C., Bacik, I., Cox, J.H., Bour, S., Bennink, J.R. *et al.* (1998). CD4 glycoprotein degradation induced by human immunodeficiency virus type 1 Vpu protein requires the function of proteasomes and the ubiquitin-conjugating pathway. *J. Virol.* **72**, 2280–2288.

Schubert, U., Bour, S., Ferrer-Montiel, A. V., Montal, M., Maldarell, F., and Strebel, K. (1996a). The two biological activities of human immunodeficiency virus type 1 Vpu protein involve two separable structural domains. *J. Virol.* **70**, 809–819.

Schubert, U., Bour, S., Willey, R.L., and Strebel, K. (1999). Regulation of virus release by the macrophage-tropic human immunodeficiency virus type 1 AD8 isolate is redundant and can be controlled by either Vpu or Env. *J. Virol.* **73**, 887–896.

Schubert, U., Ferrer-Montiel, A.V., Oblatt-Montal, M., Henklein, P., Strebel, K., and Montal, M. (1996b). Identification of an ion channel activity of the Vpu transmembrane domain and its involvement in the regulation of virus release from HIV-1-infected cells. *FEBS Lett.* **398**, 12–18.

Schubert, U., Henklein, P., Boldyreff, B., Wingender, E., Strebel, K., and Porstmann, T. (1994). The human immunodeficiency virus type 1 encoded Vpu protein is phosphorylated by casein kinase-2 (CK-2) at positions Ser52 and Ser56 within a predicted alpha-helix-turn-alpha-helix-motif. *J. Mol. Biol.* **236**, 16–25.

Schubert, U. and Strebel, K. (1994). Differential activities of the human immunodeficiency virus type 1-encoded Vpu protein are regulated by phosphorylation and occur in different cellular compartments. *J. Virol.* **68**, 2260–2271.

Schwartz, M.D., Geraghty, R.J., and Panganiban, A.T. (1996). HIV-1 particle release mediated by Vpu is distinct from that mediated by p6. *Virology* **224**, 302–309.

Schwartz, S., Felber, B.K., Fenyo, E.M., and Pavlakis, G.N. (1990). Env and Vpu proteins of human immunodeficiency virus type 1 are produced from multiple bicistronic mRNAs. *J. Virol.* **64**, 5448–5456.

Singh, D.K., McCormick, C., Pacyniak, E., Lawrence, K., Dalton, S.B., Pinson, D.M. *et al.* (2001). A simian human immunodeficiency virus with a nonfunctional Vpu (deltavpuSHIV(KU-1bMC33)) isolated from a macaque with neuroAIDS has selected for mutations in env and nef that contributed to its pathogenic phenotype. *Virology* **282**, 123–140.

Spevak, W., Keiper, B.D., Stratowa, C., and Castanon, M.J. (1993). *Saccharomyces cerevisiae* cdc15 mutants arrested at a late stage in anaphase are rescued by Xenopus cDNAs encoding N-ras or a protein with beta-transducin repeats. *Mol. Cell Biol.* **13**, 4953–4966.

Stephens, E.B., McCormick, C., Pacyniak, E., Griffin, D., Pinson, D.M., Sun, F. *et al.* (2002). Deletion of the vpu sequences prior to the env in a simian-human immunodeficiency virus results in enhanced Env precursor synthesis but is less pathogenic for pig-tailed macaques. *Virology* **293**, 252–261.

Stephens, E.B., Mukherjee, S., Sahni, M., Zhuge, W., Raghavan, R., Singh, D.K. *et al.* (1997). A cell-free stock of simian-human immunodeficiency virus that causes AIDS in pig-tailed macaques has a limited number of amino acid substitutions in both SIVmac and HIV-1 regions of the genome and has offered cytotropism. *Virology* **231**, 313–321.

Strebel, K., Klimkait, T., and Martin, M.A. (1988). A novel gene of HIV-1, vpu, and its 16-kilodalton product. *Science* **241**, 1221–1223.

Strebel, K., Klimkait, T., Maldarelli, F., and Martin, M.A. (1989). Molecular and biochemical analyses of human immunodeficiency virus type 1 vpu protein. *J. Virol.* **63**, 3784–3791.

Torres, J., Kekal, A., and Arkin, I.T. (2001). Mapping the energy surface of transmembrane helix-helix detections. *Biophys. J.* **81**, 2681–2692.

Turner, B.G. and Summers, M.F. (1999). Structural biology of HIV. *J. Mol. Biol.* **285**, 1–32.

Vondrasek, J., van Bukirk, C.P., and Wlodawer, A. (1997). Database of three-dimensional structures of HIV proteinases. *Nat. Struct. Biol.* **4**, 8.

Veglia, G. and Opella, S.J. (2000). Lanthanide ion binding to adventitious sites aligns membrane proteins in micelles for solution NMR spectroscopy. *J. Amer. Chem. Soc.* **47**, 11733–11734.

Vincent, M.J., Raja, N.U., and Jabbar, M. A. (1993). Human immunodeficiency virus type 1 Vpu protein induces degradation of chimeric envelope glycoproteins bearing the cytoplasmic and anchor domains of CD4: Role of the cytoplasmic domain in Vpu-induced degradation in the endoplasmic reticulum. *J. Virol.* **67**, 5538–5549.

Wang, J., Denny, J., Tian, C., Kim, S., Mo, Y., Kovacs, F. *et al.* (2000). Imaging membrane protein helical wheels. *J. Magn. Reson.* **144**, 162–167.

Waugh, J.S. (1976). Uncoupling of local field spectra in nuclear magnetic resonance: Determination of atomic positions in solids. *Proc. Natl. Acad. Sci. USA* **73**, 1394–1397.

Wiertz, E.J., Jones, T.R., Sun, L., Bogyo, M., Geuze, H.J., and Ploegh, H.L. (1996). The human cytomegalovirus US11 gene product dislocates MHC class I heavy chains from the endoplasmic reticulum to the cytosol. *Cell* **84**, 769–779.

Willbold, D., Hoffman, S., and Rosch, P. (1997). Secondary structure and tertiary fold of the human immunodeficiency virus protein U (Vpu) cytoplasmic domain in solution. *Eur. J. Biochem.* **245**, 581–588.

Willey, R.L., Maldarelli, F., Martin, M. A., and Strebel, K. (1992). Human immunodeficiency virus type 1 Vpu protein regulates the formation of intracellular gp160-CD4 complexes. *J. Virol.* **66**, 226–234.

Wlodawer, A. (2002). Rational approach to AIDS drug design through structural biology. *Annu. Rev. Med.* **53**, 595–614.

Wlodawer, A., Miller, M., Jaskolski, M. *et al.* (1989). Conserved folding in retroviral proteases: Crystal structure of a synthetic HIV-1 protease. *Science* **245**, 616–621.

Wray, V., Federau, T., Henklein, P., Klabunde, S., Kunert, O., Schomburg, D. *et al.* (1995). Solution structure of the hydrophilic region of HIV-1 encoded virus protein U (Vpu) by CD and 1H NMR spectroscopy. *Int. J. Pept. Protein Res.* **45**, 35–43.

Wray, V., Kinder, R., Federau, T., Henklein, P., Bechinger, B., and Schubert, U. (1999). Solution structure and orientation of the transmembrane anchor domain of the HIV-1-encoded virus protein U by high-resolution and solid-state NMR spectroscopy. *Biochem.* **38**, 5272–5282.

Wu, C.H., Ramamoorthy, A., and Opella, S.J. (1994). High resolution heteronuclear dipolar solid-state NMR spectroscopy. *J. Magn. Reson.*, **A109**, 270–272.

Yaron, A., Hatzubai, A., Davis, M., Lavon, I., Amit, S., Manning, A.M. *et al.* (1998). Identification of the receptor component of the IkappaBalpha-ubiquitin ligase. *Nature* **396**, 590–594.

12

Structure, Phosphorylation, and Biological Function of the HIV-1 Specific Virus Protein U (Vpu)

Victor Wray and Ulrich Schubert

Knowledge describing the structure and function of the small regulatory human immunodeficiency virus type 1 (HIV-1) viral protein U (Vpu) has increased significantly over the last decade. Vpu is an 81 amino acid class I oligomeric integral-membrane phosphoprotein that is encoded exclusively by HIV-1. It can therefore be anticipated, that Vpu might contribute to the increased pathogenic potential of HIV-1 when compared with HIV-2 that has so far had a lower impact on the acquired immune deficiency syndrome (AIDS) pandemic. Various biological functions have been ascribed to Vpu: first, in the endoplasmic reticulum (ER) Vpu induces degradation of CD4 in a process involving the ubiquitin–proteasome pathway and phosphorylation of its cytoplasmic tail. In addition, there is also evidence that Vpu interferes with major histocompatibility complex (MHC) class I antigen presentation and regulates Fas mediated apoptosis. Second, Vpu augments virus release from a post ER compartment by a cation-selective ion channel activity mediated by its transmembrane (TM) anchor. The phosphorylation of the molecule is mediated by the ubiquitous protein kinase caseinkinase 2 (CK-2) within a central conserved dodecapeptide at positions Ser^{52} and Ser^{56} located in a flexible hinge region between two helical domains. Structural information, provided experimentally mainly by solution- and solid-state nuclear magnetic resonance (NMR) spectroscopy and made possible through the availability of synthetic and recombinant material, have shown that the biological activities of Vpu are localized in two distinct domains that are mainly confined to the C-terminal cytoplasmic and N-terminal TM domains, respectively. Similar to other small viral proteins that interact with membranes Vpu is a very flexible molecule whose structure is exceptionally environment dependent. It assumes it's most structured form in the hydrophobic environment in or at the membrane. An initial 20–23 residue α-helix in the N-terminus adopts a TM alignment while the cytoplasmic tail forms an α-helix-flexible-α-helix-turn motif, of which at least a part is bound parallel to the membrane surface. Details of the arrangement of oligomeric forms of the molecule that are presumably

---

Victor Wray • Department of Structural Biology, German Research Centre for Biotechnology, Mascheroder Weg 1, D-38124 Braunschweig, Germany. **Ulrich Schubert** • Institute for Clinical and Molecular Virology, University of Erlangen-Nürnberg, Schlossgarten 4, D-91054 Erlangen, Germany.

Viral Membrane Proteins: Structure, Function, and Drug Design, edited by Wolfgang Fischer.
Kluwer Academic / Plenum Publishers, New York, 2005.

required for the ion channel activity, are emerging from recent theoretical calculations, while this particular function is currently the area of pharmaceutical interest.

1. Introduction

Since the first description of the acquired immune deficiency syndrome (AIDS) in 1981 and the identification of human immunodeficiency Virus (HIV) as the causative ethnological agent (Barre-Sinoussi *et al.*, 1983) enormous progress has been made in understanding the pathogenesis of human lentiviruses. Nonetheless, our knowledge of the contribution of particular virus factors to the induction of immunodeficiency is still far from complete. In general, all replication competent retroviruses contain *gag, pol*, and *env* genes encoding structural proteins and viral enzymes. Beside these retrovirus typical genes Lentiviruses, Spumaviruses, and the human T-cell leukemia virus (HTLV) and its relatives encode small additional gene products with regulatory functions in the viral life cycle. Unraveling the molecular structure of these regulatory HIV proteins has proven essential for understanding and manipulating the molecular mechanism of these viral factors (reviewed in Miller and Sarver, 1997).

Human immunodeficiency virus type 1 contains at least six regulatory genes, with the Vpu unique to HIV-1. No structural homolog has been detected in primate Lentiviruses even in closely related species such as HIV-2 or simian immunodeficiency virus (SIV), except for the HIV-1 related isolate SIV_{CPZ} (Huet *et al.*, 1990). Depending on the particular HIV-1 isolate, Vpu is an 80- to 82-residue long type I anchored amphipathic membrane phosphoprotein with a functional and structural discernible domain architecture: first, Vpu augments virus release from a post ER compartment by a cation-selective ion channel activity mediated by its TM anchor (Ewart *et al.*, 1996; Schubert *et al.*, 1996a, b; for review see Lamb and Pinto, 1997). Second, it affects the cell surface expression of several glycoproteins involved in host immune response: well characterized is the Vpu induced degradation of the primary virus receptor CD4 in the ER. This process requires the CK-2 dependent phosphorylation of two conserved serine residues within the cytoplasmic tail of Vpu (Schubert *et al.*, 1992, 1994; Friborg *et al.*, 1995; Paul and Jabbar, 1997) and the formation of multiprotein complexes containing cellular factors such as h-βTrCP and Skp1p (Margottin *et al.*, 1998) that presumably link the ubiquitin conjugating machinery to the cytoplasmic tail of CD4 leading to ubiquitination and finally proteolysis of CD4 by the 26S proteasome in the cytosol (Fujita *et al.*, 1997; Schubert *et al.*, 1998). In addition, there is also evidence that Vpu decreases the tendency for syncytia formation (Yao *et al.*, 1993), reduces the transport of certain glycoproteins (Vincent and Jabbar, 1995), interferes with an early step in the biosynthesis of MHC class I molecules (Kerkau *et al.*, 1997), and increases the susceptibility to CD95 (Fas) mediated apoptosis (Casella *et al.*, 1999) potentially by regulating the levels of Fas receptor at the cell surface. While the importance of the Vpu induced interference of MHC-I and Fas pathways for the HIV-1 replication cycle are rather obscure at the moment there are indications that the Vpu induced CD4 degradation ensures high infectivity of HIV-1 by preventing incorporation of envelope glycoproteins into budding virions, a function that is also supported by the accessory protein Nef (Lama *et al.*, 1999).

Like other accessory HIV gene products Vpu is not essential for virus replication in tissue culture. However, it is conceivable that the two major biological functions of Vpu, down-regulation of CD4 and augmentation of virus particle release, may contribute to the enhanced pathogenic potential of HIV-1 when compared to its close relative, HIV-2 (Kanki *et al.*, 1994;

Marlink *et al.*, 1994). Such a hypothesis is supported by *in vivo* studies indicating that Vpu indeed increases HIV-1 pathogenicity in a SCID-hu mice model system (Aldrovandi and Zack, 1996) or virus load and disease progression in a chimeric simian–human immunodeficiency viruses (SHIV)/monkey model (Li *et al.*, 1995).

Clearly, if Vpu is going to serve as a new target for anti-retroviral therapy and, thus, play any role in the combinatory treatment of AIDS, a prerequisite is the understanding of the relationship between its function and the molecular structure. Over the last few years, both aspects of Vpu have been intensely investigated such that these details are currently furnishing areas of pharmaceutical interest (Ewart *et al.*, 2002).

2. Structure and Biochemistry of Vpu

Shortly after the discovery of Vpu (Cohen *et al.*, 1988; Strebel *et al.*, 1988), the protein was characterized as a membrane phosphoprotein that is co-translationally integrated into the ER membrane with the suspected class I topology and a reported molecular mass of 16 kDa (Strebel *et al.*, 1989). Besides theoretical prediction of an amphipathic protein sequence (Strebel *et al.*, 1989) nothing was known about the molecular structure of the protein at the time of its first biochemical and functional characterization. To shed more light onto these topics, we were interested in the determination of its atomic structure, the biochemical characterization of the Vpu phosphorylation, and ultimately in the development of *in vitro* assays for Vpu functions with the help of chemical synthesis of Vpu peptides.

A substantial amount of biochemical data exists that characterizes Vpu as a type I oriented integral oligomeric membrane phosphoprotein composed of an amphipathic sequence of 81 amino acids comprising a hydrophobic N-terminal TM anchor proximal to a polar C-terminal cytoplasmic domain (Strebel *et al.*, 1989; Klimkait *et al.*, 1990; Maldarelli *et al.*, 1993). The latter contains a highly conserved dodecapeptide from Glu-47 to Gly-58 that is conserved among all Vpu sequences of known HIV-1 isolates (Huet *et al.*, 1990; Chen *et al.*, 1993) and contains two seryl residues in positions 52 and 56 which are phosphorylated by CK-2 in a positive cooperative manner in HIV-1 infected cells (Schubert *et al.*, 1992, 1994). Earlier studies by proteinase K digestion of membrane integrated Vpu indicated that the membrane anchor is located at the N-terminus and contains less than 30 residues (Maldarelli *et al.*, 1993). Reconstitution of the synthetic membrane anchor of Vpu in planar lipid bilayers identified a cation-selective ion channel activity (Schubert *et al.*, 1996b) which was also demonstrated for full-length Vpu in bilayers (Ewart *et al.*, 1996) and in amphibian oocytes (Schubert *et al.*, 1996). The idea that Vpu functions as an ion channel was furthermore supported by the observation that Vpu enhances membrane permeability when expressed in pro- and eukaryotic cells (Gonzales and Carrasco, 1998).

Our structural studies on Vpu involved the establishment of a general NMR approach for the study of short membrane-associated peptides using a combination of high-resolution NMR in solution and evaluation of the resulting structures by solid-state NMR in oriented bilayers. These techniques in combination with solid phase peptide synthesis afford a very powerful and straightforward approach to determining structural details of membrane-bound Vpu peptides. This strategy has the advantages that peptide synthesis provides a ready source of customized NMR probes, which can be easily and selectively labeled with ^{15}N. Alternatively, the availability of such hydrophobic peptides is usually limited when recombinant DNA techniques are applied, due to the low yield of mostly toxic and insoluble

TM peptides when expressed in bacteria, or to their general tendency to form high order aggregates.

Over the last 10 years, considerable effort has been expended on determining the structure of Vpu as a prerequisite for understanding the molecular mechanism of its biological activities. Initial difficulties in producing recombinant material, and the realization that the molecule represents a type I integral membrane protein (Maldarelli *et al.*, 1993) whose structure is environment-dependent, led to early studies being conducted on synthetic material (Henklein *et al.*, 1993). Our early attempts to produce recombinant Vpu failed as the protein, when expressed in its authentic form, exhibited a high degree of cytotoxicity, probably because of its activity on biological membranes. In addition, it formed high ordered aggregates (Schubert *et al.*, 1992). Further, we demonstrated that the recombinant Vpu could be isolated from bacteria in a membrane-bound state, which was *a bona fide* substrate for *in vitro* phosphorylation studies. However, the total yield of expressed protein in *Escherichia coli* was far too low for production of purified material for subsequent NMR experiments (Schubert *et al.*, 1992). Later studies by others where successful in producing recombinant Vpu as a fusion protein that formed inert inclusion bodies in *E. coli* (Ma *et al.*, 2002). The failure of such peptides, either produced synthetically or by recombinant techniques, to crystallize precluded X-ray crystallographic studies and hence structural studies has relied on both solution- and solid-state NMR spectroscopy as the main approach for producing models of the structural domains in the molecule.

Biochemical investigations revealed that Vpu is an amphiphatic class I oligomeric integral membrane phosphoprotein that has 27 hydrophobic amino acids at its N-terminus which function as a membrane anchor and a C-terminus of 54 hydrophilic and charged amino acids that comprise the cytoplasmic domain. This latter domain was the first to receive attention. Despite considerable differences in the amino acid sequences among Vpu proteins from different HIV-1 isolates, the prediction of secondary structure suggested a strong conservation of a supposed α-helix-flexible-α-helix-turn motif for this domain (Schubert *et al.*, 1994). Initial circular dichroism data of a series of nine overlapping peptides corresponding to the cytoplasmic domain (Wray *et al.*,1995) and of larger fragments (Henklein *et al.*, 1993) indicated the presence of only transitory amounts of stable structure in aqueous solution alone while addition of trifluoroethanol (TFE), a solvent that tends to favor secondary structure through stabilization of intramolecular interactions and simulates a membrane-like milieu in solution (Buck, 1998), afforded experimental evidence of the presence of limiting structures with two helices in regions 28–52 and 58–72 of the cytoplasmic domain (Wray *et al.*, 1995) separated by a loop containing the two phosphorylation sites at Ser^{52} and Ser^{56} (Coadou *et al.*, 2002).

The exact nature of the solution structure of the cytoplasmic domain (Federau *et al.*, 1996) was established using standard two-dimensional homonuclear ^{1}H NMR techniques (Wüthrich, 1986) in combination with restrained molecular dynamics (MD) and energy minimization (EM) calculations (Brünger, 1992). The combined experimental data for the cytoplasmic domain of Vpu indicate Vpu^{32-81} and a mutant in which the phosphoacceptors were exchanged for Asn, in 50% aqueous TFE at pH 3.5, are predominantly monomeric and adopt similar well-defined helix-interconnection-helix-turn conformations in which the four regions are bounded by residues 37–51, 52–56, 57–72, and 73–78 (Federau *et al.*, 1996). Identical regions of secondary structure were determined from the ^{13}C and αH chemical shifts of ^{2}H-/^{13}C-/^{15}N-labeled peptides analyzed in lipid micelles by multidimensional heteronuclear NMR spectroscopy and indicated the start of the first cytoplasmic helix was at or near residue 31 (Marassi *et al.*, 1999; Ma *et al.*, 2002). Both helices are amphiphatic in character, but show different charge distributions. In general, the cytoplasmic region is N-terminally positively charged, passes through a region of alternating charges in the first helix, and then becomes negatively charged. The flexibility of the interconnecting hinge region permits orientational freedom of the helices and comprises a highly conserved dodecapeptide (Schubert *et al.*, 1992). The presence of *cis–trans* isomerism of Pro^{75} manifests itself as a doubling of cross-peaks of Ala^{74} and Trp^{76} in the 2D ^{1}H spectra.

A ^{1}H NMR investigation of a peptide related to Vpu^{37-81} in an organic-free high-salt aqueous solution (Willbold *et al.*, 1997) showed two helical secondary structures similar to those in TFE and a less well-defined helix from 75–79 that was a turn in TFE. In order to be compatible with the CD data in water these structures must be less stable than those in TFE. Interestingly a small number of long-range Nuclear Overhauber Effects (NOEs) provided evidence of a tertiary fold. Clearly this feature is very susceptible to the solution conditions and is absent in TFE. The appearance of such weak tertiary structure in membrane-like environments still requires verification in the context of the full-length molecule.

Attention was then focussed on the N-terminal hydrophobic domain of Vpu, which is primarily associated with an ion channel activity either by itself (Schubert *et al.*, 1996b) or in the context of full-length Vpu (Ewart *et al.*, 1996; Ma *et al.*, 2002), and the orientation of the various secondary structure elements of the full-length protein with respect to the membrane. MD/EM calculations using NOE data generated as above for the soluble synthetic peptide Vpu^{1-39} in 50% TFE indicated a compact well-defined U-shaped tertiary structure involving a short helix (residues 10–16) on the N-terminal side and a longer helix (22–36) on the C-terminal side. The side chains of the aromatic residues, Trp^{22} and Tyr^{29}, in the latter helix are directed toward the center of the molecule around which the hydrophobic core of the folded molecule is positioned (Wray *et al.*, 1999). In contrast to helices of the cytoplasmic region, the helix region in the N-terminus is present at the lowest TFE concentrations and may even be present in the absence of membrane mimetic although of limited solubility (Wray *et al.*, 1999). The tertiary structure however was inconsistent with the formation of ion-conductive membrane pores in planar lipid bilayers (Schubert *et al.*, 1996b) as this U-folded N-terminus is unlikely to be able to span the bilayer. Consequently proton-decoupled ^{15}N cross-polarization solid-state NMR spectroscopy has been employed to investigate full-length Vpu and its isolated domains oriented in phospholipid bilayers using either synthetic discretely ^{15}N-labeled amino acids (Wray *et al.*, 1999; Henklein *et al.*, 2000) or uniformly ^{15}N-labeled recombinant material (Marassi *et al.*, 1999; Ma *et al.*, 2002; Ho Park *et al.*, 2003). Details of these solid-state NMR approaches are reviewed more comprehensively in Chapter 13 by Bechinger and Chapter 11 by Opella, respectively, in this volume. In brief the ^{15}N chemical shift data and line widths of N-terminal domain molecules are consistent with a TM alignment of a helical polypeptide, implying that the nascent helices in the folded solution structure reassemble to form a linear α-helix involving residues 6–29 that lies parallel to the bilayer normal with a tilt angle of $\leq 30°$ (Wray *et al.*, 1999) and placing Trp^{22} near to the Glu^{28}-Tyr-Arg-motif (important for helix termination, anchorage and pore selectivity, Sramala *et al.*, 2003) at the membrane-cytoplasm interface. Detailed simulations of the novel ^{1}H–^{15}N dipolar coupling/^{15}N chemical shift distribution afford an average tilt angle of approximately 13° (Ho Park *et al.*, 2003). A somewhat smaller value of $6.5 \pm 1.7°$ has been concluded from independent site-specific Fourier transform Infrared dichroism data for ^{13}C-labeled Vpr^{1-31} peptides (Kukol and Arkin, 1999).

For the cytoplasmic domain both solid-state NMR approaches (Marassi *et al.*, 1999; Henklein *et al.*, 2000) agree that the first helical cytoplasmic domain, residues 31–51, interacts strongly with the bilayer and assumes an orientation parallel to the membrane surface. In contrast, data for the second C-terminal helix in an isolated fragment (Vpu^{51-81}) showed little interaction with the membrane (Henklein *et al.*, 2000), while interpretation of data from a comparison of uniformally ^{15}N-labeled cytoplasmic fragments for this helix is ambiguous (Marassi *et al.*, 1999). Interestingly synchrotron radiation-based X-ray reflectivity methods applied to Langmuir monolayers of mixtures of Vpu and an appropriate phospholipid (Zheng *et al.*, 2001) offers significant insight into structural changes that occur at the membrane upon changing protein concentration and surface pressure. Thus the tilt angle of the TM domain decreases with increasing pressure and at medium pressure both cytoplasmic helices lie on the surface of the phospholipid headgroups. At low pressure the increased tilt angle of the membrane domain disrupts the position of the cytoplasmic helices and forces them into the bulk water phase with presumably concurrent loss in structure. In contrast at the highest pressures measured where there was insufficient space for both helices to lie on the surface the second helix is forced off the surface to form a two-helix bundle. Although these experimental parameters may not reflect physiological conditions in a comprehensive way, they do emphasize the various possibilities that are inherent in the system and that are presumably strongly influenced by environmental conditions, protein phosphorylation state, membrane composition, and protein–protein interactions.

3. Biochemical Analysis of Vpu Phosphorylation

Following the first report that Vpu is post-translationally modified by phosphorylation (Strebel *et al.*, 1989) we searched for consensus sequences for eukaryotic protein kinases within the Vpu protein and identified the seryl residues in positions 52 and 56 as two potential phosphorylation sites that correspond to the consensus $^{S}/_{T}XX^{D}/_{E}$, the minimal sequence recognized by the ubiquitous casein kinase-2 (CK-2) (Schubert *et al.*, 1992). The two CK-2 phosphorylation sites are conserved in all known Vpu sequences and represent the consensus S^{52}GN(E/D)S(E/D)G(E/D)^{59}. The assumption that Vpu is a substrate for CK-2 *in vivo* was first supported by our observation that phosphorylation of Vpu in HIV-1 infected T-cells can be blocked by inhibitors specific for CK-2 (Schubert *et al.*, 1992). Furthermore, using bacterial expressed recombinant Vpu as a substrate for *in vitro* phosphorylation studies we were able to show that membrane-bound full-length Vpu can be phosphorylated by purified CK-2 or by CK-2 containing extracts from mammalian cells. For identification of phosphoacceptor sites in Vpu and for biochemical characterization of the kinase reaction we employed synthetic peptides of the cytoplasmic tail of Vpu. Initially, we investigated phosphorylation of synthetic peptides comprising the hydrophilic, polar C-terminal domain of Vpu from position I^{32} to L^{81}. We were able to demonstrate that a peptide, Vpu^{32-81}, containing the wild-type sequence was phosphorylated *in vitro* by purified recombinant enzyme CK-2 or by whole cell extract of mammalian cells, and that inhibitors specific for CK-2 can block the *in vitro* phosphorylation. In contrast, a corresponding mutant peptide, $Vpum^{2/6}$, was spared by CK-2 confirming that phosphorylation occurs at both predicted phosphoacceptor sites. Serine residues in positions 52 and 56 were replaced by asparagine in the sequence of the mutated peptide $Vpum^{2/6}$. Since both, serine and asparagine, have similar effects on secondary structure according to the "structure derived correlation matrix (SCM)" described by Niefind and Schomburg (1991), these replacements should not influence the structure of the protein backbone, a prediction which was later confirmed by ^{1}H NMR spectroscopy (Wray *et al.*, 1995; Federau *et al.*, 1996). Direct evidence was provided that CK-2 targets the phosphoacceptor sites individually in both positions. For this purpose, the K_m values of CK-2 to three 54 amino acid peptides comprising the entire hydrophilic part and containing single serine to asparagine transitions in either position 52 or 56 were established (Vpu*52*, Vpu*56*, and Vpu*wt* (Schubert *et al.*, 1994)). The 3-fold higher K_m value of CK-2 to Vpu*56* revealed a preferential phosphorylation of S^{56} over S^{52}. This would be in accordance with a positive cooperative mechanism of CK-2-phosphorylation of Vpu in which phosphorylation of the residue S^{56} occurs first. Subsequently, phosphoserine in position 56 stimulates phosphorylation of the second phosphoacceptor S^{52}. This model of sequential CK-2 phosphorylation was recently supported by the findings from Paul and Jabber (1997) that dual phosphorylation of serine residues in both positions, 52 and 56, but not individual phosphorylation of either one of these residues are sufficient to induce Vpu mediated proteolysis of CD4. These *in vitro* analyses provided the basis for follow up studies demonstrating that in the context of HIV-1 the Vpu induced degradation of CD4 is strictly dependent on the phosphorylation of both acceptor sites, serine 52 and 56 (Schubert and Strebel, 1994). In contrast, the virus release function of Vpu, is not controlled by phosphorylation, and furthermore, this activity depends on the ability of the membrane anchor of Vpu to form cation-selective ion channels. This observation led to the model of the two distinct functional and structural domain architecture of Vpu (Schubert *et al.*, 1996a).

Although phosphorylation is of crucial importance for the regulation of Vpu function its structural consequences have received scant attention. Solution NMR studies have been restricted to monitoring changes that occur on phosphorylation in the fragment 41–62 in water and 50% TFE at pH 3.5 and 7.2 (Coadou *et al.*, 2002). Distinct changes are observed in the chemical shift and NOE patterns which correspond to some loss of helix propensity in the region 42–49 as well as a change toward a more β-strand-like structure for residues 50–62 with a corresponding displacement of the C-terminal helix. The relevance of these to the membrane-bound protein are difficult to assess, particularly as the sensitivity of current solid-state NMR methods did not disclose any significant changes in the orientation of the first cytoplasmic helix upon phosphorylation (Henklein *et al.*, 2000) while the consequences for the entire domain awaits investigation.

Thus the experimental evidence demands a dynamic two domain model for the membrane-associated structure of monomeric full-length Vpu shown in Figure 12.1 that corresponds in many aspects to those already reported in the literature by us (Wray *et al.*, 1999; Henklein *et al.*, 2000) and others (Marassi *et al.*, 1999). Most recently attention has been focused on the ion channel activity of Vpu and its similarity to other viral ion channels (for review see Fischer and Sansom, 2002). Although direct experimental evidence of the structure of the ion channel are unavailable a considerable number of MD investigations of various Vpu fragments embedded in octane/water and lipid bilayer systems (for recent results see Cordes *et al.*, 2002 and references therein and Lopez *et al.*, 2002) combined with conductance measurements provide evidence for water-filled five-helix bundles, the relative orientations of their monomeric units and rationale of the weak cation selectivity. Progress in this area is considered in Chapter 14 by Lemaitre *et al.* (this volume).

In summary, although the monomeric model provides details of the topology and positions of secondary structure in the bound state it is clear that considerable more experimental and theoretical work is required if a meaningful rationalization is to be achieved of Vpu in its functional forms *in vivo* where it exists as a phosphoprotein in multiprotein

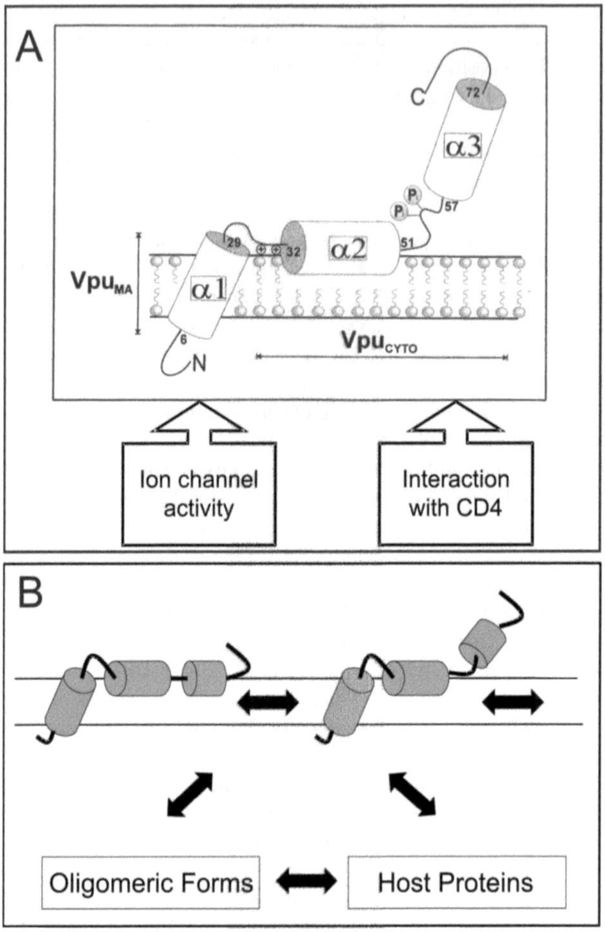

Figure 12.1. Model of the membrane-bound structures of monomeric full-length Vpu (A) and its dynamic forms (B).

complexes involving CD4, β-TrCp, and Skp 1 (Margottin *et al.*, 1998) or in homo-oligomeric non-phosphorylated forms required for its ion channel activity in the cell membrane. Improvements in experimental technologies, particularly solid-state NMR, X-ray diffraction, FTIR dichroism and cryoelectron microscopy, as well as improvements in computational technology should provide the keys to our future understanding and facilitate approaches to antiviral drug design.

References

Aldrovandi, G.M., and Zack, J.A. (1996). Replication and pathogenicity of human immunodeficiency virus type 1 accessory gene mutants in SCID-hu Mice. *J. Virol.* **70**, 1505–1511.

Barre-Sinoussi, F., Cherman, J.C., Rey, F., Nugeyre, M.T., Chamaret, S., Gruest, J. *et al.* (1983). Isolation of a T-lymphotropic retrovirus from a patient at risk for acquired immunodeficiency syndrome (AIDS). *Science* **220**, 868–870.

Buck, M. (1998). Trifluoroethanol and colleagues: Cosolvents come of age. Recent studies with peptides and proteins. *Q. Rev. Biophys.* **31**, 297–355.

Brünger, A.T. (1992). *X-PLOR, Version 3.1, A system for X-ray crystallography and NMR.* Yale University Press, New Haven and London.

Casella, C.R., Rapaport, E.L., and Finkel, T.H. (1999). Vpu increases susceptibility of human immunodeficiency virus type 1-infected cells to fas killing. *J. Virol.* **73**, 92–100.

Chen, M.Y., Maldarelli, F., Karczewski, M.K., Willey, R.L., and Strebel, K. (1993). Human immunodeficiency virus type 1 Vpu protein induces degradation of CD4 *in vitro*: The cytoplasmic domain of CD4 contributes to Vpu sensitivity. *J. Virol.* **67**, 3877–3884.

Coadou, G., Evrard-Todeschi, N., Gharbi-Benarous, J., Benarous, R., and Girault, J.P. (2002). HIV-1 encoded virus protein U (Vpu) solution structure of the 41–62 hydrophilic region containing the phosphorylated sites Ser52 and Ser56. *Int. J. Biol. Macromol.* **30**, 23–40.

Cohen, E.A., Terwilliger, E.F., Sodroski, J.G., and Haseltine, W.A. (1988). Identification of a protein encoded by the vpu gene of HIV-1. *Nature* **334**, 532–534.

Cordes, F.S., Tustian, A.D., Sansom, M.S., Watts, A., and Fischer, W.B. (2002). Bundles consisting of extended transmembrane segments of Vpu from HIV-1: Computer simulations and conductance measurements. *Biochemistry* **41**, 7359–7365.

Ewart, G.D., Mills, K., Cox, G.B., and Gage, P.W. (2002). Amiloride derivatives block ion channel activity and enhancement of virus-like particle budding caused by HIV-1 protein Vpu. *Eur. Biophys. J.* **31**, 26–35.

Ewart, G.D., Sutherland, T, Gage, P.W., and Cox, G.B. (1996). The Vpu protein of human immunodeficiency virus type 1 forms cation-selective ion channels. *J. Virol.* **70**, 7108–7115

Federau, T., Schubert, U., Flossdorf, J., Henklein, P., Schomburg, D., and Wray, V. (1996). Solution structure of the cytoplasmic domain of the human immunodeficiency virus type 1 encoded virus protein U (Vpu). *Int. J. Pept. Protein Res.* **47**, 297–310.

Fischer, W.B. and Sansom, M.S. (2002). Viral ion channels: Structure and function. *Biochim. Biophys. Acta* **1561**, 27–45.

Friborg, J., Ladha, A. Goettlinger, H., Haseltine, W.A., and Cohen, E.A. (1995). Functional analysis of the phosphorylation sites on the human immunodeficiency virus type 1 Vpu protein. *J. Acquired Immune Def. Syndr. Hum. Retrovir.* **8**, 10–22.

Fujita, K., Omura, S., and Silver, J. (1997). Rapid degradation of CD4 in cells expressing HIV-1 Env and Vpu is blocked by proteasome inhibitors. *J. Gen. Virol.* **78**, 619–625.

Gonzalez, M.E. and Carrasco, L. (1998). The human immunodeficiency virus type 1 Vpu protein enhances membrane permeability. Biochemistry **37**, 13710–13719.

Henklein, P., Schubert, U., Kunert, O., Klabunde, S., Wray, V., Kloppel, K.D. *et al.* (1993). Synthesis and characterization of the hydrophilic C-terminal domain of the human immunodeficiency virus type 1-encoded virus protein U (Vpu). *Pept. Res.* **6**, 79–87.

Henklein, P., Kinder, R., Schubert, U., and Bechingerm, B. (2000). Membrane interactions and alignment of structures within the HIV-1 Vpu cytoplasmic domain: Effect of phosphorylation of serines 52 and 56. *FEBS Lett.* **482**, 220–224.

Ho Park, S., Mrse, A.A., Nevzorov, A.A., Mesleh, M.F., Oblatt-Montal, M., Montal, M. *et al.* (2003). Three-dimensional structure of the channel-forming transmembrane domain of virus protein "u" (Vpu) from HIV-1. *J. Mol. Biol.* **333**, 409–424.

Huet, T., Cheynier, R., Meyerhans, A., Roelants, G., and Wain-Hobson, S. (1990). Genetic organization of a chimpanzee lentivirus related to HIV-1. *Nature* **345**, 356–359.

Kanki, P.J., Travers, K.U., MBoup, S., Hsieh, C.C., Marlink, R.G., Gueye-Ndiaye, A. *et al.* (1994). Slower heterosexual spread of HIV-2 than HIV-1. Lancet **343**, 943–946.

Klimkait, T., Strebel, K., Hoggan, M.D., Martin, M.A., and Orenstein, J.M. (1990). The human immunodeficiency virus type 1-specific protein vpu is required for efficient virus maturation and release. *J. Virol.* **64**, 621–629.

Kerkau, T., Bacik, I., Bennink, J.R., Yewdell, J.W., Hunig, T. *et al.* (1997). The human immunodeficiency virus type 1 (HIV-1) Vpu protein interferes with an early step in the biosynthesis of major histocompatibility complex (MHC) class I molecules. *J. Exp. Med.* **185**, 1295–1305.

Kukol, A. and Arkin, I.T. (1999). Vpu transmembrane peptide structure obtained by site-specific fourier transform infrared dichroism and global molecular dynamics searching. *Biophys J.* **77**, 1594–1601.

Lama, J., Mangasarian, A., and Trono, D. (1999). Cell-surface expression of CD4 reduces HIV-1 infectivity by blocking Env incorporation in a Nef- and Vpu-inhibitable manner. *Curr. Biol.* **9**, 622–631.

Lamb, R.A. and Pinto, L.H. (1997). Do Vpu and Vpr of human immunodeficiency virus type 1 and NB of influenza B virus have ion channel activities in the viral life cycles? *Virology* **229**, 1–11.

Li, J.T., Halloran, M., Lord, C.I., Watson, A., Ranchalis, J., Fung, M. *et al.* (1995). Persistent infection of macaques with simian-human immunodeficiency viruses. *J. Virol.* **69**, 7061–7067.

Lopez, C.F., Montal, M., Blasie, J.K., Klein, M.L., and Moore, P.B. (2002). Molecular dynamics investigation of membrane-bound bundles of the channel-forming transmembrane domain of viral protein U from the human immunodeficiency virus HIV-1. *Biophys. J.* **83**, 1259–1267.

Ma, C., Marassi, F.M., Jones, D.H., Straus, S.K., Bour, S., Strebel, K. *et al.* (2002). Expression, purification, and activities of full-length and truncated versions of the integral membrane protein Vpu from HIV-1. *Protein Sci.* **11**, 546–557.

Maldarelli, F., Chen, M.-Y., Willey, R.L., and Strebel, K. (1993). Human immunodeficiencyvirus type 1 Vpu protein is an oligomeric type 1 integral membrane protein. *J. Virol.* **67**, 5056–5061

Marassi, F.M., Ma, C., Gratkowski, H., Straus, S.K., Strebel, K., Oblatt-Montal, M. *et al.* (1999). Correlation of the structural and functional domains in the membrane protein Vpu from HIV-1. *Proc. Natl. Acad. Sci. USA* **96**, 14336–14341.

Margottin, F., Bour, S.P., Durand, H., Selig, L., Benichou, S., Richard, V. *et al.* (1998). A novel human WD protein, h-beta TrCp, that interacts with HIV-1 Vpu connects CD4 to the ER degradation pathway through an F-box motif. *Mol. Cell* **1**, 565–574.

Marlink, R., Kanki, P., Thior, I., Travers, K., Eisen, G., Siby, T. *et al.* (1994). Reduced rate of disease development after HIV-2 infection as compared to HIV-1. *Science* **265**, 1587–1590.

Miller, R. and Sarver, N. (1997). HIV accessory proteins as therapeutic targets. *Nature Med.* **3**, 389–394.

Niefind, K. and Schomburg, D. (1991). Amino acid similarity coefficients for protein modelling and sequence alignment derived from main chain folding angles. *J. Mol. Biol.* **219**, 481–497.

Paul, M. and Jabbar, M.A. (1997). Phosphorylation of both phosphoacceptor sites in the HIV-1 Vpu cytoplasmic domain is essential for Vpu-mediated ER degradation of CD4. *Virology* **232**, 207–216.

Schubert, U., Anton, L.C., Bacik, I., Cox, J.H., Bour, S., Bennink, J.R. *et al* (1998). CD4 glycoprotein degradation induced by human immunodeficiency virus type 1 Vpu protein requires the function of proteasomes and the ubiquitin conjugating pathway. *J. Virol.* **72**, 2280–2288.

Schubert, U., Bour, S., Ferrer-Montiel, A.V., Montal, M., Maldarelli, F., and Strebel, K. (1996a). The two biological activities of human immunodeficiency virus type 1 Vpu protein involve two separable structural domains. *J. Virol.* **70**, 809–819.

Schubert, U., Ferrer-Montiel, A.V., Oblatt-Montal, M., Henklein, P., Strebel, K., and Montal, M. (1996b). Identification of an ion channel activity of the Vpu transmembrane domain and its involvement in the regulation of virus release from HIV-1-infected cells. *FEBS Lett.* **398**, 12–8.

Schubert, U., Henklein, P., Boldyreff, B., Wingender, E., Strebel, K., and Porstmann, T. (1994). The human immunodeficiency virus type 1 encoded Vpu protein is phosphorylated by casein kinase-2 (CK-2) at positions Ser52 and Ser56 within a predicted alpha-helix-turn-alpha-helix-motif. *J. Mol. Biol.* **236**, 16–25.

Schubert, U., Schneider, T., Henklein, P., Hoffmann, K., Berthold, E., Hauser, H. *et. al.* (1992). Human-immunodeficiency-virus-type-1-encoded Vpu protein is phosphorylated by casein kinase II. *Eur. J .Biochem.* **204**, 875–883.

Schubert, U. and Strebel, K. (1994). Differential activities of the human immunodeficiency virus type-1 encoded Vpu protein are regulated by phosphorylation and occur in different cellular compartments. *J. Virol.* **68**, 2260–2271.

Sramala, I., Lemaitre, V., Faraldo-Gomez, J.D., Vincent, S., Watts, A., and Fischer, W.B. (2003). Molecular dynamics simulations on the first two helices of Vpu from HIV- 1. *Biophys. J.* **84**, 3276–3284.

Strebel, K., Klimkait, T., and Martin, M.A. (1988). A novel gene of HIV-1, vpu, and its 16-kilodalton product. *Science* **241**, 1221–1223.

Strebel, K., Klimkait, T., Maldarelli, F., and Martin, M.A. (1989). Molecular and biochemical analyses of human immunodeficiency virus type 1 vpu protein. *J. Virol.* **63**, 3784–3791.

Vincent, M. J. and Jabbar, M.A. (1995). The human immunodeficiency virus type 1 Vpu protein: A potential regulator of proteolysis and protein transport in the mammalian secretory pathway. *Virology* **213**, 639–649.

Willbold, D., Hoffmann, S. and Rosch, P. (1997). Secondary structure and tertiary fold of the human immunodeficiency virus protein U (Vpu) cytoplasmic domain in solution. *Eur. J. Biochem.* **245**, 581–588.

Wray, V., Federau, T., Henklein, P., Klabunde, S., Kunert, O., Schomburg, D. *et al.* (1995). Solution structure of the hydrophilic region of HIV-1 encoded virus protein U (Vpu) by CD and ^{1}H NMR spectroscopy. *Int. J. Pept. Protein Res.* **45**, 35–43.

Wray, V., Kinder, R., Federau, T., Henklein, P., Bechinger, B., and Schubert, U. (1999). Solution structure and orientation of the transmembrane anchor domain of the HIV-1-encoded virus protein U by high-resolution and solid-state NMR spectroscopy. *Biochemistry* **38**, 5272–5282.

Wüthrich, K. (1986). *NMR of Proteins and Nucleic Acids*. Wiley, New York.

Yao, X.J., Garzon, S., Boisvert, F., Haseltine, W.A., and Cohen, E.A. (1993). The effect of vpu on HIV-1-induced syncytia formation. *J. Acq. Immune. Defic. Syndr.* **6**, 135–141.

Zheng, S., Strzalka, J., Ma, C., Opella, S.J., Ocko, B.M., and Blasie, J.K. (2001). Structural studies of the HIV-1 accessory protein Vpu in langmuir monolayers: Synchrotron X-ray reflectivity. *Biophys. J.* **80**, 1837–1850.

13

Solid-State NMR Investigations of Vpu Structural Domains in Oriented Phospholipid Bilayers: Interactions and Alignment

Burkhard Bechinger and Peter Henklein

The accessory 81-residue protein viral protein U (Vpu) of human immunodeficiency virus-1 (HIV-1) fulfills two important functions during the viral life cycle. First, its intact N-terminus is required during viral release, a function that is correlated to its capability to form channels in planar lipid bilayers. Second, its cytoplasmic domain is involved in the binding and degradation of viral receptors including CD4 and major histocompatibility complex-I (MHC-I). In order to develop a structural model of Vpu in phospholipid bilayers, various Vpu polypeptides have been prepared by chemical synthesis and labeled with ^{15}N at selected backbone amide sites. After reconstitution of the peptides into oriented phospholipid bilayers, proton-decoupled ^{15}N solid-state nuclear magnetic resonance (NMR) spectra were recorded. The measured ^{15}N chemical shifts are indicative of the interactions and alignment of structural domains of Vpu within the lipid membrane. Whereas the hydrophobic helix 1, which is located at the N-terminus, adopts a transmembrane tilt angle of approximately 20°, the amphipathic helix 2 in the center of the polypeptide is oriented parallel to the membrane surface. No major changes in the topology of this membrane-associated amphipathic helix were observed upon phosphorylation of serine residues 52 and 56, although this modification regulates biological function of the Vpu cytoplasmic domain. The alanine-62 position of Vpu^{51-81} exhibits a pronounced ^{15}N chemical shift anisotropy suggesting that interactions with the lipid bilayer of the C-terminal part of the protein are weak.

1. Introduction

The HIV-1 encodes structural as well as several small regulatory proteins (Frankel and Young, 1998). Among the latter ones, the Vpu, fulfills important accessory functions during the life

Burkhard Bechinger • Université Louis Pasteur, Faculté de chimie, ILB, 4 rue Blaise Pascal, 67070 Strasbourg, France. **Peter Henklein** • Universitätsklinikum Charité Humboldt Universität, Institut für Biochemie, Monbijoustr. 2, 10117 Berlin, Germany.

Viral Membrane Proteins: Structure, Function, and Drug Design, edited by Wolfgang Fischer.
Kluwer Academic / Plenum Publishers, New York, 2005.

cycle of HIV-1. Biochemical experiments indicate that Vpu is an integral oligomeric membrane protein of 81 amino acids. Structural and functional studies show that the protein consists of two distinct domains. Whereas the N-terminus serves as a hydrophobic membrane anchor, the polar C-terminus is located in the cytoplasm (Maldarelli *et al.*, 1993).

On the one hand, the hydrophobic N-terminus is involved in cation-selective channel-formation and virus-release (Ewart *et al.*, 1996, 2002; Schubert *et al.*, 1996b; Lamb and Pinto, 1997). Reconstitution into planar lipid bilayers of chemically prepared sequences of the N-terminal portion of the protein results in cation selective ion-channel activities (Schubert *et al.*, 1996b). Membrane permeabilization was also characterized for full-length Vpu in lipid bilayers (Ewart *et al.*, 1996) as well as in amphibian oocytes (Schubert *et al.*, 1996b). In order to represent the channel activity of Vpu, oligomeric structures of the membrane anchor domain have been modeled using computational methods (Grice *et al.*, 1997; Cordes *et al.*, 2002; Fischer and Sansom, 2002; Lopez *et al.*, 2002).

On the other hand, processes that occur in the ER are tightly connected to the cytoplasmic C-terminus (Schubert *et al.*, 1996; Lamb and Pinto, 1997; Tiganos *et al.*, 1997). The best characterized of these is the downregulation of host cell receptor proteins such as CD4, the major cellular receptor for HIV-1, as well as MHC class I molecules (Willey *et al.*, 1992; Kerkau *et al.*, 1997). The degradation of CD4 requires regulatory control by caseine kinase 2-mediated phosphorylation of serines 52 and 56 (Paul and Jabbar, 1997; Tiganos *et al.*, 1997), the formation of multiprotein complexes containing CD4, Vpu, hβTrCp, and Skp1 (Margottin *et al.*, 1998), and the function of the ubiquitin-proteasome pathway (Fujita *et al.*, 1997; Schubert *et al.*, 1998).

In order to better understand the functional mechanisms of this polypeptide, structural data in membrane environments are required. Whereas x-ray diffraction techniques and solution NMR spectroscopy can provide high-resolution conformations from crystalline samples or of proteins that exhibit fast isotropic reorientation, they fail in lipid bilayer environments. In such cases, valuable information has been obtained using solid-state NMR spectroscopy. The latter technique has provided accurate answers to specific questions such as distances between selected sites within membrane proteins (Griffin, 1998; Helmle *et al.*, 2000). Furthermore, the alignments of helical domains relative to the membrane surface have been determined using this technique (Cross, 1997; Bechinger *et al.*, 1999). In order to develop their full strength, NMR techniques require combinations of specific, selective, and uniform labeling of the biomolecules (McIntosh and Dahlquist, 1990; Cross, 1997; Griffin, 1998; Hong and Jakes, 1999; Castellani *et al.*, 2002; Vogt *et al.*, 2003).

Solid-state NMR spectroscopy has a proven record to provide important structural information of polypeptides in membrane environments (reviewed in Cross, 1997; Griffin, 1998; Davis and Auger, 1999; Watts *et al.*, 1999; de Groot, 2000; Bechinger and Sizun, 2003). Membrane-associated polypeptides, which are oriented with respect to the magnetic field, exhibit chemical shifts as well as dipolar and quadrupolar interactions that are dependent on the alignment of chemical bonds and molecular domains (Cross, 1997; Bechinger *et al.*, 1999). Therefore, from these parameters, important structural and topological information is extracted from polypeptides that are associated with phospholipid bilayers. In particular, proton-decoupled ^{15}N solid-state NMR spectroscopy provides valuable information on the alignment of α-helical polypeptides in oriented phospholipid membranes. The technique has also been used to analyze the secondary structure of membrane-associated peptides (Cross, 1997; Bechinger *et al.*, 1999; Bechinger and Sizun, 2003). Whereas at orientations of the helix parallel to the magnetic field direction, the ^{15}N peptide backbone amides exhibit

^{15}N chemical shift values approaching 230 ppm, ^{15}N chemical shifts <100 ppm are indicative of in-plane oriented peptide helices (Bechinger and Sizun, 2003). When all orientations in space are present at random in nonoriented samples, the NMR resonances add up to broad "powder pattern" line shapes.

Due its hydrophobic nature, the biochemical expression of Vpu remains difficult and hydrophobic proteins of this type can sometimes not be obtained at the purity desired. These problems are enhanced when large quantities of protein are needed for structural investigations. In addition, selective or specific labeling schemes are required for NMR assignment purposes. This remains difficult to achieve, although uniform labeling of proteins in bacterial cell cultures has become routine. The preparation of samples labeled at a single of only a few sites thus imposes the most stringent requirements on the biochemical preparation method.

Solid-phase peptide synthesis offers alternative means to prepare polypeptides of considerable size. This technique works best for short sequences albeit proteins of more than 100 residues have been prepared (Kochendorfer and Kent, 1999). The technique allows the modification of residues or the incorporation of isotopic labels at single or a few sites almost at will. In contrast, due to high costs, the technique is at present not suitable for the labeling of larger polypeptides in a uniform manner.

The synthesis and the purification of long, hydrophobic polypeptides remains a challenge and is far from routine. Concomitantly, it can be difficult to obtain preparative amounts of pure polypeptides, which are required for NMR spectroscopic investigations. Fortunately, various domains of Vpu as well as the full-length polypeptide could be prepared in quantities and qualities sufficient for structural investigations including solution and solid-state NMR spectroscopy (Henklein *et al.*, 1993; Willbold *et al.*, 1997; Wray *et al.*, 1999).

The primary amino acid sequence of Vpu indicates that this protein exhibits an amphipathic character (Schubert *et al.*, 1994): a hydrophobic membrane anchor is followed by a hydrophilic and charged C-terminus. The polar region of Vpu exhibits the potential to form two amphipathic α-helices, one being zwitter, the other being anionic. They are joined by a strongly acidic turn that contains the two phosphoacceptor sites Ser52 and Ser56.

2. Peptide Synthesis

For structural and biochemical investigations, various peptides have been synthesized including the full-length peptide Vpu^{1-81} and fragments thereof. Peptides encompassing the hydrophobic membrane anchor are the fragment Vpu^{1-59}, the shortest membrane anchor prepared, Vpu^{1-27}, and a peptide including parts of the cytoplasmic region, Vpu^{1-39}. Furthermore, the hydrophilic sequence Vpu^{28-81} and modifications of this sequence, where the serine residues 52 and 56 were replaced by asparagines, were also prepared.

All of these peptides were synthesized by automated solid-phase peptide synthesis using Fmoc chemistry (Atherton *et al.*, 1981; Carpino and Han, 2003). To assemble the peptides, an automatic 433A Peptide synthesizer (Applied Biosystems, Darmstadt, FRG) and TentaGel resins with a capacity from 0.17 to 0.2 mmol/g (Rapp Polymere, Tübingen, FRG) were used. The syntheses were carried out on a 0.1 mmol scale with 10 times excess of the Fmoc-protected amino acids (Orpegen, Heidelberg, FRG). *N*-Methylpyrrolidon (NMP) was purchased from Biosolve (Holland), and the condensation reagents HBTU and HATU from Applied Biosystems (Darmstadt, FRG).

To cleave the peptide from the resin and to remove the side chain protecting groups at the same time, the product was triturated for 3 hr by application of a mixture of 89% TFA (trifluroacetic acid), 5% water, 3% cresole, and 3% triisopropylsilane. The peptides were purified by reversed phase HPLC using a preparative column (40 × 300) filled with VYDAC 218 TPB 1520 material.

A major problem during the handling of the peptides is their poor solubility in most common solvents, in particular when peptide fragments containing the membrane anchor moiety are investigated. The membrane anchor Vpu^{1-27} was completely insoluble in common organic solvents (DMF, DMSO, acetonitrile, methanol). This peptide was purified by trituration with methanol and acetonitrile to remove soluble byproducts. Among all the solvents tested, pure (TFA) worked best to dissolve this peptide. On the other hand, we found that a mixture of 50% trifluoroethanol (TFE) in water was suitable to dissolve and purify the elongated peptides Vpu^{1-39}, Vpu^{1-59} as well as the full-length peptide Vpu^{1-81}. All peptides were characterized by analytical HPLC and by mass spectrometric analysis using an Applied Biosystems Voyager Biospectrometry Workstation (Foster City, USA). Typical mass spectra are shown in Figure 13.1.

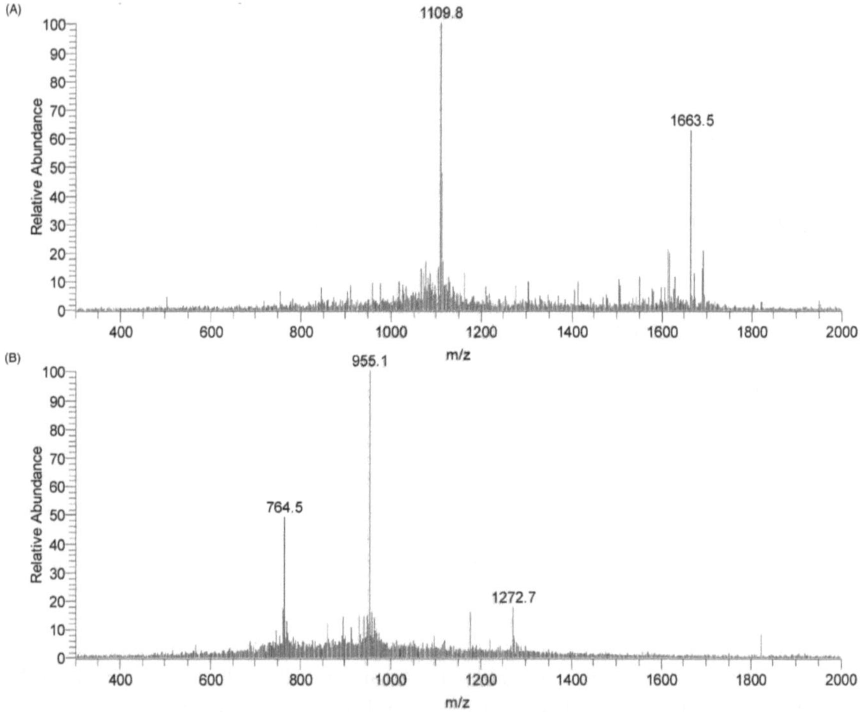

Figure 13.1 Electrospray ionization mass spectra of the products. (A) Vpu^{51-81}, labeled with ^{15}N at the alanine-62 position, theoretical mass 3327. (B) Vpu^{27-57}, labeled with ^{15}N at the leucine-45 position, theoretical mass 3818.

3. Results and Discussion

A substantial amount of biochemical data exists that characterize Vpu as type I oriented integral oligomeric membrane phosphoprotein composed of a hydrophobic N-terminal membrane anchor (Vpu_{MA}) and a polar C-terminal domain (Vpu_{CYTO}) (Maldarelli *et al.*, 1993). Hydrophobicity analysis and protease K digestion studies suggest that the membrane anchor of Vpu is located within the first N-terminal 27 amino acids (Strebel *et al.*, 1988; Maldarelli *et al.*, 1993). This sequence has previously been identified to exhibit cation selective ion channel activity, that regulates virus release (Ewart *et al.*, 1996; Schubert *et al.*, 1996).

CD spectroscopic analysis indicates that the N-terminal region of Vpu adopts stable α-helical conformations in water/TFE mixtures, at concentrations of the organic solvent as low as 10% (Wray *et al.*, 1999). This result suggests that the peptide exhibits a high propensity to form α-helical structures not only in the presence of membranes but also in their absence. More structural details of the peptides have been obtained using solution NMR spectroscopy in combination with restrained molecular dynamics calculations. In TFE/water mixtures, Vpu^{1-39} exhibits a well-defined tertiary structure. A turn at the N-terminus (Met-1 to Ile-6) is followed by a linker (Ala-7 to Val-9) that leads into a short helix (Ala-10 to Ile-16). Thereafter, a short loop region connects into a second longer helix that stretches from Trp-22 to Arg-36 (Wray *et al.*, 1999). Several unambiguously identified long-range NOEs found between the side chains of Trp-22 and Ile-26 with Ala-7 define a stable U-type tertiary structure. The compact U-structure establishes close helix–helix contacts with the side chains of the two aromatic amino acids buried in the interior of the structure.

However, such a U-folded solution structure is unlikely to be able to cross the phospholipid bilayer. It is thus difficult to accommodate with the finding that the hydrophobic N-terminus of Vpu forms an ion-conductive membrane pore (Schubert *et al.*, 1996b). In order to investigate the reasons for this apparent discrepancy, the polypeptide has been studied in lipid bilayer environments using solid-state NMR spectroscopy. The solid-state NMR studies presented in this chapter, therefore, provide an ideal complement to the solution-NMR investigations. The latter are presented in more detail by our collaborators Victor Wray and Ulrich Schubert in Chapter 12, this volume.

In previous electrophysiological investigations, channel formation of the peptide Vpu^{1-27} was characterized (Schubert *et al.*, 1996b). Therefore, synthetic Vpu^{1-27} peptides were prepared and reconstituted into oriented phospholipid bilayers and investigated by proton-decoupled ^{15}N cross-polarization solid-state NMR spectroscopy. The spectra of Vpu^{1-27} selectively labelled with ^{15}N at either position Ala-10, Ala-14, and Ala-18 and oriented in phospholipid bilayers were recorded. All of these helical polypeptides exhibit ^{15}N chemical shifts between 210 and 220 ppm indicative of transmembrane alignments (Wray *et al.*, 1999).

A quantitative analysis confirms that the experimental ^{15}N chemical shifts measured in oriented lipid bilayers cannot be accommodated with a single orientation of the U-shaped structure of Vpu^{1-39}, which is prevalent in water/TFE solutions. Care was taken to account for the uncertainties in determining the chemical shift values or in the description of the chemical shift interactions with the magnetic field (Wray *et al.*, 1999). In contrast, a linear α-helical model structure when tilted by approximately 20° with respect to the bilayer normal is in excellent agreement with the experimental chemical shift values. Transmembrane alignments were also observed by the newly developed method of site-directed FTIR spectroscopy (tilt angle 6.5° ±1.7°; Kukol and Arkin, 1999). Thus, major structural rearrangements of the polypeptide backbone involving the loss of the tertiary fold occur during the insertion from aqueous solutions into the membrane. Deuterium NMR spectra of $^{2}H_3$-alanine labeled Vpu^{1-39} in oriented membranes indicates the presence of a variety of peptide alignments (Aisenbrey and Bechinger, unpublished). This observation suggests that equilibria between different conformations, oligomerization states, and/or considerable asymmetry within oligomers exist.

Multidimensional solution-state NMR and CD spectroscopies indicate that in the presence of TFE/water, the cytoplasmic domain of Vpu consists of helix 2 (residues 37–51), a loop region involving the two phosphorylation sites at serines 52 and 56, helix 3 (residues 57–72) and a C-terminal turn (74–78) (Federau *et al.*, 1996). Investigations in TFE/water of Vpu^{41-62} with the serines 52 and 56 phosphorylated suggest that the loop region opens up upon phosphorylation to make accessible interactions sites for other proteins (Coadou *et al.*, 2002). In aqueous buffer of 620 mM salt concentration, these C-terminal helices are shortened and helix 3 appears irregular (Willbold *et al.*, 1997). At the same time, the short loop at the C-terminus is characterized by Ramachandran angles in the α-helical region. A few additional long-range contacts between the loop regions and the C-terminus indicate an antiparallel alignment of helices 2 and 3 (Willbold *et al.*, 1997). CD and NMR structural data obtained with peptide fragments in TFE/water are in good agreement with the conformations of the full-length cytoplasmic domain suggesting that, to first approximation, the individual helical domains of Vpu act as independent units (Wray *et al.*, 1995; Federau *et al.*, 1996; Willbold *et al.*, 1997). The membrane alignments of the hydrophobic and cytoplasmic peptide domains were also tested by recent molecular modeling calculations (Sramala *et al.*, 2003; Sun, 2003).

Also, for the cytoplasmic domain, ^{15}N-labelled peptides were prepared by solid-phase peptide synthesis and reconstituted into oriented phosphatidylcholine membranes. Thereby the interactions of helix 2 and helix 3 could be investigated independently of one another (Henklein *et al.*, 2000). Nitrogen-15 labels were introduced into the central regions of helix 2, or in the domain that had been assigned "helix 3" in previous investigations (Federau *et al.*, 1996; Willbold *et al.*, 1997). Site-directed mutagenesis indicates that both of these structural domains are essential for CD4 degradation (Tiganos *et al.*, 1997). In oriented 1-palmitoyl-2-oleoyl-*sn*-glycero-3-phosphocholine (POPC) membranes, a narrow resonance at 68 ppm is observed for [^{15}N-Leu45]-Vpu^{27-57} at orientations of the bilayer normal parallel to the magnetic field direction (Henklein *et al.*, 2000). The leucine-45 site is positioned in the central portion of helix 2 (Federau *et al.*, 1996; Willbold *et al.*, 1997). A quantitative analysis of the solid-state NMR data indicates an orientation of helix 2 where the C_α position of D39 is positioned slightly deeper within the membrane when compared to that of R48 (Figure 13.2A). This orientation is in excellent agreement with the size and the direction of the hydrophobic moment calculated for this helix encompassing residues K37–D51 using the algorithm of (Eisenberg *et al.*, 1984). The range of possible helix orientations is slightly increased when potential deviations of the chemical shift measurement, or of the main tensor element σ_{11} and σ_{22} are taken into consideration (Henklein *et al.*, 2000). When the published model of the aqueous solution structure of the cytoplasmic domain (Willbold *et al.*, 1997) was oriented relative to the membrane interface in such a manner that the alignment of helix 2 agrees with the solid-state NMR data, many charged and polar amino acids beyond residue 64 become immersed in the membrane interior. However, a few minor modifications within the flexible regions C-terminal of helix 2 overcome this difficulty and allow solvation of most, if not all, charged amino acids of the C-terminal domain in an aqueous environment (Figure 13.3).

Phosphorylation of serines 52 and 56 leaves the ^{15}N chemical shift position of the main peak of ^{15}N-leucine-45 unaltered indicating the absence of major changes in conformation

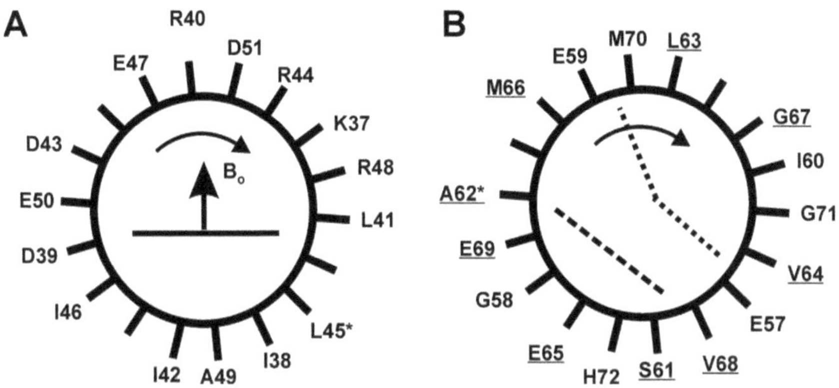

Figure 13.2. (A) Edmundson helical wheel diagram of residues 37–51 of Vpu (Federau *et al.*, 1996). The ^{15}N labeled position is marked with a star. The wheel is oriented to agree with the ^{15}N chemical shift of Leu45 when the membrane normal is oriented parallel to the magnetic field direction B_0. The solid line represents the suggested delineation of the hydrophilic–hydrophobic interface of the lipid bilayer. (B) Edmundson helical wheel diagram of residues 57–72 of Vpu. Different structures have been obtained depending on the chemical environment. Whereas in water random coil conformations are observed, helical structures extending from Glu57 to Glu69/His72 have been obtained in TFE/water mixtures. Somewhat shorter helices are observed for example in high-salt solutions, in the presence of high detergent concentrations, or during molecular modeling calculations in water/1,2-C24-PC monolayers (Sun, 2003). The separation of hydrophilic and hydrophobic residues is shown for the longest and shortest helices, respectively, that is, when including residues 57–72 (dotted line; Federau *et al.*, 1996) or when considering solely the underlined residues 61–69 (hatched line (Sun, 2003)).

and topology of the membrane-associated domain. However, an additional broad component between 50 and 200 ppm represents a wide range of other amide orientations suggesting that the interactions of helix 2 with the membrane are weakened (Henklein *et al.*, 2000). This is probably a result of electrostatic repulsion of some of the peptide, due to the accumulation of negative charges at the membrane surface that arises from the phosphoserines. Furthermore, it has been shown that phosphorylation results in major conformational changes in the loop region (Coadou *et al.*, 2002) and thus probably also in the structure and/or alignment of residues further C-terminal (Henklein *et al.*, 2000).

When the Vpu^{51-81} peptide is labeled with ^{15}N at the alanine 62 position and reconstituted into oriented lipid bilayers, the proton-decoupled ^{15}N solid-state NMR spectra show a wide distribution of molecular orientations. The spectra cover the full ^{15}N chemical shift anisotropy typically observed for backbone labeled alanine residues (Oas *et al.*, 1987; Shoji *et al.*, 1989; Lazo *et al.*, 1995; Lee *et al.*, 1998, 2001). At the same time, the ^{31}P solid-state NMR spectra of the samples indicate that the phospholipid membranes are well aligned. This combination of results is suggestive of the absence of strong interactions between "helix 3" and the POPC membranes (Figure 13.3). Similar spectral line shapes are also obtained from peptides labeled at position 62, which comprise the full-length cytoplasmic domain (unpublished result). The widths of the ^{15}N chemical shift resonances from this site indicate that immobilized structures are formed. Notably, in the absence of TFE only, a few residues of the C-terminal region of Vpu exhibit instable and irregular secondary structures (Wray *et al.*, 1995; Federau *et al.*, 1996; Willbold *et al.*, 1997). Therefore, both side-chain and backbone functional groups are available for intermolecular interactions.

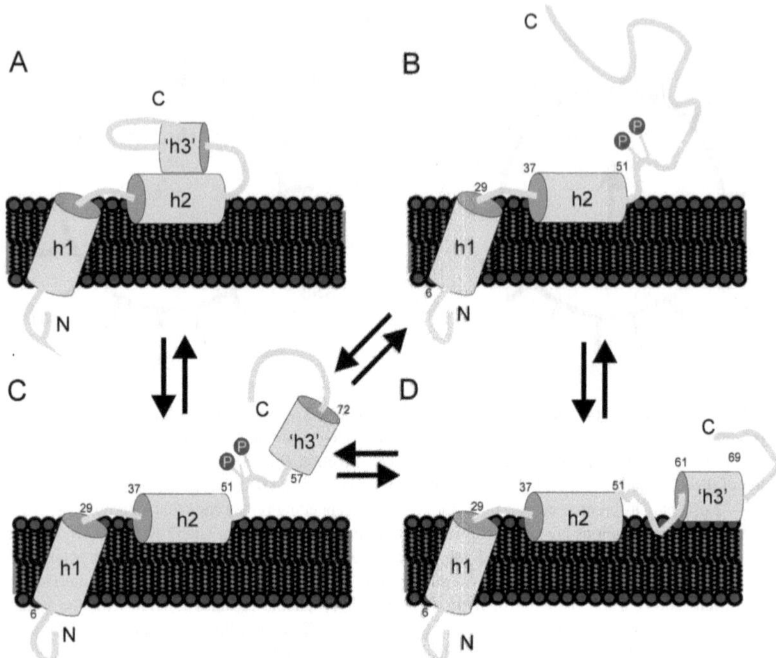

Figure 13.3. Some models of Vpu in phospholipid bilayers, taking into consideration part of the conformational space experimentally observed for the region C-terminal of helix 2. The approximate positions of the Ser52 and Ser56 phosphorylation sites are indicated in panels B and C.

Based on the structural data presented, the following working model of membrane-associated Vpu has been established. The first, most N-terminal helix, anchors the polypeptide in the membrane. The tilt angle of this hydrophobic helix is approximately 20° (Figure 13.3). The amphipathic helix 2 is located at the membrane interface at orientations parallel to the surface. This alignment of helix 2 does not change upon phosphorylation of serines 52 and 56; however, the association of helix 2 is weakened in the presence of two phosphates. The solid-state NMR data preclude a strong interaction of helix-3 with the membrane, a finding, which has been confirmed by recent molecular dynamics simulations (Sun, 2003). Overall, neither the structure nor the alignment of the region C-terminal of helix 2 seem well defined (Figure 13.3). Helix formation has been observed experimentally in the presence of high detergent concentrations (residues 58–70), TFE/water (57–69), or in high salt solutions (59–67), but not in the presence of water (reviewed in Sun, 2003, table 1). These residues are also characterized by the highest RMSD values in molecular modeling calculations (Sun, 2003). Concomitantly, the amphipathic character of this region is strongly dependent on the exact outline of the helix. Figure 13.2B illustrates that inclusion of Glu57 in a helical wheel analysis considerably reduces the hydrophobic angle subtended by the peptide (Federau *et al.*, 1996; Henklein *et al.*, 2000) when compared with shorter helical domains (e.g., Willbold *et al.*, 1997; Sun, 2003). Indeed different structures have been observed for the C-terminal domain (Figure 13.3) indicating that the membrane interactions of this region are dependent on the detailed experimental conditions (Zheng *et al.*, 2001). More stable membrane association of parts of the cytoplasmic region are observed in the presence of the hydrophobic membrane anchor (Henklein *et al.*, 2000; Ma *et al.*, 2002; Sun, 2003). It is thus possible that anchoring the cytoplasmic domain to the membrane involves alterations in entropic energy contributions, which are sufficient to shift the membrane insertion equilibrium of residues C-terminal of Glu57 (Figure 13.3C,D). Alternatively, molecular modeling calculations indicate that tight interactions exist within the loop region connecting helices 1 and 2 (Sramala *et al.*, 2003). If the interactions between helices 2 and 3 are stronger than the experimental results obtained so far suggest, a connection between helices 1 and 3 might also be established.

Notably, it has been shown that membrane insertion and helix formation are often tightly connected processes (Deber and Li, 1995) and this might apply for the C-terminal region of Vpu as well. Furthermore, a model in which helix 2 adopts antiparallel arrangements with other structured regions further toward the C-terminus (Willbold *et al.*, 1997) also in the presence of phospholipid bilayers remains to be considered (Figure 13.2A, Henklein *et al.*, 2000).

Although the ^{15}N chemical shift of the membrane-associated helix 2 (Leu45) is unaltered upon phosphorylation, it can be expected that the association equilibrium of helix 2 with the bilayer is modified due to electrostatic interactions. This is compatible with the appearance of an additional broad component in the ^{15}N spectrum after phosphorylation (Henklein *et al.*, 2000). This effect should be more pronounced in the presence of acidic phospholipids. In view of the conformational and topological flexibility of residues C-terminal of Glu57, even modest alterations can lever considerable structural changes of the cytoplasmic Vpu domain making accessible interaction sites of Vpu (Coadou *et al.*, 2002). Phosphorylation has also been shown to affect association of membrane polypeptides (Cornea *et al.*, 1997). It remains to be shown that such conformational rearrangements are indeed involved in the regulation of Vpu's activity.

Acknowledgments

We are grateful to Barbara Brecht-Jachan for technical help during peptide synthesis, to Rudolf Kinder for early solid-state NMR measurements, as well as to Victor Wray and Ulrich Schubert for valuable discussion contributions. The financial support by the *Deutsche Forschungsgemeinschaft* (DFG) and the *Agence Nationale de Recherches sur le Sida* (ANRS) are gratefully acknowledged.

References

Atherton, E., Logan, C.J., and Sheppart, R.C. (1981). *J. Chem. Soc. Perkin Trans.* I, 538–546.

Bechinger, B., Kinder, R., Helmle, M., Vogt, T.B., Harzer, U., and Schinzel, S. (1999). *Biopolymers* **51**, 174–190.

Bechinger, B. and Sizun, C. (2003). *Concepts Magn. Reson.* **18A**, 130–145.

Carpino, L.A. and Han, G.Y., *J. Org. Chem.* **37**, 3404.

Castellani, F., van Rossum, B., Diehl, A., Schubert, M., Rehbein, K., and Oschkinat, H. (2002). *Nature* **420**, 98–102.

Coadou, G., Evrard-Todeschi, N., Gharbi-Benarous, J., Benarous, R., and Girault, J.P. (2002). *Int. J. Biol. Macromol.* **30**, 23–40.

Cordes, F.S., Tustian, A.D., Sansom, M.S., Watts, A., and Fischer, W.B. (2002). *Biochemistry* **41**, 7359–7365.

Cornea, R.L., Jones, L.R., Autry, J.M., and Thomas, D.D. (1997). *Biochemistry* **36**, 2960–2967.
Cross, T.A. (1997). *Meth. Enzymol.* **289**, 672–696.
Davis, J.H. and Auger, M. (1999). *Prog. NMR Spectosc.* **35**, 1–84.
de Groot, H.J.M. (2000). *Curr. Opin. Struct. Biol.* **10**, 593–600.
Deber, C.M. and Li, S.C. (1995). *Biopolymers* **37**, 295–318.
Eisenberg, D., Weiss, R.M., and Terwilliger, T.C. (1984). *Proc. Natl. Acad. Sci. USA* **81**, 140–144.
Ewart, G.D., Mills, K., Cox, G.B., and Gage, P.W. (2002). *Eur. Biophys. J. JID-8409413* **31**, 26–35.
Ewart, G.D., Sutherland, T., Gage, P.W., and Cox, G.B. (1996). *J. Virol.* **70**, 7108–7115.
Federau, T., Schubert, U., Flossdorf, J., Henklein, P., Schomburg, D., and Wray, V. (1996). *Int. J. Pept. Protein Res.* **47**, 297–310.
Fischer, W.B. and Sansom, M.S. (2002). *Biochim. Biophys. Acta* **1561**, 27–45.
Frankel, A.D. and Young, J.A.T. (1998). *Ann. Rev. Biochem.* **67**, 1–25.
Fujita, K., S. Omura, and Silver, J. (1997). *J. Gen. Virol.* **78**, 619–625.
Grice, A.L., Kerr, I.D., and Sansom, M.S.P. (1997). *FEBS Lett.* **405**, 299–304.
Griffin, R.G., (1998). *Nat. Struct. Bio. NMR Suppl.* **5**, 508–512.
Helmle, M., Patzelt, H., Gärtner, W., Oesterhelt, D., and Bechinger, B. (2000). *Biochemistry* **39**, 10066–10071.
Henklein, P., Kinder, R., Schubert, U., and Bechinger, B. (2000). *FEBS Lett.* **482**, 220–224.
Henklein, P., Schubert, U., Kunert, O., Klabunde, S., Wray, V., Kloppel, K.D. *et al.* (1993). *Pept. Res.* **6**, 79–87.
Hong, M. and Jakes, K. (1999). *J. Biomol. NMR* **14**, 71–74.
Kerkau, T., Bacik, I., Bennink, J.R., Yewdell, J.W., Hunig, T., Schimpl, A. *et al.* (1997). *J. Exp. Med.* **185**, 1295–1305.
Kochendorfer, G.G. and Kent, S.B.H. (1999). *Curr. Opin. Chem. Biol.* **3**, 665–671.
Kukol, A. and Arkin, I.T. (1999). *Biophys. J.* **77**, 1594–1601.
Lamb, R.A. and Pinto, L.H. (1997). *Virology* **229**, 1–11.
Lazo, N.D., Hu, W., and Cross, T.A. (1995). *J. Magn. Res.* **107**, 43–50.
Lee, D.K., Wei, Y. and Ramamoorthy, A. (2001). *J. Phys. Chem. B* **105**, 4752–4762.
Lee, D.K., Wittebort, R.J. and Ramamoorthy, A. (1998). *J. Am. Chem. Soc.* **120**, 8868–8874.
Lopez, C.F., Montal, M., Blasie, J.K., Klein, M.L., and Moore, P.B. (2002). *Biophys. J.* **83**, 1259–1267.
Ma, C., Marassi, F.M., Jones, D.H., Straus, S.K., Bour, S., Strebel, K. *et al.* (2002). *Protein Sci.* **11**, 546–557.
Maldarelli, F., Chen, M.Y., Willey, R.L., and Strebel, K. (1993). *J. Virol.* **67**, 5056–5061.
Margottin, F., Bour, S.P., Durand, H., Selig, L., Benichou, S., Richard, V. *et al.* (1998). *Mol. Cell* **1**, 565–574.
McIntosh, L. and Dahlquist, F.W. (1990). *Q. Rev. Biophys.* **23**, 1–38.
Oas, T.G., Hartzell, C.J., Dahlquist, F.W., and Drobny, G.P. (1987). *J. Am. Chem. Soc.* **109**, 5962–5966.
Paul, M., and Jabbar, M.A. (1997). *Virology* **232**, 207–216.
Schubert, U., Antón, L.C., Bacik, I., Cox, J.H., Bour, S., Bennink, J.R. *et al.* (1998). *J. Virol.* **72**, 2280–2288.
Schubert, U., Bour, S., Ferrer-Montiel, A.V., Montal, M., Maldarell, F., and Strebel, K. (1996a). *J. Virol.* **70**, 809–819.
Schubert, U., Ferrer-Montiel, A.V., Oblatt-Montal, M., Henklein, P., Strebel, K., and Montal, M. (1996b). *FEBS Lett.* **398**, 12–18.
Schubert, U., Henklein, P., Boldyreff, B., Wingender, E., Strebel, K., and Porstmann, T. (1994). *J. Mol. Bio.* **236**, 16–25.
Shoji, A., Ozaki, T., Fujito, T., Deguchi, K., Ando, S., and Ando, I. (1989). *Macromolecules* **22**, 2860–2863.
Sramala, I., Lemaitre, V., Faraldo-Gomez, J.D., Vincent, S., Watts, A., and Fischer, W.B. (2003). *Biophys. J.* **84**, 3276–3284.
Strebel, K., Klimkait, T., and Martin, M.A. (1988). *Science* **241**, 1221–1223.
Sun, F., (2003). *J. Mol. Model (Online)* **9**, 114–123.
Tiganos, E., Yao, X.J., Friborg, J., Daniel, N., and Cohen, E.A. (1997). *J. Virol.* **71**, 4452–4460.
Vogt, T.C.B., Schinzel, S., and Bechinger, B. (2003). *J. Biomol. NMR* **26**, 1–11.
Watts, A., Burnett, I.J., Glaubitz, C., Grobner, G., Middleton, D.A., Spooner, P.J.R. *et al.* (1999). *Nat. Prod. Rep.* **16**, 419–423.
Willbold, D., Hoffmann, S., and Rosch, P. (1997). *Eur. J. Biochem.* **245**, 581–588.
Willey, R.L., Maldarelli, F., Martin, M.A., and Strebel, K. (1992). *J. Virol.* **66**, 226–234.
Wray, V., Federau, T., Henklein, P., Klabunde, S., Kunert, O., Schomburg, D. *et al.* (1995). *Int. J. Pept. Protein Res.* **45**, 35–43.
Wray, V., Kinder, R., Federau, T., Henklein, P., Bechinger, B., and Schubert, U. (1999). *Biochemistry* **38**, 5272–5282.
Zheng, S., Strzalka, J., Ma, C., Opella, S.J., Ocko, B.M., and Blasie, J.K. (2001). *Biophys. J.* **80**, 1837–1850.

14

Defining Drug Interactions with the Viral Membrane Protein Vpu from HIV-1

V. Lemaitre, C.G. Kim, D. Fischer, Y.H. Lam, A. Watts, and W.B. Fischer

The replication of HIV-1 is strongly enhanced by a small membrane protein called virus protein U (Vpu). Vpu achieves its task by (a) interacting with CD4, the HIV-1 receptor, and (b) by amplifying particle release at the site of the plasma membrane. While the first role is due to interactions of the cytoplasmic site of Vpu with CD4, the second role may be due to ion channel activity caused by the self-assembly of the protein. Recently, a blocker has been proposed which abolishes channel activity. In this chapter, the mechanism of blocking is described using computational methods, including a brief overview of other viral ion channel blockers.

1. Introduction

The discovery of the structure and function of a protein goes mostly in parallel with the wish and also the need to find potential modulators inhibiting the discovered function, especially in the case of viral proteins. Several textbook discoveries are described in the literature (for a general review, see Blundell *et al.*, 2002) in a solely rational approach, and potential drugs have been design for the HIV-1 protease (Lam *et al.*, 1994; for a review see Wlodawer and Erickson, 1993; Hodge *et al.*, 1997) and influenza neuraminidase (von Itzstein *et al.*, 1993) based on their respective crystal structures (Colman *et al.*, 1983; Varghese *et al.*, 1983; Baumeister *et al.*, 1991).

Even more recent example comprises the discovery of an inhibitor of the main protease M^{pro} of the SARS virus from an application of a structure-based computational approach (Yang *et al.*, 2003). This protein is involved in the proteolytic processing of transcribed

V. Lemaitre, C.G. Kim, D. Fischer, Y.H. Lam, A. Watts, and W.B. Fischer • Biomembrane Structure Unit, Department of Biochemistry, Oxford University, South Parks Road, Oxford OX1 3QU, UK.

W.B. Fischer • Bionanotechnology Interdisciplinary Research Consortium, Clarendon Laboratory, Department of Physics, Oxford University, Parks Road, Oxford OX1 3SU, UK.

V. Lemaitre • Nestec S.A., BioAnalytical Science Department, Vers-Chez-Les-Blanc, CH-1000 Lausanne 26, Switzerland.

Viral Membrane Proteins: Structure, Function, and Drug Design, edited by Wolfgang Fischer.
Kluwer Academic / Plenum Publishers, New York, 2005.

proteins essential for the viral life cycle. High-resolution data (around 2 Å) obtained in the presence of the hexapeptidyl inhibitor CbZ-Val-Asn-Ser-Thr-Leu-Gln-CMK revealed the binding conformation of the drug. In addition, structures recorded at different pHs reveal large structural arrangements of this protein upon changes in ionization states.

Out of these examples, one protein, neuraminidase, is a membrane protein (see Chapter 17 by Garman and Laver, this book). Up to date, antiviral drugs have been found against two other membrane proteins, including virus adsorption inhibitors (e.g., gp120 from HIV-1) and virus-cell fusion inhibitors like the 5-helix peptide (Root *et al.*, 2001; for a review, see De Clercq, 2002). However, resistance development against these drugs (e.g., in the case of HIV-1 protease, Condra *et al.*, 1995), necessitates the hunt for other targets such as the smaller, so-called accessory proteins, which include also membrane proteins (for HIV-1, see Miller and Sarver, 1997).

1.1. Short Viral Membrane Proteins

Compared to the larger soluble proteins, which can be seen as highly structured candidates with large surface areas on which putative binding motifs can be discovered, membrane proteins comprise a more difficult target. This is due to the difficulties in obtaining enough structural information for accessible regions (loops), which is a desirable prerequisite for fast drug discovery. Recent attempts to derive relatively large quantities of the full length of short viral membrane proteins, with up to 100 amino acids by applying standard solid phase peptide synthesis (SPPS), have been successfully achieved for M2 from influenza A (Kochendörfer *et al.*, 1999). The two 50mer ends of the protein have been synthesized and chemically ligated to obtain the full-length protein. It is claimed that with this method, even noncoded amino acids can be incorporated, which would mean in case of Vpu, two phosphorylated serines (Ser-52 and Ser-56) would need to be added. Despite the still missing experimental verification the method will play an important role in the future in obtaining sufficient quantities of small proteins in general (Kochendoerfer *et al.*, 2004).

However, so far structural information has emerged gradually by investigating parts of the whole, for which Vpu from HIV-1 may serve as an example. CD, FTIR, and NMR spectroscopy (see Chapters 11 and 13 by Opella *et al.* and Bechinger and Henklein, respectively, this book) have been carried out on parts of Vpu, for example, the transmembrane (TM) part in various lengths and solely the cytoplasmic part. Thereby, the fragments have been produced either by SPPS or expression methodology. The structural information is sufficient to enable the generation of computational models. These models, especially the models representing the TM segment of Vpu, have been embedded in hydrophobic slabs, surrounded by octane to mimic a lipid bilayer, or in fully hydrated lipid bilayers on an all atom basis (reviewed in Fischer, 2003). Recently, even a full-length model based on the structural information of all these fragments has been produced and embedded in a lipid monolayer (Sun, 2003). The models can be used for docking approaches and molecular dynamics (MD) simulations enabling detailed structural analysis, until high-resolution data are available. Once a high-resolution structure is available, these methods may furthermore be used to refine the static structure, predict protein mechanics and putative drug–protein interactions.

In this chapter, the efforts to obtain information about drug–protein interactions of Vpu are summarized and compared with investigations on other short viral membrane proteins. The emphasis is to outline the computational methods used for this enterprise and to demonstrate their potential for drug discovery especially in the case of Vpu, for which a high-resolution structure is not yet available.

1.2. The Vpu Protein

Vpu is an 81 amino acid protein encoded by HIV-1 with a high degree of sequence conservation (Willbold *et al.*, 1997). Its role and structure are reported in detail in other chapters of this book. In brief, its function in the life cycle of HIV-1 is 2 fold (for reviews, see Fischer and Sansom, 2002; Bour and Strebel, 2003; Fischer, 2003; Montal, 2003): (a) to interact with CD4 in the endoplasmic reticulum to initiate the ubiquinine mediated degradation of the CD4–Vpu complex, and (b) to enhance particle release at the site of the plasma membrane altering the electrochemical gradient via ion channel formation by homo oligomerization in the lipid membrane or interacting with other ion channels (Hsu *et al.*, 2004). While the first function is fairly established, the second is still open to debate (Lamb and Pinto, 1997). For example, it is not known whether ion channel activity is an intrinsic part of Vpu's function. However, Vpu, either with its TM domain synthesized using SPPS or as full-length protein from expression and reconstituted into lipid bilayers, shows channel activity (Ewart *et al.*, 1996; Schubert *et al.*, 1996; Marassi *et al.*, 1999; Cordes *et al.*, 2002; Park *et al.*, 2003). Recently, a drug has been described which blocks channel activity of Vpu *in vitro* (Ewart *et al.*, 2002) and also of p7 from hepatitis C virus (Premkumar *et al.*, 2004). The molecule causing this effect is a derivative of amiloride, cyclohexamethylene amiloride.

Spectroscopic studies of Vpu incorporated into membranes (or mimic membranes) using CD (Wray *et al.*, 1995, 1999), FTIR (Kukol and Arkin, 1999), and NMR spectroscopy (Federau *et al.*, 1996; Willbold *et al.*, 1997; Ma *et al.*, 2002) have identified structural elements allowing the following description of the Vpu structure: a helical TM segment is followed by a larger cytoplasmic domain with a helix–loop helix–helix/turn motif. The cytoplasmic domain seems to be in contact with the membrane (Marassi *et al.*, 1999; Henklein *et al.*, 2000; Zheng *et al.*, 2003). The tilt angle of the TM segment with respect to the membrane normal using different techniques and Vpu constructs has been found to range from approximately 6° to 30° (Kukol and Arkin, 1999; Marassi *et al.*, 1999; Wray *et al.*, 1999; Henklein *et al.*, 2000). The most recent NMR spectroscopic study indicates a kink of the TM segment of about 12–15° around residue Ile-17 (Park *et al.*, 2003). Based on these experimental results, several models of Vpu have been proposed from computer simulations (Grice *et al.*, 1997; Cordes *et al.*, 2002; Lopez *et al.*, 2002; Sramala *et al.*, 2003; Sun, 2003).

For further details of Vpu, the reader may refer to Chapters 11, 12, 13, and 15 by S.J. Opella *et al.*, Wray and Schubert, B. Bechinger and P. Henklein, and P. Gage *et al.*, respectively, in this book.

2. The Methods

2.1. Docking Approach

Docking approaches have been proven to be valuable tools in identifying not only the drug-binding sites on proteins but also to screen large databases for other potential drugs, if the binding site is known (Blundell *et al.*, 2002; Glick *et al.*, 2002). In docking methods, the protein and the ligand are transferred to a point on a grid. Smaller spacing of the grid increases the accuracy of the method, which proceeds in parallel with an increase in computer time for the calculations. Pioneering work in this field has been achieved by Peter Goodford (Goodford, 1985; Boobyer *et al.*, 1989; Wade *et al.*, 1993). In this method, putative docking sites and the potency of different conformations of a drug have been assessed based on calculating electrostatic interactions. To date, several other docking software are available differing in the function implemented to evaluate the fit (scoring) of a drug to the binding site (scoring function). In principle, both the protein and the drug are held rigid during the calculation to save computer time. More advanced methods allow for the ligand to be flexible (Wang *et al.*, 1999; Kua *et al.*, 2002) and even the protein (Carlson and McCammon, 2000). Another improvement of the docking approach is the introduction of the energy of solvation of the ligand for evaluation of binding affinities and consequently the scoring (Shoichet *et al.*, 1999). In the present study, AUTODOCK (Morris *et al.*, 1998) was used, in which the Lamarckian genetic algorithm (GA) is applied. The ligand explores randomly translational, orientational, and conformational space with respect to a rigid protein. To avoid extreme CPU time due to the endless number of possible combinations in the search, GAs are combined with a local search (LS) protocol, which allows for energy minimization at low temperatures. However, a general drawback for the docking software is the inability to include the electrostatic contributions of a lipid bilayer environment.

2.2. Molecular Dynamics (MD) Simulations

MD simulations *per se* describe the changes of positions, velocities, and orientations of a system with time, based on Newtonian principles. From the trajectories, time averages of macroscopic properties can be deduced. Any screening of a protein surface by a drug or ligand would cost almost endless MD simulation time, which would impose software failures and consequently inaccurate results. Moving a ligand around a protein surface might be envisaged by the use of "artificial" force on the ligand. Steered MD simulation is a step in this direction used till date for protein folding (Lu *et al.*, 1998; Gao *et al.*, 2002) and to propose pathways of molecules into or out of known biding sites (Kosztin *et al.*, 1999; Isralewitz *et al.*, 2001; Shen *et al.*, 2003). A workaround is the combined use of a docking approach followed by a MD simulation or molecular mechanics calculations (Wang *et al.*, 2001; Beierlein *et al.*, 2003). However, using solely MD simulations for the exploration of the drug and the protein is becoming an increasingly valuable tool. Recent investigations on the effect of general anesthetics on the ion channel gramicidin A have been investigated (Tang and Xu, 2002). In this study, the missing local changes on the protein caused by the drugs have been correlated with a low affinity of the general anesthetics to their target. In another approach, effects of mutations on the structure of the HIV-1 integrase have been analyzed using MD simulations (Barreca *et al.*, 2003). Simulations have also been done in the presence of an inhibitor (5CITEP) to address drug–protein interactions. In the present chapter, the putative binding site of amiloride (Am) and cyclohexamethylene amiloride (Hma) with the TM domain of Vpu has been derived from the docking approach. In the docking approach, a pentameric bundle was used to discover the binding site. Simulations were then run with the protein bundle—drug complex for 12 ns. The topology of the drug and its partial atomic charges have been determined by PRODRUG (van Aalten *et al.*, 1996) and adapted for the force field GROMOS43a2.

3. Analysis of Drug–Protein Interactions of Vpu with a Potential Blocker

According to the results shown experimentally (Ewart *et al.*, 2002), Hma and, to a lesser extent, dimethyl amiloride block channel activity of full-length Vpu, reconstituted into a lipid membrane, and a peptide representing the TM domain of the protein. Am itself has almost negligible effect. With these data at hand, and the large amount of structural data available (see Fischer, 2003, and references therein; Park *et al.*, 2003), computational methods can be approached to address the following questions: (a) the location of a putative binding site and the affinity of the blocker; (b) the dynamics of the blocker at the binding site and the blocking mechanism; and (c) the entry to the binding site.

3.1. Using the Docking Approach

The question of where the blocker binds can be addressed using docking programs, even though they are not designed to search within pore-like structures. In the present study, the program AUTODOCK is used for Am and the most potent blocker Hma, both in protonated (AM^+, HMA^+)—the most likely form under physiological conditions—and unprotonated (AM, HMA) forms (C. Kim, V. Lemaitre, A. Watts, W.B. Fischer, in preparation). The Vpu models used are single-stranded Vpu, corresponding to the TM helix of Vpu, Vpu_{1-32} (Figure 14.1) and the extended kinked model Vpu_{1-52}. Models with assembled helices forming pentameric and hexameric bundles are generated and also used in the docking approach. The single peptides are rationalized by the idea of a possible binding mechanism in which the blocker binds prior to the formation of ion conducting bundles. The results can be summarized in such a way that the putative binding site is toward the C terminal end of the peptides and, in case of the bundles, within the pore. The specific residue to which the blockers establish contact is Ser-23. The type of interaction is via hydrogen bonding. Calculated binding constants are found to be of the same range as those derived from experiments

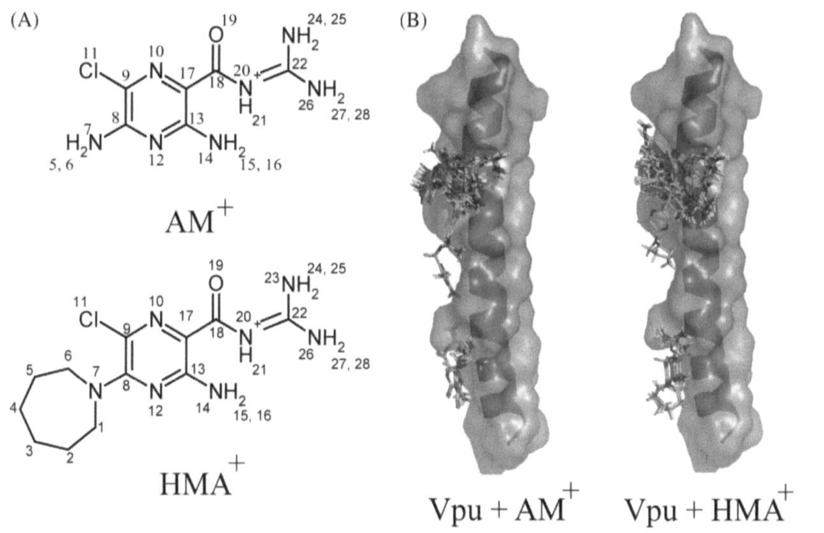

Figure 14.1. (A) Structures and atom number used for AM^+ (top) and HMA^+ (bottom) for the docking experiments. (B) Docking result of both blockers on a single strand of Vpu (QPIPIVAIVA[10] LVVAIIIAIV[20] VWSIVIIEYR[30] KI, HV1H2, shown in gray with the helix highlighted) corresponding to the TM segment. The TM segment is shown with the results for AM^+ (left) and HMA^+ (right).

(Fischer, Lam, Watts, Fischer, unpublished results). Based on these findings MD simulations may follow to address the dynamics of the blocker and the protein in an almost realistic environment of a hydrated lipid bilayer (Lemaitre *et al.*, 2004). Studies of membrane proteins embedded in a hydrated lipid bilayer have been highly successful in describing the mechanism of function of ion channels on an atomic level (Tieleman *et al.*, 1997; Sansom *et al.*, 1998; Shrivastava and Sansom, 2000; Berneche and Roux, 2001; de Groot and Grubmüller, 2001; Im and Roux, 2002; Zhu *et al.*, 2002; Böckmann and Grubmüller, 2002).

3.2. Applying MD Simulations

Based on the findings from the docking approach, cyclohexamethylene amiloride and amiloride, both in their protonated and deprotonated sates, have been placed within a pentameric bundle (Figure 14.2). The whole system has then been placed into a hydrated lipid bilayer. The analysis gives an insight into the effect of the blockers on the protein structure.

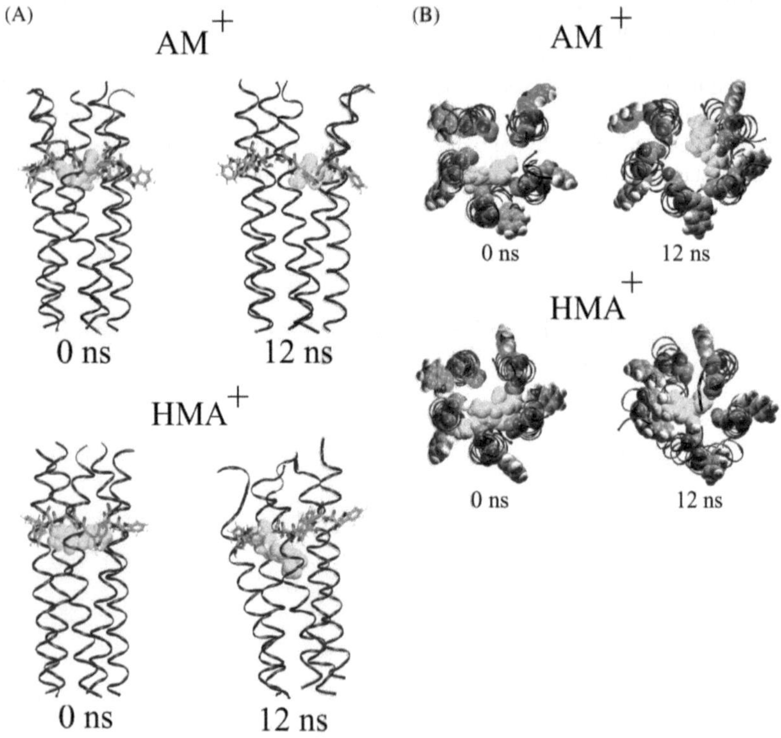

Figure 14.2. (A) Side view of the blockers (in light gray) within a pentameric bundle consisting of the circular assembly of the helical TM segments. AM^+ is shown in the top bundles at 0 ns (left) and 12 ns (right), HMA^+ in the lower bundles. The lipid bilayer and the water molecules present during the simulations are omitted for clarity. Tryptophans and serines are indicated by sticks. The C-terminal end is pointing up, while the N-terminal end is pointing down. The same structures as in (A) are shown with a view down into the pore from the C-terminal end (B).

The analysis of the data reveals no major structural rearrangements of the helices. For a more detailed analysis of the pore, the bundle is subdivided into three regions: an N-terminal (residues 1–11), a middle (residues 12–22), and a C-terminal section (residues 23–32). The data uncover small structural sorting which allows the following interpretation: while amiloride induces widening of the pore, cyclohexamethylene induces a more funnel-like shape, with the narrow part at the N-terminal end.

Principal Component Analysis (PCA) has been applied to determine concerted motion taking place in the bundles during the simulation (Figure 14.3). PCA (Amadei *et al.*, 1993), sometimes called "Essential Dynamics," allows finding correlated motions within an object, for example, a protein (Garcia, 1992; Wong *et al.*, 1993; van Aalten *et al.*, 1995; Yang *et al.*, 2001; Grottesi and Sansom, 2003). These correlated motions or principal components have been shown to describe motions which are relevant for the function of proteins. The technique involves the removal of the overall rotation and translation to isolate the internal motion only. This is achieved by a least square fitting to a reference structure and computation of the covariance matrix C of the atomic coordinates. The principal components are then obtained by diagonalizing the matrix C with an orthonormal transformation (Amadei *et al.*, 1993).

The only atoms that have been considered for the analysis of the Vpu bundles are the $C\alpha$ atoms. This enables noise arising from the random motion of the side chains to be removed and to reduce the size of the matrices which needs to be diagonalized. Furthermore, the analysis is performed on the equilibrated part of the trajectory, discarding the first nanosecond. Figure 14.3 shows the first principal motions which have the largest amplitude in each of the bundles indicated by arrows. The motions describe a correlated change in the twist or in the kink of individual helices forming the pore for the Vpu bundle without blocker (Figure 14.3, (1)) and the Vpu bundle in the presence of AM^+ (Figure 14.3, (3)). The main correlated motion in the presence of HMA^+ (Figure 14.3, (2)) describes a concerted change in the tilt angle of two helices, indicating a closure of the pore which finally explains the funnel-like shape.

The mode of blocking also includes occlusion. AM and AM^+ are within the pore but are not occluding it completely, while HMA and HMA^+ almost completely occupy the space within the pore (Figure 14.3). When in the vicinity of at least one of the serines, the apparent minimum pore radius left within the middle section is about 1.42 ± 0.46 Å for AM^+ and 0.63 ± 0.40 Å for HMA^+. The latter value is too small for even allowing a single filed water (Roux and Karplus, 1991; Woolley and Wallace, 1992; Chiu *et al.*, 1999; de Groot *et al.*, 2002) to pass the blocker. Water molecules in the peptide antibiotic gramicidin are assumed to pass the pore in a single file way, which means that one water molecule has only neighbors in front and in the back.

The root-mean-square deviation (RMSD) values for all blockers remain below 0.2 nm with slightly higher values for HMA and HMA^+, and the largest fluctuations for the latter (data not shown). The larger values and spread is indicative of the amiloride derivative for the flexible cyclohexamethylene ring. The root mean square fluctuations (RMSF) of the individual atoms of the blockers indicate the central body of Am (a 3,4,6-substituted pyrazine ring) and Hma, remain fairly rigid (RMSF < 0.1 nm) independent of the protonation state of the blockers. Only the hydrogen atoms of the amino groups of the pyrazine ring and the guanidinium group show the largest fluctuations (≥ 0.1 nm). The curves albeit very similar for both blockers, when compared with similar atoms, adopt slightly lower values for all atoms in the protonated blocker. The cyclohexamethylene ring in Hma fluctuates around 0.1 nm, independent of the protonation state. Thus, the hexamethylene ring adds a mobile part to the rigid pyrazine ring.

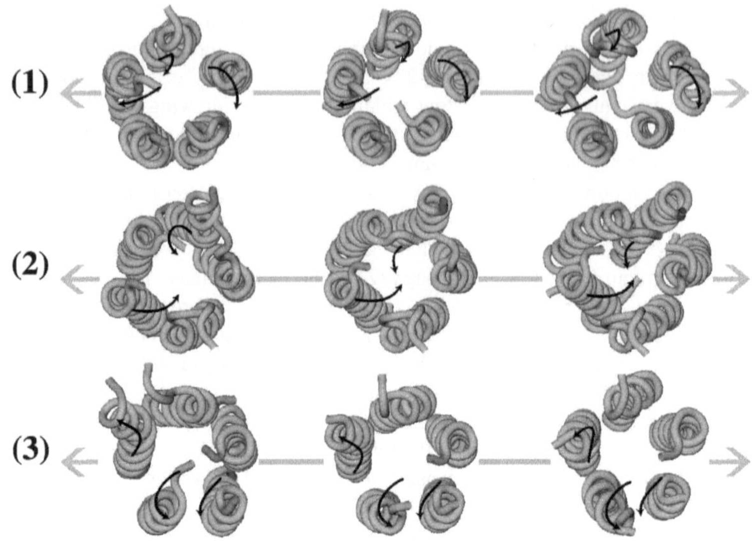

Figure 14.3. First principal components, indicating the correlated motion with the largest amplitude, on the Cα atoms of the Vpu bundle without any blocker (1), Vpu bundle in the presence of HMA^+ (2), and AM^+ (3). The small arrows indicate the motion calculated for the Cα atoms of the helices. The large horizontal arrows symbolize the protein oscillation between the two conformations on either side. The small arrows highlight the section of the bundle where the motion takes place.

The different conformations generated by the MD simulations for the blockers tested were clustered using a full linkage algorithm, using a cut-off of 0.03 nm for Am and 0.04 nm for Hma. In the case of Am, this means that molecules with an RMSD smaller than 0.03 nm relative to all the existing members of a cluster will belong to this cluster. The mostly populated conformation for AM^+ with 96.0% is shown in Figure 14.4A, the most frequently adopted conformation for HMA^+ with 95.1%, in Figure 14.4B. The carbonyl group linked to the guanidinium group in the protonated blocker is pointing toward the primary amine group of the pyrazine ring. Deprotonation of the blockers reveals a conformational change in the blocker so that the amine part of the guanidinium group is pointing to the primary amine group of the pyrazine ring (data not shown).

The blockers, independent of their protonation state, interact with the protein via hydrogen bonding. The hydrogen bond partners are the serine side chains (Ser-23) which point into the pore. Therefore, the most prominent difference between AM^+ and HMA^+ is that, in addition to these hydrogen bonds, HMA^+ also interacts with one of the tryptophans (Trp-22) which comes to reside at the helix–helix and helix–lipid interface. The cause of this interaction might be due to the lower average velocity of HMA^+ within the site. In addition to this dynamic affect, HMA^+ orients with its hydrophobic cyclomethylene ring stronger to the hydrophobic part of the pore toward the N-terminal end. This is reflected by an average angle of $43.0 \pm 11.8°$ for HMA^+ and $11.3 \pm 12.7°$ for HMA with respect to the membrane plane. AM^+ and AM adopt angles of $5.3 \pm 23.3°$ and $8.6 \pm 51.8°$, respectively. The larger standard deviation reflects the higher flexibility of the Am within the pore.

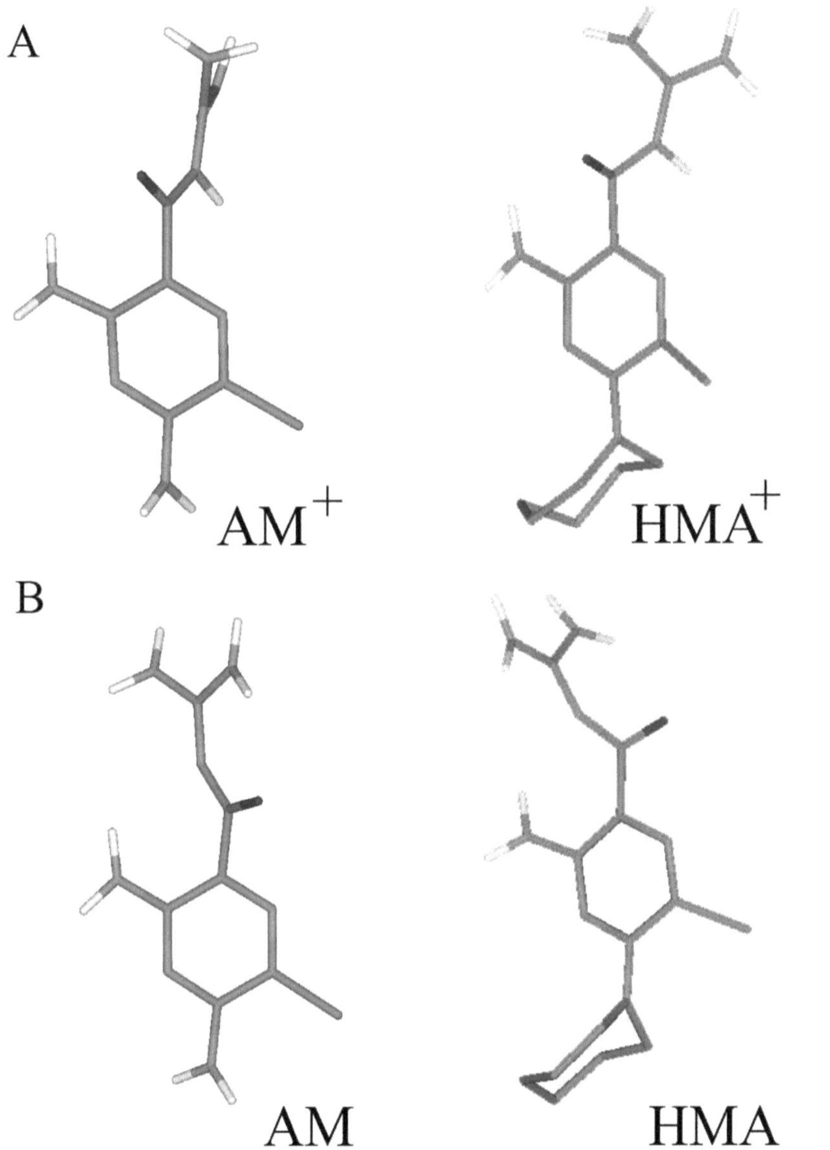

Figure 14.4. Most populated structures during the MD simulations for the (A) protonated and the (B) deprotonated blockers.

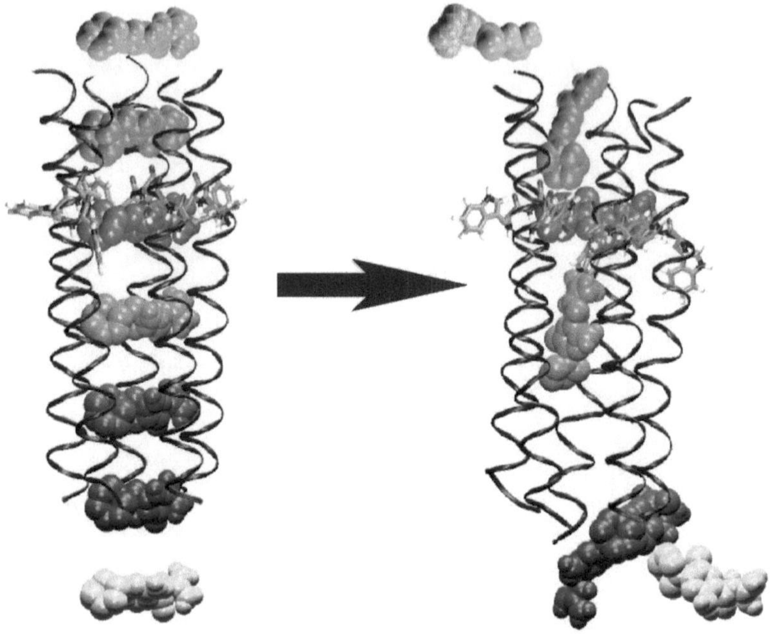

Figure 14.5. A pentameric bundle of the TM segment of Vpu with the individual HMA^+ superimposed at different positions along the pore axis at the beginning (left structure) of the simulation and after 1 ns (right structure). The tryptophans and the serines are highlighted by sticks.

The MD simulations have been based on results obtained from a docking approach performed with the AUTODOCK software. The blocker was placed at a site suggested by AUTODOCK. The question arises, would MD simulations be able to deliver the same result? In order to address this question, a series of 1 ns simulation have been performed with a pentameric bundle and HMA^+ at different position along the *z*-axis within the pore of a pentameric bundle (Figure 14.5). The results in terms of the stability of the bundle and the overall structure remain similar to the results of the long simulations here (data not shown). The integrity of the bundle is not destroyed, independent of the position of the blocker. It is interesting to note that in all of the positions, other than the one used for the longer simulations (position at the Ser-23 site), does the blocker leaves its starting position after 1 ns. To put the individual pictures in context, any position of the HMA^+ toward the C- or N-terminal end of the bundle may force the blocker to orientate its long axis parallel to the pore axis, and beyond a certain threshold position even to escape from the pore.

4. How Realistic is the Protein Model?

Computational approaches are always judged by the relevance of their results which stand and fall with the model assumed. In this case, the question is allowed: how realistic are the bundles? Experimental studies propose that Vpu exists as an oligomer (Maldarelli *et al.*, 1993). Computational experiments on solely the TM segments have been undertaken to relate the experimentally derived cation selectivity of Vpu to the number of segments forming the bundle (Grice *et al.*, 1997). In these studies, the low energy for a K^+ to pass the pore compared to the higher value obtained for Cl^- for the pentameric bundle supports the idea of a pentameric model. In a combined experimental and computational approach, the tetrameric bundle was excluded as a putative pore (Cordes *et al.*, 2002). However, direct structural evidence is still lacking. Thus, Vpu could also be a tetramer unless proven by, for example, X-ray, spectroscopy or otherwise, and thus, investigations are carried out with the pentameric and hexameric bundles. The overall orientation of the helices, having the serines facing the lumen of the pore, is based on findings for other ion channels (Leonard *et al.*, 1988; Galzi *et al.*, 1991; Pebay-Peyroula *et al.*, 1997; Pautsch and Schulz, 2000; Sass *et al.*, 2000). Most recent solid-state NMR spectroscopic investigations propose a kink around Ile-17 (Park *et al.*, 2003). The computer models are generated without an explicit kink; however, the analysis of the bundles using residues Ile-16 to Ala-18 as center point also reveal on average a kink of the helices. Differences in the absolute values may derive from the different lipids used and the length of the peptides, including additional residues at the C-terminal end to enable facilitated purification.

5. The Putative Binding Site

The site of binding is at the moment pure computationally based speculation. However, we may bring the site into context with experimental evidence.

Am (3,5-diamino-6-chloro-*N*-(diaminomethylene)pyrazinecarboxamide) is a moderately strong base with a pK_a of 8.7 (Schellenberg *et al.*, 1985; Kleyman and Cragoe, 1988). At physiological pH at least 95%, Am exists in a positively charged form due to the protonated guanidinium group (Cragoe *et al.*, 1967). In the case of epithelial Na^+ channels, this protonated form is essential for blocking, indicative of a pH-dependent activity (Kleyman and Cragoe, 1988). Am is soluble in water up to 16 mM (Benos, 1982). The deprotonated form is lipid soluble and can easily cross the cell membrane with the consequence of accumulating within the cell and altering a number of cellular processes (Benos, 1982; Kleyman and Cragoe, 1988; Grinstein *et al.*, 1989). Am cannot be metabolized in the body, so it is eliminated intact in urine and has a biological half-life of 9.6 ± 1.8 hr (Smith and Smith, 1973). Am and its derivatives are a class of potent blockers of Na^+ transporters such as the epithelial Na^+ channels (ENaC), Na^+/H^+ exchanger and, to a lesser extent, the Na^+/Ca^{2+} exchanger (Kleyman and Cragoe, 1988, 1990). Cyclohexamethylene amiloride was reported to be especially selective for the Na^+/H^+ exchanger (Kleyman and Cragoe, 1988). Hydrophobic interactions with parts of the channel adjacent to the Am binding site increases the binding affinity of the blocker (Garty and Palmer, 1997). For Am, a putative binding site in the Na^+ channel has been proposed to be approximately 20% within the TM electrical field (Li *et al.*, 1987; Fyfe and Canessa, 1998) on the extracellular side (Benos, 1982). The orientation is proposed to allow the guanidinium group to penetrate into the pore (reviewed in Alvarez de la Rosa *et al.*, 2000). Using anti-Am antibodies, a six amino acid sequence of WYRFHY (extracellular loop of α-rENaC), WYKLHY (β- and γ-ENaC subunit) and WYHFHY (δ-ENaC subunit) (Kieber-Emmos *et al.*, 1995; Waldmann *et al.*, 1995; Ismailov *et al.*, 1997; Schild *et al.*, 1997) has been identified as a putative binding site. Replacement of the second His-282 in the δ-ENaC subunit by glutamate abolishes the blocking while replacement by arginine increases the blocking. Another point mutation within the TM region replacing a crucial serine residue also abolishes affinity for Am (Waldmann *et al.*, 1995). This indicates that the sequence, including an aromatic residue, a positively charged residue, and a serine in the vicinity of a hydrophobic site, is a key feature for binding Am and its derivatives. Consequently, the binding site proposed, with the guanidinium group in the vicinity of the serines of the putative pore of Vpu in the present study, is in agreement with findings on the ENaC.

Also, local anesthetics such as QX-222 and QX-314, which reflect amiloride in their overall structure, are penetrating the pore of the nicotinic acetylcholine receptor and induce blocking via occlusion (Neher and Steinbach, 1978). The binding site for these blockers, and also procaine (Adams, 1977), is proposed to be at a site within the pore which should be near to the ring of serines and threonines found in the pore of the receptor (Leonard *et al.*, 1988). Based on experimental findings, computational experiments, using the Monte Carlo minimization method, have been performed on a model of the nAChR based on its TM helices of M2 (Tikhonov and Zhorov, 1998). The data analyzed the binding geometry of, for example, QX-222 and Chlorpromazin (CPZ). It is found that the charged groups interact with the side chains of the serines and threonines (Thr-4 or Ser-8), and the uncharged groups interact with the ring of hydrophobic residues (Leu-11, Ala-12). The blockers orient their long axis along the axis of the bundle due to the amphiphatic character of the drugs.

These results are indicative of a generalized pattern of binding behavior and underline the relevance of the results obtained for the Am and its derivative in the present study.

One of the important factors for the discovery of potential blocker and drugs is also the knowledge of how the blocker reaches its binding site. The diffusion of the drug to (on rate) the site and away (off rate) from it is involved in defining the binding constant (Lüdemann *et al.*, 2000). Also, it is essential to know the correct description of the energetics within the narrow geometry of the pore. Indirect methods like Brownian dynamics or steered MD, and calculation of free energies and reaction path methods may give a reasonable picture of the diffusion of the blocker into the pore. These methods need to be taken into account to improve the understanding of the mechanism of blocking and the success of virtual drug screening for pore blockers.

6. Water in the Pore

During the generation of the bundles and finally the hydration of the bilayer system, water molecules are found within the pore. However, during the 300 ps equilibration step, they totally escape and do not re-enter the pore during the simulation. During the entire duration of the simulations, the hydrophilic C-terminal end is engulfed by water reaching the serines and the blocker. Further toward the N-terminal end, no water molecule remains within the pore.

Simulations on bundles of Vpu consisting of peptides representing the TM sequence IAIVA^{10} LVVAIIIAIV^{20} VWSIVII indicate a reduced number of water molecules toward the C-terminal end during the early stage of the simulation (Lopez *et al.*, 2002). The pentameric and hexameric bundles are embedded in an octane slab which mimics the lipid bilayer. The long-range electrostatic interactions have been treated with Ewald sums. These findings are indicative for a fluctuating number of water molecules within the pore. Recent investigations on theoretical hydrophobic pore models show that the presence of water in these pores is strongly dependent on the pore radius (Beckstein and Sansom, 2003). Below 0.4 nm, the water density found decreases to almost zero. Water molecules, if bigger than 0.4 nm, traverse the pore in an "avalanche-like fashion" (Beckstein and Sansom, 2003). For very small radii, simulations on a very long timescale (>50 ns) are needed to assess the proper behavior of the water molecules in confined geometry. Thus, the Vpu bundle model presented here shows a proper bundle with a temporary state of low water content.

7. MD Simulations for Drug Screening?

The use of MD simulations for drug screening of membrane proteins still needs further testing. Up to now simulations are still time consuming, which imposes a major drawback when compared to docking approaches. However, the possibility of implementing more realistic conditions upon the models, such as for membrane proteins within a bilayer in an all-atom representation including an explicit representation of solvent, will, in the future, overcome the existing disadvantages through the advancement of both software and hardware (Mangoni *et al.*, 1999). In the meantime, the combination of MD simulations with docking approaches has been proven to be a powerful tool for lead discovery (Wang *et al.*, 2001; Kua *et al.*, 2002). MD simulations with the drug will increasingly play an important role in the exploration of the dynamics of the ligand and its impact on the protein structure (Tang and Xu, 2002). Here, if Vpu is identified as a potential drug target, then identification of residues involved in blocking action will help in further rational drug design.

8. Other Viral Ion Channels and Blockers

One of the earliest antiviral drugs, amantadine (amantadine-1), and also remantadine (α-methyl-1-adamantanemethanamine) (Davies *et al.*, 1964; Hoffmann, 1973) are known to target the viral proton channel M2 from influenza A (Wang *et al.*, 1993), with the latter especially used in chemoprophylaxis and therapy of influenza A. Its site of blocking is assumed to be within the lumen of the pore with residues such as Val-27, Ala-30, and Ser-31 involved (Duff and Ashley, 1992; Pinto *et al.*, 1992; Duff *et al.*, 1993). For a detailed review, see Chapter 8 by Y. Tang *et al.*, this book. Also NB, an influenza B membrane protein forming ion channels, can be blocked by amantadine (Sunstrom *et al.*, 1996; Fischer *et al.*, 2001). However, the binding constants proposed for blocking are too high to propose NB as a potential target. The short viral membrane protein p7 from Hepatitis C virus also exhibits ion channel activity (Griffin *et al.*, 2003; Pavlovic *et al.*, 2003). For this channel, amantadine seems to block channel activity (Griffin *et al.*, 2003). However, it could be shown that this channel can also be blocked by long-alkyl-chain iminosugar derivatives (Pavlovic *et al.*, 2003). Structural and, consequently, computational data, still need to be produced to get more insight about the putative binding site of these blockers. With Vpu as another viral ion channel forming protein, the number of HIV protein targets has been enlarged by the discovery of Am derivatives as potential channel blockers (Ewart *et al.*, 2002; see also Chapter 15 by P. Gage *et al.*, this book). Derivatives of amantadine, spiro 5- and 6-membered analogs, have been identified for weak HIV-1 activity (Kolocouris *et al.*, 1996). Since they have not been active against HIV-2, which does not encode Vpu, it is tempting to assume blocking of either one or both of the two roles of Vpu in the infected cell. Possibly other viral ion channel proteins and viroporins may become highly important future drug targets and we might be just at the beginning of a fruitful time in this respect.

9. Speculation of Binding Sites in the Cytoplasmic Site

It remains speculative of that the potential ligand-binding site of the AM derivatives might be in the cytoplasmic part of Vpu. The beauty of this approach would be to also inhibit Vpu's function of initiating CD4 degradation. Also, in this particular case of Vpu and in general, future drug design has to take into consideration the immediate proximity of the lipid membrane.

10. Conclusions

Small viral membrane proteins which are forming ion channels are increasingly in the line of fire as drug targets. Some of these proteins, like Vpu, are highly conserved. However, for some of the viruses, antiviral drugs against other proteins are already available. This is mostly because these other proteins are globular, relatively easily accessible and, with moderate effort, crystallizable. Membrane proteins still remain a difficult target in this respect. On the other hand, these proteins, like Vpu, need to assemble to fullfill their task which comprises another challenge in finding the proper target structure. It seems likely that, especially if functioning as ion channels, these small proteins will show their Achilles' heel and may become potential targets.

Acknowledgments

WBF thanks the E.P. Abraham Research Fund for financial support of this work. The Bionanotechnology IRC, the Engineering and Physical Science Research Council (EPSRC), Medical Research Council (MRC), the Biological Science Research Council (BBSRC, studentship for VL), and the CJ Corporation (for financial support to CGK) are all acknowledged for grant support to AW.

References

Adams, P.R. (1977). Voltage jump analysis of procaine action at frog end-plate. *J. Physiol. (London)* **268**, 291–318.

Alvarez de la Rosa, D., Canessa, C.M., Fyfe, G.K., and Zhang, P. (2000). Structure and regulation of amiloride-sensitive sodium channels. *Annu. Rev. Physiol.* **62**, 573–594.

Amadei, A., Linssen, A.B., and Berendsen, H.J.C. (1993). Essential dynamics of proteins. *Protein Struct. Func. Genet.* **17**, 412–425.

Barreca, M.L., Lee, K.W., Chimirri, A., and Briggs, J.M. (2003). Molecular dynamics studies of the wild-type and double mutant HIV-1 integrase complexed with the 5CITEP inhibitor: Mechanism for inhibition and drug resistance. *J. Biophys.* **84**, 1450–1463.

Baumeister, W.P., Ruigrok, R.W. and Cusack, S. (1991). The 2.2Å resolution crystal structure of influenza B neuraminidase and its complex with sialic acid. *EMBO J.* **11**, 49–56.

Beckstein, O. and Sansom, M.S.P. (2003). Liquid-vapor oscillations of water in hydrophobic nanopores. *Proc. Natl. Acad. Sci. USA* **100**, 7063–7068.

Beierlein, F., Lanig, H., Schuerer, G., Horn, A.H.C., and Clark, T. (2003). Quantum mechanical/molecular mechanical (QM/MM) docking: An evaluation for know test systems. *Mol. Phys.* **101**, 2469–2480.

Benos, D.J. (1982). Amiloride: A molecular probe of sodium transport in tissues and cells. *Am. J. Physiol.* **242**, C131–C145.

Berneche, S. and Roux, B. (2001). Energetics of ion conductance through the K^+ channel. *Nature* **414**, 73–77.

Blundell, T.L., Jhoti, H., and Abell, C. (2002). High-throughput crystallography for lead discovery in drug design. *Nat. Rev. Drug Discov.* **1**, 45–54.

Böckmann, R. and Grubmüller, H. (2002). Nanosecond molecular dynamics simulation of primary mechanical steps in F1-ATP synthase. *Nat. Struct. Biol.* **9**, 198–202.

Boobyer, D.N.A., Goodford, P.J., McWhinnie, P.M., and Wade, R.C. (1989). New hydrogen-bond potentials for use in determining energetically favourable binding sites on molecules of known structure. *J. Med. Chem.* **32**, 1083–1094.

Bour, S. and Strebel, K. (2003). The HIV-1Vpu protein: A multifunctional enhancer of viral particle release. *Microb. Infect.* **5**, 1029–1039.

Carlson, H.A. and McCammon, J.A. (2000). Accommodating protein flexibility in computational drug design. *Mol. Pharm.* **57**, 213–218.

Chiu, S.W., Subramaniam, S., and Jakobsson, E. (1999). Simulation study of a gramicidin/lipid bilayer system in excess water and lipid. I. Structure of the molecular complex. *J. Biophys.* **76**, 1929–1938.

Colman, P.M., Varghese, J.N., and Laver, W.G. (1983). Structure of the catalytic and antigenic sites in influenza virus neuraminidase. *Nature* **303**, 41–44.

Condra, J.H., Schleif, W.A., Blahy, O.M., Gabryelski, L.J., Graham, D.J., Quintero, J.C. *et al.* (1995). In vivo emergence of HIV-1 variants resistant to multiple protease inhibitors. *Nature* **374**, 569–571.

Cordes, F.S., Tustian, A., Sansom, M.S.P., Watts, A., and Fischer, W.B. (2002). Bundles consisting of extended transmembrane segments of Vpu from HIV-1: Computer simulations and conductance measurements. *Biochemistry* **41**, 7359–7365.

Cragoe, E.J., Jr., Woltersdorf, O.W., Jr., Bicking, J.B., Kwong, S.F., and Jones, J.H. (1967). Pyrazine diuretics. II. N-amidino-3-amino-5-substituted 6-halopyrazinecarboxamides. *J. Med. Chem.* **10**, 66–75.

Davies, W.L., Grunert, R.R., Haff, R.F., McGahen, J.W., Neumayer, E.M., Paulshock, M. *et al.* (1964). Antiviral activity of 1-adamantanamine (amantadine). *Science* **144**, 862–863.

De Clercq, E. (2002). Strategies in the design of antiviral drugs. *Nat. Rev. Drug Discov.* **1**, 13–25.

de Groot, B.L. and Grubmüller, H. (2001). Water permeation across biological membranes: Mechanism and dynamics of aquaporin-1 and GlpF. *Science* **294**, 2353–2357.

de Groot, B.L., Tieleman, D.P., Pohl, P., and Grubmüller, H. (2002). Water permeation through gramicidin A: Desformylation and the double helix: A molecular dynamics study. *J. Biophys.* **82**, 2934–2942.

Duff, K.C. and Ashley, R.H. (1992). The transmembrane domain of influenza A M2 protein forms amantadine-sensitive proton channels in planar lipid bilayers. *Virology* **190**, 485–489.

Duff, K.C., Gilchrist, P.J., Saxena, A.M., and Bradshaw, J.P. (1993). The location of amantadine hydrochloride and free base within phospholipid multilayers: A neutron and X-ray diffraction study. *Biochim. Biophys. Acta* **1145**, 149–156.

Ewart, G.D., Mills, K., Cox, G.B., and Gage, P.W. (2002). Amiloride derivatives block ion channel activity and enhancement of virus-like particle budding caused by HIV-1 protein Vpu. *Eur. J. Biophys.* **31**, 26–35.

Ewart, G.D., Sutherland, T., Gage, P.W., and Cox, G.B. (1996). The Vpu protein of human immunodeficiency virus type 1 forms cation-selective ion channels. *J. Virol.* **70**, 7108–7115.

Federau, T., Schubert, U., Floßdorf, J., Henklein, P., Schomburg, D., and Wray, V. (1996). Solution structure of the cytoplasmic domain of the human immunodeficiency virus type 1 encoded virus protein U (Vpu). *Int. J. Pept. Protein Res.* **47**, 297–310.

Fischer, W.B. (2003). Vpu from HIV-1 on an atomic scale: Experiments and computer simulations. *FEBS Lett.* **552**, 39–46.

Fischer, W.B., Pitkeathly, M., and Sansom, M.S. (2001). Amantadine blocks channel activity of the transmembrane segment of the NB protein from influenza B. *Eur. J. Biophys.* **30**, 416–420.

Fischer, W.B. and Sansom, M.S. (2002). Viral ion channels: Structure and function. *Biochim. Biophys. Acta* **1561**, 27–45.

Fyfe, G.K. and Canessa, C.M. (1998). Subunit composition determines the single channel kinetics of the epithelial sodium channel. *J. Gen. Physiol.* **112**, 423–432.

Galzi, J.L., Revah, F., Bessis, A., and Changeux, J.P. (1991). Functional architecture of the nicotinic acetylcholine receptor: From electric organ to brain. *Annu Rev Pharmacol. Toxicol.* **31**, 37–72.

Gao, M., Wilmanns, M., and Schulten, K. (2002). Steered molecular dynamics studies of titin i1 domain unfolding. *J. Biophys.* **83**, 3435–3445.

Garcia, A.E. (1992). Large-amplitude nonlinear motions in proteins. *Phys. Rev. Lett.* **68**, 2696–2699.

Garty, H. and Palmer, L.G. (1997). Epithelial sodium channels: Function, structure, and regulation. *Physiol. Rev.* **77**, 359–396.

Glick, M., Robinson, D.D., Grant, G.H., and Richards, W.G. (2002). Identification of ligand binding sites on proteins using a multi-scale approach. *J. Am. Chem. Soc.* **124**, 2337–2344.

Goodford, P.J. (1985). A computational procedure for determining energetically favourable binding sites on biologically important macromolecules. *J. Med. Chem.* **28**, 849–857.

Grice, A.L., Kerr, I.D., and Sansom, M.S.P. (1997). Ion channels formed by HIV-1 Vpu: A modelling and simulation study. *FEBS Lett.* **405**, 299–304.

Griffin, S.D.C., Beales, L.P., Clarke, D.S., Worsfold, O., Evans S.D., Jäger J. *et al.* (2003). The p7 protein of hepatitis C virus forms an ion channel that is blocked by the antiviral drug, amantadine. *FEBS Lett.* **535**, 34–38.

Grinstein, S., Rotin, D., and Mason, M.J. (1989). Na^+/H^+ exchange and growth factor-induced cytosolic pH changes. Role in cellular proliferation. *Biochim. Biophys. Acta* **988**, 73–97.

Grottesi, A. and Sansom, M.S.P. (2003). Molecular dynamics simulations of a K^+ channel blocker: Tc1 toxin from *Tityus cambridgei. FEBS Lett.* **535**, 29–33.

Henklein, P., Kinder, R., Schubert, U., and Bechinger, B. (2000). Membrane interactions and alignment of structures within the HIV-1 Vpu cytoplasmic domain: Effect of phosphorylation of serines 52 and 56. *FEBS Lett.* **482**, 220–224.

Hodge, C.N., Straatsma, T.P., McCammon, J.A., and Wlodawer, A. (1997). Rational design of HIV protease inhibitors. In S. Chiu, R.M. Burnett, and R.L. Garcea, (eds), *Structural Biology of Viruses*, Oxford University Press, Oxford, pp. 451–473.

Hoffmann, C.E. (1973). Amantadine. HCl and related compounds. In W.A. Carter (ed.), *Selective Inhibitors of Viral Functions*, CRC Press, Cleveland, OH, pp 199–211.

Hsu, K., Seharaseyon, J., Dong, P., Bour, S., Marbon E. (2004). Mutual functional destruction of HIV-1 Vpu and host TASK-1 channel. *Mol.cell* **14**, 259-267.

Im, W. and Roux, B. (2002). Ions and counterions in a biological channel: A molecular dynamics simulation of OmpF porin from *Escherichia coli* in an explicit membrane with 1 M KCl aqueous salt solution. *J. Mol. Biol.* **319**, 1177–1197.

Ismailov, I.I., Kieber-Emmos, T., Lin, C., Berdiev, B.K., Shlyonsky, V.G., Patton, H.K. *et al.* (1997). Identification of an amiloride binding domain within the alpha-subunit of the epithelial Na^+ channel. *J. Biol. Chem.* **272**(34), 21075–21083.

Israelewitz, B., Gao, M., and Schulten, K. (2001). Steered molecular dynamics and mechanical functions of proteins. *Curr. Opin. Struct. Biol.* **11**, 224–230.

Kieber-Emmos, T., Lin, C., Prammer, K.V., Villalobos, A., Kosari, F., and Kleyman, T.R. (1995). Defining topological similarities among ion transport proteins with anti-amiloride antibodies. *Kidney Int.* **48**, 956–964.

Kleyman, T.R. and Cragoe, E.J., Jr. (1988). Amiloride and its analogs as tools in the study of ion transport. *J. Membr. Biol.* **105**, 1–21.

Kleyman, T.R. and Cragoe, E.J., Jr. (1990). Cation transport probes: amiloride series. *Meth. Enzymol.* **191**, 739–755.

Kochendörfer, G.G., Salom, D., Lear, J.D., Wilk-Orescan, R., Kent, S.B.H., and DeGrado, W.F. (1999). Total chemical synthesis of the integral membrane protein influenza A virus M2: Role of its C-terminal domain in tetramer assembly. *Biochemistry* **38**, 11905–11913.

Kochendörfer, G.G., Jones, D. H., Lee, S., Oblatt-Montal, M., Opella, S.J., Montal, M. (2004). Functional characterization and NMR spectroscopy on full-length Vpu from HIV-1 prepared by total chemical synthesis. *J. Am. Chem. Soc* **126**, 2439–2446.

Kolocouris, N., Kolocouris, A., Foscolos, G.B., Fytas, G., Neyts J., Padalko, E. *et al.* (1996). Synthesis and antiviral activity evaluation of some new aminoadamantane derivatives. 2. *J. Med. Chem.* **39**, 3307–3318.

Kosztin, D., Izrailev, S., and Schulten, K. (1999). Unbinding of retinoic acid from its receptor studied by steered molecular dynamics. *Biophys. J.* **76**(1 Pt 1), 188–197.

Kua, J., Zhang, Y., and McCammon, J.A. (2002). Studying enzyme binding specificity in acetylcholinesterase using a combined molecular dynamics and multiple docking approach. *J. Am. Chem. Soc.* **124**, 8260–8267.

Kukol, A. and Arkin, I.T. (1999). Vpu transmembrane peptide structure obtained by site-specific fourier transform infrared dichroism and global molecular dynamics searching. *Biophys. J.* **77**, 1594–1601.

Lam, P.Y.S., Jadhav, P.K., Eyermann, C. J., Hodge, C.N., Ru, Y., Bacheler, L.T. *et al.* (1994). Rational design of potent, bioavailable, nonpeptide cyclic ureas as HIV protease inhibitors. *Science* **263**, 380–384.

Lamb, R.A. and Pinto, L.H. (1997). Do Vpu and Vpr of human immunodeficiency virus type 1 and NB of influenza B virus have ion channel activities in the viral life cycles? *Virology* **229**, 1–11.

Lemaitre, V., Ali, R., Kim, C.G., Watts, A., Fischer, W.B. (2004). Interaction of amiloride and one of its derivation with Vpu from HIV-1: a molecular dynamics simulation. *FEBS Lett.* **563**, 75-81.

Leonard, R.J., Labarca, C.G., Charnet, P., Davidson, N., and Lester, H.A. (1988). Evidence that the M2 membrane-spanning region lines the ion channel pore of the nicotinic receptor. *Science* **242**, 1578–1581.

Li, J.H., Cragoe, E.J., Jr., and Lindemann, B. (1987). Structure-activity relationship of amiloride analogs as blockers of epithelial Na channels: II. Side-chain modifications. *J. Membr. Biol.* **95**, 171–185.

Lopez, C.F., Montal, M., Blasie, J.K., Klein, M.L., and Moore, P.B. (2002). Molecular dynamics investigation of membrane-bound bundles of the channel-forming transmembrane domain of viral protein U from the human immunodeficiency virus HIV-1. *Biophys. J.* **83**, 1259–1267.

Lu, H., Isralewitz, B., Krammer, A., Vogel, V., and Schulten, K. (1998). Unfolding of titin immunoglobulin domains by steered molecular dynamics simulation. *Biophys. J.* **75**, 662–671.

Lüdemann, S.K., Lounnas, V., and Wade, R.C. (2000). How do substrates enter and products exit the buried active site of cytochrome P450? 1. Random expulsion molecular dynamics investigation of ligand access channels and mechanisms. *J. Mol. Biol.* **303**, 797–811.

Ma, C., Marassi, F.M., Jones, D.H., Straus, S.K., Bour, S., Strebel, K. *et al.* (2002). Expression, purification, and activities of full-length and truncated versions of the integral membrane protein Vpu from HIV-1. *Protein Sci.* **11**, 546–557.

Maldarelli, F., Chen, M.Y., Willey, R.L., and Strebel, K. (1993). Human immunodeficiency virus type 1 Vpu protein is an oligomeric type I integral membrane protein. *J. Virol.* **67**, 5056–5061.

Mangoni, M., Roccatano, D., and Di Nola, A. (1999). Docking of flexible ligands to flexible receptors in solution by molecular dynamics simulation. *Protein Struct, Func. Genet.* **35**, 153–162.

Marassi, F.M., Ma, C., Gratkowski, H., Straus, S.K., Strebel, K., Oblatt-Montal, M. *et al.* (1999). Correlation of the structural and functional domains in the membrane protein Vpu from HIV-1. *Proc. Natl. Acad. Sci. USA* **96**, 14336–14341.

Miller, R.H. and Sarver, N. (1997). HIV accessory proteins as therapeutic targets. *Nat. Med.* **3**, 389–394.

Montal, M. (2003). Structure–function correlates of Vpu, a membrane protein of HIV-1. *FEBS Lett.* **552**, 47–53.

Morris, G.M., Goodsell, D.S., Huey, R., Hart, W.E., Halliday, S., Belew, R. *et al.* (1998). Automated docking using a Lamarckian genetic algorithm and and empirical binding free energy function. *J. Comp. Chem.* **19**, 1639–1662.

Neher, E. and Steinbach, J.H. (1978). Local anesthetics transiently block current through single acetylcholine-receptor channels. *J. Physiol.* **277**, 153–176.

Park, S.H., Mrse, A.A. Nevzorov, A.A., Mesleh, M.F., Oblatt-Montal, M., Montal, M. *et al.* (2003). Three-dimensional structure of the channel-forming transmembrane domain of virus protein "u" (Vpu) from HIV-1. *J. Mol. Biol.* **333**, 409–424.

Pautsch, A. and Schulz, G.E. (2000). High-resolution structure of the OmpA membrane domain. *J. Mol. Biol.* **298**, 273–282.

Pavlovic, D., Neville, D.C.A., Argaud, O., Blumberg, B., Dwek, R.A., Fischer, W.B. *et al.* (2003). The hepatitis C virus p7 protein forms an ion channel that is inhibited by long-alkyl-chain iminosugar derivatives. *Proc. Natl. Acad. Sci. USA* **100**, 6104–6108.

Pebay-Peyroula, E., Rummerl, G., Rosenbusch, J.P., and Landau, E.M. (1997). X-ray structure of bacteriorhodopsin at 2.5 Angstroms from microcrystals grown in lipid cubic phases. *Science* **277**, 1676–1681.

Pinto, L.H., Holsinger, L.J., and Lamb, R.A. (1992). Influenza virus M2 protein has ion channel activity. *Cell* **69**, 517–528.

Premkumar, A., Wilson, L., Ewart, G.D., and Gage, P.W. (2004). Cation-selective ion channels formed by p7 of hepatitis C virus are blocked by hexamethylene amiloride. *FEBS Lett.* **557**, 99–103.

Root, M.J., Kay, M.S., and Kim, P.S. (2001). Protein design of an HIV-1 entry inhibitor. *Science* **291**, 884–888.

Roux, B. and Karplus, M. (1991). Ion transport in a model gramicidin channel: Structure and thermodynamics. *Biophys. J.* **59**, 961–981.

Sansom, M.S., Tieleman, D.P., Forrest, L.R., and Berendsen, H.J. (1998). Molecular dynamics simulations of membranes with embedded proteins and peptides: Porin, alamethicin and influenza virus M2. *Biochem. Soc. Trans.* **26**, 438–443.

Sass, H., Buldt, G., Gessenich, R., Hehn, D., Neff, D., Schlesinger, J. *et al.* (2000). Structural alterations for proton translocation in the M state of wild-type bacteriorhodopsin. *Nature* **40**, 649–653.

Schellenberg, G.D., Anderson, L., Cragoe, E.J., Jr., and Swanson, P.D. (1985). Inhibition of synaptosomal membrane Na^+-Ca^{2+} exchange transport by amiloride and amiloride analogues. *Mol. Pharmacol.* **27**, 537–543.

Schild, L., Schneeberger, E., Gautschi, I., and Firsov, D. (1997). Identification of amino acid residues in the alpha, beta, and gamma subunits of the epithelial sodium channel (ENaC) involved in amiloride block and ion permeation. *J. Gen. Physiol.* **109**, 15–26.

Schubert, U., Ferrer-Montiel, A.V., Oblatt-Montal, M., Henklein, P., Strebel K., and Montal, M. (1996). Identification of an ion channel activity of the Vpu transmembrane domain and its involvement in the regulation of virus release from HIV-1-infected cells. *FEBS Lett.* **398**, 12–18.

Shen, L., Shen J., Luo, X., Cheng, F., Xu Y., Chen, K. *et al.* (2003). Steered molecular dynamics simulation on the binding of NNRTI to HIV-1 RT. *Biophys. J.* **84**, 3547–3563.

Shoichet, B.K., Leach, A.R., and Kuntz, I.D. (1999). Ligand solvation in molecular docking. *Protein. Struct. Func. Genet.* **34**, 4–16.

Shrivastava, I.H. and Sansom, M.S.P. (2000). Simulations of ion permeation through a potassium channel: Molecular dynamics of KcsA in a phospholipid bilayer. *Biophys. J.* **78**, 557–570.

Smith, A.J. and Smith, R.N. (1973). Kinetics and bioavailability of two formulations of amiloride in man. *Br. J. Pharmacol.* **48**, 646–649.

Sramala, I., Lemaitre, V., Faraldo-Gomez, J.D., Vincent, S., Watts, A., and Fischer, W.B. (2003). Molecular dynamics simulations on the first two helices of Vpu from HIV-1. *Biophys. J.* **84**, 3276–3284.

Sun, F. (2003). Molecular dynamics simulation of human immunodeficiency virus protein U (Vpu) in lipid/water Langmuir monolayer. *J. Mol. Mod.* (Online) **9**, 114–123.

Sunstrom, N.A., Prekumar, L.S., Prekumar, A., Ewart, G., Cox, G.B., and Gage, P.W. (1996). Ion channels formed by NB, an influenza B virus protein. *J. Membr. Biol.* **150**, 127–132.

Tang, P. and Xu, Y. (2002). Large-scale molecular dynamics simulations of general anesthetic effects on the ion channel in the fully hydrated membrane: The implication of molecular mechanisms of general anesthesia. *Proc. Natl. Acad. Sci. USA* **99**, 16035–16040.

Tieleman, D.P., Marrink, S.J., and Berendsen, H.J. (1997). A computer perspective of membranes: Molecular dynamics studies of lipid bilayer systems. *Biochim. Biophys. Acta* **1331**, 235–270.

Tikhonov, D.B. and Zhorov, B.S. (1998). Kinked-helices model of the nicotinic acetylcholine receptor ion channel and its complexes with blockers: Simulation by Monte Carlo minimization method. *Biophys. J.* **74**, 242–255.

van Aalten, D.M.F., Amadei, A., Linssen, A.B., Eijsink, V., Vriend, G., and Berendsen, H.J.C. (1995). The essential dynamics of thermolysin: Confirmation of the hinge-bending motions and comparison of simulations in vacuum and water. *Proteins* **22**, 45–54.

van Aalten, D.M.F., Bywater, R., Findlay, J.B., Hendlich, M., Hooft, R.W., and Vriend, G. (1996). PRODRG, a program for generating molecular topologies and unique molecular descriptors from coordinates of small molecules. *J. Comp. Aid. Mol. Des.* **10**, 255–262.

Varghese, J.N., Laver, W.G., and Colman, P.M. (1983). Structure of the influenza virus glycoprotein antigen neuraminidase at 2.9 Å resolution. *Nature* **303**, 35–40.

von Itzstein, M., Wu, W.-Y., Kok, G.B., Pegg, M.S., Dyason, J.C., Jin, B. *et al.* (1993). Rational design of potent sialidase-based inhibitors of influenza virus replication. *Nature* **363**, 418–423.

Wade, R.C., Clark, K.J., and Goodford, P.J. (1993). Further developments of hydrogen bond formations for use in determining energetically favourable binding sites on molecules of known structure. 1. Ligand probe groups with the ability to form two hydrogen bonds. *J. Med. Chem.* **36**, 140–146.

Waldmann, R., Champigny, G., and Lazdunski, M. (1995). Functional degenerin-containing chimeras identify residues essential for amiloride-sensitive Na^+ channel function. *J. Biol. Chem.* **270**, 11735–11737.

Wang, C., Takeuchi, K., Pinto, L.H., and Lamb, R.A. (1993). Ion channel activity of influenza A virus M_2 protein: Characterization of the amantadine block. *J. Virol.* **67**, 5585–5594.

Wang, J., Kollman, P.A., and Kuntz, I.D. (1999). Flexible ligand docking: a multistep strategy approach. *Protein Struct. Func. Genet.* **36**, 1–19.

Wang, J., Morin, P., Wang, W., and Kollman, P.A. (2001). Use of MM-PBSA in reproducing the binding free energies to HIV-1 RT of TIBO derivatives and predicting the binding mode to HIV-1 RT of efavirenz by docking and MM-PBSA. *J. Am. Chem. Soc.* **123**, 5221–5230.

Willbold, D., Hoffmann, S., and Rösch, P. (1997). Secondary structure and tertiary fold of the human immunodeficiency virus protein U (Vpu) cytoplasmatic domain in solution. *Eur. J. Biochem.* **245**, 581–588.

Wlodawer, A. and Erickson, J.W. (1993). Structure-based inhibitors of HIV-1 protease. *Annu. Rev. Biochem.* **62**, 543–585.

Wong, C.F., Zheng, C., Shen J., McCammon, J.A., and Wolynes, P.G. (1993). Cytochrome c: a molecular proving ground for computer simulations. *J. Phys. Chem.* **97**, 3100–3110.

Woolley, G.A. and Wallace, B.A. (1992). Model ion channels: Gramicidin and alamethicin. *J. Membr. Biol.* **129**, 109–136.

Wray, V., Federau, T., Henklein, P., Klabunde, S., Kunert, O., Schomburg, D. *et al.* (1995). Solution structure of the hydrophilic region of HIV-1 encoded virus protein U (Vpu) by CD and ^{1}H NMR-spectroscopy. Int. *J. Pept. Protein Res.* **45**, 35–43.

Wray, V., Kinder, R., Federau, T., Henklein, P., Bechinger, B., and Schubert, U. (1999). Solution structure and orientation of the transmembrane anchor domain of the HIV-1-encoded virus protein U by high resolution and solid-state NMR spectroscopy. *Biochemistry* **38**, 5272–5282.

Yang, C., Jas, G.S., and Kuczera, K. (2001). Structure and dynamics of calcium-activated calmodulin in solution. *J. Biomol. Struct. Dyn.* **19**, 247–271.

Yang, H., Yang, M., Ding, Y., Liu, Y., Lou, Z., Zhou, Z. *et al.* (2003). The crystal structure of severe acute respiratory syndrome virus main protease and its complex with an inhibitor. *Proc. Natl. Acad. Sci. USA* **100**, 13190–13195.

Zheng, S., Strzalka, J., Jones, D.H., Opella, S.J., and Blasie, J.K. (2003). Comparative structural studies of Vpu peptides in phospholipid monolayers by X-ray scattering. *Biophys. J.* **84**, 2393–2415.

Zhu, F., Tajkhorshid, E., and Schulten, K. (2002). Pressure-induced water transport in membrane channels studied by molecular dynamics. *Biophys. J.* **83**, 154–160.

15

Virus Ion Channels Formed by Vpu of HIV-1, the 6K Protein of Alphaviruses and NB of Influenza B Virus

Peter W. Gage, Gary Ewart, Julian Melton, and Anita Premkumar

1. Virus Ion Channels

One of the features of cytolytic viral infection of cells is an increase in plasma membrane permeability. The loss of cellular metabolites and osmotic effects eventually lead to rupture of the cell membrane and lysis of affected cells. It has been suggested that virus-encoded pores, or "viroporins" are responsible for this phenomenon (Carrasco, 1977). Defined as "small proteins that form pores in lipid bilayers," viroporins were thought to be nonselective pores that could become increasingly larger as infection proceeded (Carrasco, 1995). Pores are passive and nonspecific conduits across cell membranes, allowing water, ions, and low molecular weight molecules (e.g., translation inhibitors such as hygromycin) to pass through. In contrast, ion channels are selective for particular ions. Most virus ion channels studied so far are cation- or proton-selective. For this reason, we use the term "virus ion channels" rather than "viroporins" for the channels that show ion selectivity.

The precedent establishing that some viruses encode membrane proteins capable of forming ion channels came from research some years ago on the influenza A M2 protein showing that M2 forms pH-activated proton channels that are essential for normal replication of influenza A virus. The drug, amantadine, blocks the channel formed by M2 and thereby stops virus replication. While amantadine, and an analog, rimantadine, have proven to be useful antiviral agents during influenza A epidemics, they are not perfect anti-influenza drugs because resistant strains can arise by mutation of amino acid residues in the M2 transmembrane domain that contribute to the channel pore. Despite this, the principle was established that a drug capable of blocking a virus-encoded ion channel could inhibit virus replication.

Evidence that M2 ion channels are not functioning as mere pores or viroporins is provided by experiments that show their proton-selectivity to be unchanged, even when channels have very high conductances (Mould *et al.*, 2000). Many virus genomes encode small, hydrophobic proteins, similar to M2 in basic characteristics, though with very little, if any,

Peter W. Gage, Gary Ewart, Julian Melton, and Anita Premkumar • Division of Molecular Bioscience, John Curtin School of Medical Research, Australian National University, Canberra, Australia.

Viral Membrane Proteins: Structure, Function, and Drug Design, edited by Wolfgang Fischer.
Kluwer Academic / Plenum Publishers, New York, 2005.

amino acid sequence similarity. The Vpu and 6K proteins, among others, were initially suggested as possible virus ion channels on the basis of their hydrophobicity and small size (Carrasco *et al.*, 1993). Other possible virus ion channels identified by hydrophobicity and size included the 2B protein from Coxsackie virus, and the 2B and 3A proteins of poliovirus. The 2B and 3A proteins have been shown to modify membrane permeability when expressed in *Escherichia coli* (Lama and Carrasco, 1992). The Coxsackie virus 2B protein is membrane associated (Porter, 1993), and increases membrane permeability (van Kuppeveld *et al.*, 1997). The CM2 protein from influenza C has structural similarity to the M2 ion channel of influenza A (Pekosz and Lamb, 1997).

Virus proteins that have been shown capable of forming ion channels include M2 from influenza A virus (Duff and Ashley, 1992; Pinto *et al.*, 1992), NB and BM2 from influenza B virus (Sunstrom *et al.*, 1996; Mould *et al.*, 2000; Fischer *et al.*, 2001), Vpu and Vpr from HIV-1 (Ewart *et al.*, 1996; Piller *et al.*, 1996; Schubert *et al.*, 1996b), Kcv, from the large chlorella virus (Plugge *et al.*, 2000) and p7 from hepatitis C virus (Griffin *et al.*, 2003; Pavlovic *et al.*, 2003).

Most virus ion channels for which topological data exist have a single transmembrane domain (Fischer *et al.*, 2000a). The p7 protein from the hepatitis C virus is an exception in that it has two transmembrane domains (Carrere-Kremer *et al.*, 2002). However, only one of the two helices is thought to participate in pore formation. With one notable exception (the M2 protein of influenza A), the importance of virus ion channels for virus replication has not been well characterized. Information derived from the study of viral ion channels can potentially lead to new therapeutic agents, and an understanding of the mechanisms underpinning membrane fusion events within cells.

The lipid bilayer technique is commonly used to show that virus proteins form ion channels. The essential technique is described in detail elsewhere (Miller, 1986). We were initially introduced to the technique by Dr. Bob French, more than 10 years ago, and we are continually modifying how we use it. The experimental chamber consists of two plastic cups with a common, thinned wall between them. In our laboratory, we burn a small hole in the partition to a desired diameter by varying the voltage across two electrodes straddling the thinned common wall. A lipid or mixture of lipids is painted across a small (diameter of about 50 to 200 μm) hole in the side of a plastic chamber. After the lipid is painted on, it usually thins to a bilayer. We track this by applying a voltage ramp to the bilayer and monitoring the current generated as an index of the capacitance ($I = C \cdot dv/dt$). As the bilayer thins, the current (capacitance) increases to a plateau. Sometimes the formation of a bilayer is also monitored visually as the hole becomes black.

The two chambers are traditionally called *cis* and *trans* (near and far). Normally, we have 500 mM NaCl or KCl in the *cis* chamber and 50 mM NaCl or KCl in the *trans* chamber. The same solution is not used in each chamber because it would then not be possible to determine ion selectivity of channels (see below). We ground the *cis* chamber and record currents and voltages in the *trans* chamber. Protein is added to the *cis* chamber and the chamber stirred for periods by a small magnetic stirrer rotating in the bottom of the chamber. The appearance of ion channel currents is monitored between periods of stirring which is discontinued after channel activity is seen.

It is sometimes asserted that almost any peptide will form an ion channel in lipid bilayers. That is not our experience. Mutations of a channel-forming peptide could completely prevent ion channel formation. Initially, we did occasionally see "ion channels" in bilayers not intentionally exposed to any peptide or protein. We soon learned that scrupulous attention to cleanliness was essential. The lipid in a container can easily become contaminated if glass rods or brushes previously exposed to channel-forming peptide are used to obtain fresh lipid. The bridge electrode in the *cis* chamber can also retain sufficient peptide from a previous experiment to give ion channel activity. Additionally, insufficient cleaning of the plastic cups can also give unexpected ion channel activity. We routinely observe lipid bilayers for periods of 5 min or more (control) before adding peptide to the *cis* chamber.

Ion channels formed in a lipid bilayer by adding a channel-forming peptide are notoriously variable in amplitude. The reason for this has not been elucidated. Two possible reasons are that channels may increase in size as more "staves" are added to a "barrel-shaped" channel as has been proposed for alamethicin. It might then be expected that channel selectivity might become less at higher channel conductances. Alternatively, hydrophobic channels may cluster in rafts and open and close cooperatively, the rafts becoming larger as more peptide is incorporated. Thus, the size of currents recorded in bilayers depends on the amount of peptide incorporated in the bilayer, the open probability of channels, and the conductance of channels. Any of these factors can vary so that comparison of current amplitude in different solutions cannot be used to estimate single-channel conductance. On rare occasions, when very small channels are seen with a low and regular conductance, it may be that a unitary channel is active in the bilayer. Single-channel conductance is an important property of the channel but is insufficient to determine the ion selectivity of the channel. This can only be done by measuring reversal potentials in asymmetric solutions. Even if single-channel conductance were shown to be different, it could not be used to indicate relative permeabilities. Conductance does not necessarily reveal relative permeabilities of a channel. An extreme illustration of this is that a sodium channel with high selectivity for sodium ions will not produce currents if there are no sodium ions in solutions. Thus, measurement of reversal potentials in dissimilar solutions, followed by application of the Goldman–Hodgkin Katz equation, provides the only electrophysiological method of measuring relative ion permeabilities (see Vpu section also).

In this chapter, we will discuss three virus ion channels: Vpu of HIV-1, the 6K protein of alphaviruses, and NB of the influenza B virus. We will first briefly discuss the ion channel activity of Vpu, discovery of small molecule Vpu channel blockers, and ongoing work aimed at determining whether and how the channel activity is involved in HIV-1 replication.

2. Vpu of HIV-1

The genome of HIV-1 is illustrated in Figure 15.1. VPU encodes for a small protein, Vpu, with the sequence (in the HXB2 isolate) shown in Figure 15.1. Viral protein u is an integral membrane protein encoded by HIV-1. It is primarily found in the Golgi and endoplasmic reticulum (ER) membranes in infected cells. Although it can be detected in the plasma membrane of cells when artificially overexpressed, it has not been detected in the viral envelope. The protein contains from 80 to 82 amino acids (depending on the HIV-1 isolate) consisting of an N-terminal transmembrane anchor and a hydrophilic cytoplasmic C-terminal domain. The C-terminal domain contains a 12 amino acid sequence that is highly conserved and within which are two serine residues that are phosphorylated by caseine kinase-2. Vpu forms homo-oligomers, but the exact number of subunits in the native complex is not yet known. The secondary structure and tertiary fold of the cytoplasmic domain of Vpu have been determined by a combination of NMR, CD spectroscopy, and molecular dynamics

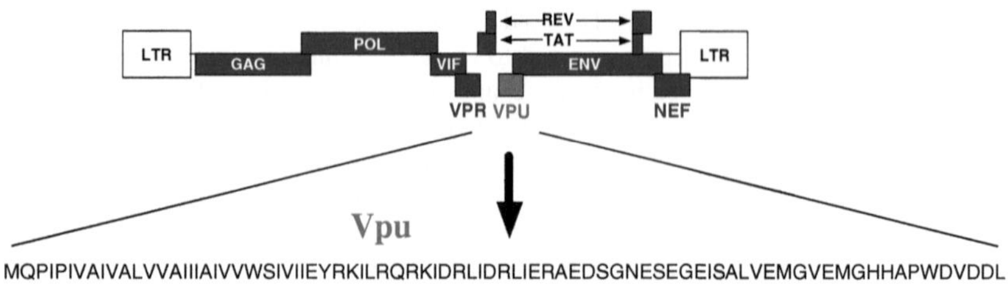

Figure 15.1. Arrangement of the HIV-1 genome. The diagram illustrates the order of the proteins known to be encoded on the HIV-1 RNA genome, highlighting the relative location of the Vpu protein with its amino acid sequence listed below.

calculations (Willbold *et al.*, 1997). There are two α-helices separated by a short flexible loop containing phosphorylated serine residues. Recent structural data for the TM domain (Wray *et al.*, 1999; Montal, 2003) supports the theoretical prediction that the region is α-helical and various tilt angles relative to the bilayer normal have been measured. A number of molecular dynamics simulation studies have been reported based on the assumption that oligomerization produces a bundle of α-helixes that spans the membrane (Grice *et al.*, 1997; Moore *et al.*, 1998; Fischer and Sansom, 2002). These studies favor formation of a pentameric complex. However, depending on the initial conditions set and restraint parameters, different conclusions about the orientation of the individual helices in the complex were reached.

2.1. Roles of Vpu in HIV-1 Replication

HIV-1, the virus responsible for the AIDS pandemic is a retrovirus belonging to the lentivirus genus. The viral genome consists of two copies of linear, positive-sense ssRNA approximately 9,700 nucleotides long, encoding 14 proteins. In the cytoplasm of infected cells, the viral ssRNA is copied to a dsDNA proviral molecule by viral reverse-transcriptase which is transported to the nucleus and integrated into the host cell genome.

The ability to initiate infection by transfecting cultured cells with dsDNA HIV-1 proviruses greatly facilitated development of experimental systems for the investigation of HIV-1 molecular biology: Mutant viruses can readily be constructed in the laboratory and their replication properties compared with the isogenic wild type. Investigation of recombinant HIV-1 constructs from which the *vpu* gene was deleted led to Vpu being labeled an "accessory" protein of HIV-1 because it is not essential for virus replication *in vitro*. Nevertheless, a number of reports have shown that Vpu significantly enhances the release of progeny virions from infected cells. The magnitude of this effect is dependent on both the virus strain and the type of cell infected. Replication of *vpu* knockout strains is only marginally decreased (0–4-fold) in cultured $CD4^+$ T-cells or cell lines derived from T-lymphocytes. However, at the other extreme, using infected macrophages, one study has reported a decrease greater than 1,000-fold in the release of new virus particles when Vpu was ablated, compared to the isogenic wild-type HIV-1 strain [see (Schubert *et al.*, 1999) and references therein]. Interestingly, it has been reported (Du *et al.*, 1993) that *vpu* can induce a significant constraint upon the host range and growth potential of HIV-1 and it was noted that adaptation of cells, such as T-cell lines, to *in vitro* replication can involve selection of viruses that have ablated *vpu* gene expression by mutation.

As revealed by more detailed probing of the effects of *vpu* deletion on HIV-1 biology, an intriguing characteristic of the Vpu protein is that it exerts a number of effects on apparently disparate cellular processes (Kerkau *et al.*, 1997; Emerman and Malim, 1998; Schubert *et al.*, 1998). Thus, in the past 10 years it has been discovered that Vpu:

1. binds to the CD4 protein in the ER and induces its degradation via the ubiquitin–proteosome pathway;
2. controls the trafficking of the HIV-1 glycoprotein gp160, and certain other glycoproteins, to the plasma membrane;
3. reduces HIV-1 cytopathicity by reducing virus-induced syncytium formation;
4. downregulates expression of MHC class 1 on the cell surface; and
5. augments the budding of newly assembled virus particles from the host-cell plasma membrane.

A major scientific challenge is to understand fully the molecular mechanisms underlying the diverse range of physiological outcomes involving Vpu. In this regard, mutagenesis studies have shown that two Vpu-mediated activities, CD4 degradation and enhancement of virion budding, reside in two separate domains of the protein. Mutations that prevent phosphorylation of the conserved serine residues in the cytoplasmic domain revealed that phosphorylation is essential for CD4 degradation, as well as inhibition of gp120 trafficking, and reduced syncytium formation. Yet, an absence of phosphorylation had only a minimal impact upon augmentation of virus release. Further, a mutant in which the sequence of the first 27 N-terminal amino acids of Vpu was randomized resulted in a protein (Vpu-RND) which had lost its ability to enhance virion budding, although the CD4 degradation activity was not affected (Schubert *et al.*, 1996a). The TM domain of Vpu expressed without the cytoplasmic domain, was still able to augment virus budding, but not degradation of CD4. Thus, it was concluded that the C-terminal hydrophilic cytoplasmic domain mediates CD4 degradation and that the N-terminal membrane-spanning domain (amino acids 1–27) modulates virus particle release (Schubert *et al.*, 1996a), though the mechanisms by which Vpu exerts these activities remain to be elucidated.

2.2. Evidence that Vpu Forms an Ion Channel

We found that Vpu can form ion channels (Ewart *et al.*, 1996). This was established experimentally using purified recombinant protein reconstituted into planar lipid bilayers. A cDNA fragment encoding Vpu was cloned into the expression plasmid p2GEX enabling IPTG-inducible expression of Vpu fused to the C-terminus of Glutathione-*S*-Transferase (GST) in *E. coli*. After purification of the expressed fusion protein by glutathione–agarose affinity chromatography, the GST tag was cleaved by digestion with thrombin and the liberated Vpu was purified to homogeneity using anion exchange and immuno-affinity chromatography steps. The zwitterionic detergent CHAPS (3-[(3-cholamidopropyl)-dimethyl-ammonio]-1-propanesulfonate) was used to maintain solubility of the hydrophobic Vpu polypeptide in aqueous buffer systems.

Purified Vpu was reconstituted into planar lipid bilayers, either directly from mixed micelles with CHAPS detergent, or from proteoliposomes prepared after dialysis of the detergent in the presence of phospholipids. Use of these techniques allows more control of the amount of channel-forming peptide incorporated into the bilayer.

Briefly, aliquots containing Vpu were added to the aqueous buffer in the *cis* chamber of the bilayer apparatus and the mixture was stirred to facilitate collision of Vpu with the bilayer. Typically, after brief periods of stirring, ion currents were detected as shown in Figure 15.2A, indicating that incorporation of Vpu into the bilayer had created a pathway for ions across the bilayer.

2.2.1. Properties of the Vpu Channel

Electrophysiological measurements of ion currents generated in the presence of Vpu showed that the reversal potential for currents was close to the equilibrium potential of monovalent cations in solutions on either side of the membrane, as illustrated in Figure 15.2B. These measurements showed that the Vpu channel is about six times more permeable to sodium ions than to chloride ions and approximately equally permeable to sodium and potassium ions.

In designing experiments to gain information about relative cation and anion permeabilities of a channel, it is necessary to use "asymmetrical" solutions so that the cation and anion equilibrium potentials are different. Use of the Goldman–Hodgkin-Katz equation then allows calculation of relative permeabilities from the reversal potential of the ion currents. In symmetrical solutions, both the cation and anion equilibrium potentials are at 0 mV and it is not possible to calculate relative permeabilities.

A minimal open state conductance of approximately 14 pS was observed, but when bilayers were exposed to higher concentrations of Vpu, larger conductance states were seen (Ewart *et al.*, 1996). This is a common observation when testing peptides for channel-forming ability in planar lipid bilayers (see above). Synthetic peptides were also used to show that Vpu forms ion channels in bilayers. A short peptide corresponding to the first

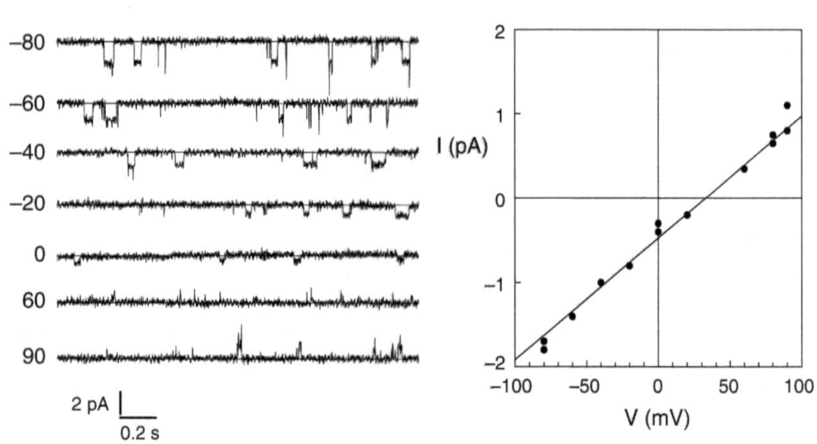

Figure 15.2. Ion channel activity of Vpu in planar lipid bilayers. Purified recombinant Vpu (approx. 7–70 ng) in proteoliposomes was added to the *cis* chamber of a planar lipid bilayer set-up which was stirred to facilitate incorporation of the protein into the bilayer. MES (2-[N-Morpholino] ethanesulfonic acid) buffer (10 mM, pH 6.0) was used and the *cis* and *trans* chambers contained 50 or 500 mM NaCl, respectively. Single-channel openings observed at the indicated holding potentials (mV) are shown in the left-hand panel and a typical current–voltage relationship generated by plotting the most frequent open state current level versus holding potential is shown in the right hand panel.

27 N-terminal residues produced ion currents with similar properties to the full-length recombinant protein and, importantly, a peptide of identical amino acid composition but with a randomized sequence did not produce channels in artificial lipid bilayers (Schubert *et al.*, 1996b). These results established that the ion channel activity of Vpu is associated with the N-terminal hydrophobic domain and that information specific to channel formation is encoded in the native amino acid sequence.

Another experimental technique has been used to confirm that Vpu forms ion channels. Patch clamping of the plasma membrane of *Xenopus* oocytes injected with cRNA encoding full-length Vpu revealed new cation currents with similar ion selectivity properties to the channels reported in the bilayer experiments (Schubert *et al.*, 1996b). Because Vpu is primarily an intracellular protein, only very small currents were generated with the wild-type Vpu cRNA and the protein could not be detected in plasma membrane fractions by Western blotting. However, site-directed mutation of two serine residues in the cytoplasmic domain, preventing phosphorylation of the protein, greatly increased both the currents detected and the level of Vpu in the plasma membrane. Injection of cRNA encoding mutant Vpu proteins, either with the TM domain deleted or with a randomized TM sequence, did not induce the new cation current.

Finally, evidence that Vpu alters membrane permeability also came from experiments involving cellular expression of Vpu. A bacterial cross-feeding assay demonstrated proline leaks out of *E. coli* cells expressing Vpu under control of a temperature-regulated promoter (Ewart *et al.*, 1996). Interestingly, the same cells were not leaky for methionine, a molecule of similar size and charge as proline, and nor were membrane vesicles prepared from these cells leaky to protons. Proline is an amino acid taken up into bacterial cells by an Na^+ symport that derives its energy from the Na^+ concentration gradient, whereas the methionine transporter is energized by ATP hydrolysis directly. Taken together, these observations indicated that, in those experiments, Vpu had not formed a large nonspecific pore in the *E. coli* plasma membrane, but rather, the proline leak was a secondary effect, most likely due to dissipation of the Na^+ ion gradient normally maintained by the cells. In contrast, when Vpu was greatly overexpressed in *E. coli* or COS cells, nonspecific permeability changes occurred and the cells became permeable to very large and/or charged molecules such as lysoyme and hygromycin B (Ewart *et al.*, 1996; Gonzalez and Carrasco, 1998). It is possible that, at high concentrations, self-association between Vpu monomers in the membrane leads to formation of larger complexes with reduced pore selectivity. This idea agrees with observations of increased channel conductance in bilayer experiments correlated to addition of higher concentrations of Vpu to the *cis* chamber (Ewart *et al.*, 1996).

2.3. The Link Between Budding Enhancement by Vpu and its Ion Channel Activity

2.3.1. Mutants Lacking Ion Channel Activity and Virus Budding

Strong circumstantial evidence to suggest a link between Vpu's virus budding enhancement and ion channel activities was first provided by Schubert *et al.* (1996b, 1998). They showed that the Vpu mutant, VpuRND, with the randomized TM sequence, when expressed in *Xenopus* oocytes, does not induce the cation-specific conductance that is observed with wild-type Vpu. Similarly, a 27-residue synthetic peptide corresponding to the randomized TM domain does not produce channels in planar lipid bilayers although the equivalent peptide with wild-type sequence clearly does. In addition, the VpuRND mutation caused a decrease in progeny virus released from cells transfected with a mutated HIV-1 proviral cDNA vector.

2.3.2. Channel-Blocking Drugs Inhibit Budding

The mechanistic association between ion channel activity and budding enhancement was further strengthened by our discovery of small molecules that block the Vpu ion channel and inhibit virus budding (Ewart *et al.*, 2002). Two derivatives of the potassium-sparing diuretic pharmaceutical amiloride were found to inhibit Vpu ion currents in planar lipid bilayer experiments. These derivatives, 5-[*N,N*′-hexamethylene]amiloride (HMA) and 5-[*N,N*′-dimethyl]amiloride (DMA), have aliphatic substituents on the N atom of the amino group at the 5 position of the pyrazine ring of amiloride and it appears that such hydrophobic substituents at this position are important for inhibition of Vpu channels, because amiloride itself is not inhibitory. The effect of DMA on ion channel activity of Vpu is illustrated in Figure 15.3. HMA was found to be a slightly more potent inhibitor of Vpu ion channels than DMA.

The effect of HMA on the budding of virus-like particles (VLP) was tested in HeLa cells co-expressing HIV-1 Gag and Vpu proteins. In those experiments, expression of Gag

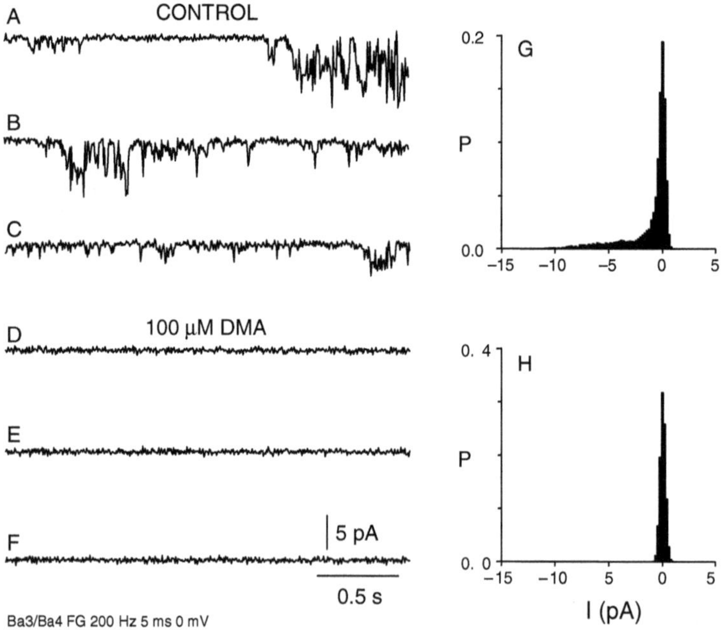

Figure 15.3. Currents generated by Vpu added to the *cis* chamber (A,B,C) are blocked following addition of 100 μM DMA to the *cis* chamber (D,E,F). The potential was 0 mV and there was 500 mM NaCl in the *cis* chamber and 50 mM NaCl in the *trans* chamber (G). An all-points histogram of a 30 s current record showing negative (downward) current before addition of drug (H). An all-points histogram of a 30 s current record after addition of 100 μM DMA showing only the baseline current.

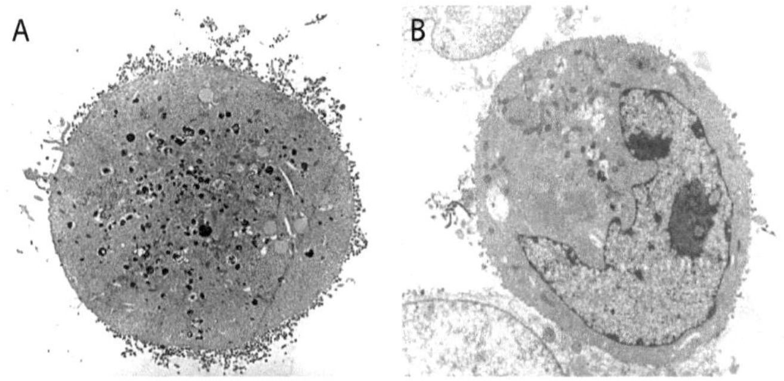

Figure 15.4. Inhibition of budding of virus-like particles (VLP) by HMA. HeLa cells at 60–70% confluence were infected with the T7 RNA polymerase expressing vaccinia virus vector vTF7.3 and then transfected with pcDNA3.1 based plasmids for expression of HIV-1 Vpu and Gag proteins under control of T7 promoters. For observation by electron microscopy, the cells were grown on glass coverslips and subsequently fixed before staining with uranyl acetate and encasing in Spurr's resin from which thin sections were cut and stained with lead citrate. The cell in micrograph A is representative of control cultures incubated in the absence of HMA and shows a distinctly "unhealthy" cell surrounded by many VLP. Micrograph B is a representative cell from a parallel culture incubated in the presence of 10 μM HMA showing a large reduction in the number of VLP associated with the cell membrane.

alone was seen to drive the formation of VLP at the plasma membrane and the presence of Vpu enhanced the release of VLP into culture medium by approximately 13-fold. HMA (10 μM) inhibited the release of VLP from Gag/Vpu expressing cells by more than 90%. It can be seen that the budding in Figure 15.4A is depressed in the presence of 10 μM HMA (Figure 15.4B).

2.3.3. HMA Inhibits HIV-1 Replication in Monocytes and Macrophages

As a consequence of the demonstrated ability to both block the Vpu ion channel and inhibit VLP budding, HMA and DMA were tested for anti-HIV-1 activity and were found to repress replication of the laboratory-adapted macrophage-tropic strain HIV_{BaL} in cultured human monocyte-derived macrophages (Ewart *et al.*, 2004). At concentrations in the range of 1–10 μM, HMA and DMA strongly inhibited production of extracellular virus as measured by p24 antigen ELISA, as illustrated in Figure 15.5.

In addition to measuring p24 in culture supernatants, the level of HIV-1 viral DNA inside the cells was assessed using semiquantitative (Q) PCR to amplify a 320 bp LTR/Gag fragment. In these experiments, some inhibition of HIV DNA accumulation in the cells was observed but it was clearly not as marked as the corresponding inhibition of p24 antigen. These findings indicated that the primary inhibitory target of HMA is at a stage of the replication cycle after reverse transcription of viral RNA and are consistent with an effect of HMA on the Vpu ion channel and virus budding.

In summary, Vpu from HIV-1, like M2 from influenza A, belongs to a growing family of virus ion channels. Recombinant and synthetic Vpu polypeptides in planar lipid bilayers form Na^+- & K^+-selective channels that show gating and have a minimal conductance of about 14 pS. The Vpu channel activity is inhibited by HMA and DMA, but not by the parent compound, amiloride. The ion channel activity of Vpu has been linked to the protein's function in augmenting virus particle release: Virion budding is strongly suppressed by either

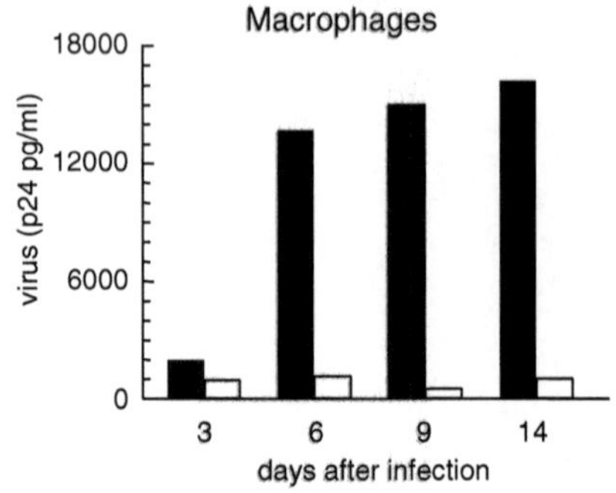

Figure 15.5. Inhibition of HIV-1 replication by HMA. Monocyte-derived macrophages (MDM) cultured from human HIV seronegative blood, were infected at a multiplicity of infection of 0.02 with the laboratory adapted R5 M-tropic HIV_{BaL} strain: 10^6 MDM per well in a 48-well tissue culture plate were incubated in the presence of virus for 1 hr, then the medium was replaced and cultures were maintained for variable periods up to 14 days after infection. Compounds were added immediately after infection of cells with the virus and media—containing the appropriate concentration of HMA—were replenished every 3 or 4 days. Culture media were sampled at the days indicated in the graph and the concentration of HIV p24 antigen in the supernatants was quantified by ELISA using the Coulter HIV-1 p24 antigen assay. The filled and open bars show the levels of HIV in the control (no drug) and HMA (10 μM)-treated cultures.

genetic mutations that ablate Vpu ion channel activity or chemical inhibitors of the channel, HMA and DMA. Though it is not currently understood how changes in cellular ion gradients contribute to the budding process, nevertheless, the rate and efficiency of budding of new virions from the plasma membrane is clearly of fundamental importance to the replication of HIV-1. Vpu plays a number of roles in the molecular biology of HIV-1, and while expression of Vpu is not essential in some *in vitro* systems, this protein does enhance the efficiency of virus replication, particularly in macrophages. At present the presumably critical roles played by Vpu in human HIV-1 infections remain elusive. However, the ability of HIV-1 to infect and efficiently replicate in macrophages is thought to be essential in AIDS pathogenesis (Schuitemaker, 1994). In fact, it has been suggested that macrophage-tropic HIV isolates may be necessary and sufficient for the development of AIDS (Mosier and Sieburg, 1994). Therefore, drugs such as HMA and DMA that block the Vpu channel and inhibit HIV-1 replication in macrophages may ultimately have potential therapeutic use as anti-AIDS agents. Clearly, there is a need for further fundamental research to define the structure of the Vpu ion channel and to understand its mechanism so that a rational approach to designing better channel-blocking drugs can be undertaken. The ultimate aim would be to design a drug of such high specificity for the Vpu channel that it would act like a magic bullet to slow or even prevent the progression from HIV-1 infection to AIDS.

3. Alphavirus 6K Proteins

Alphaviruses are members of the *Togaviridae* family, which includes flaviviruses, the smallest enveloped animal viruses. This family contains a number of important animal and human pathogens that are extremely adaptable and are found in a range of habitats on every continent except Antarctica. Alphaviruses are unusual in their ability to cross phyla in the normal course of the infection cycle (i.e., vertebrate to nonvertebrate).

3.1. Replication of Alphaviruses

Much is known about the cell biology of alphavirus replication in vertebrate cells through the study of the model viruses Sindbis virus (SINV) and Semliki Forest virus (SFV). The cellular infection process can be summarized as follows: binding and endocytosis, fusion with endosomes and release of genome to cytoplasm, transcription and translation of the viral genome, viral assembly, and budding.

3.1.2. Viral Budding

Alphavirus budding has been extensively studied, and some of the molecular mechanisms have been described. The "early" budding model was based on observations with SFV (Garoff and Simons, 1974); it was proposed that protein–protein interactions between the cytoplasmic domain of the spike proteins and the nucleocapsid drive budding. Initially, a preassembled nucleocapsid at the plasma membrane interacts with just a few spike proteins. The reduced mobility of the complex attracts more spike proteins to interact with the nucleocapsid. This causes the membrane to curve, eventually encircling the entire nucleocapsid with glycoprotein spikes. Budded virions are predominantly seen exterior to cells, appearing as electron-dense regions of diameter 40–60 nm. SFV genomes containing various deletions were used to show that budding is dependent on nucleocapsid–spike interactions. Genomes lacking either the spike protein or capsid genes were unable to generate particles individually. However, co-expression of spike proteins with capsid protein allowed nucleocapsids to assemble into particles at the plasma membrane. The molecular mechanism of the spike–C interaction have been identified using molecular modeling, and reconstructions using anti-idiotype antibodies and fitting the X-ray crystal structures of capsid and E2 into cryoelectron microscopy-derived electron density maps (Garoff *et al.*, 1998).

3.2. The 6K Protein of Alphaviruses

The importance and role(s) of the 6K protein in cellular infection are not well characterized. The 6K proteins are not highly conserved between alphaviruses (Figure 15.6A), having only 58% sequence identity but presumably have a common structure and roles.

The small size of the 6K protein (60 amino acids for Ross River virus (RRV) and 58 for Barmah Forest virus (BFV)), fatty acylation (SINV 6K is palmitoylated on cysteine residues 23, 35, 36, and 39), and hydrophobicity (Figures 15.6B and C) are typical characteristics of membrane-associated proteins.

These reasons, coupled with its toxicity when expressed in *E. coli* are probably why a purification strategy for 6K had not been published earlier (Melton *et al.*, 2002). Examination of hydrophobicity plots of 6K proteins (see Figure 15.6C) suggests two TM domains. This is supported by evidence for the topology of the polypeptide from which 6K protein is derived. However, TM prediction of BFV 6K (using a hidden Markov model) to identify residues typically found at membrane-aqueous junctions shows a single TM domain covering amino acid residues 14–32 (Sonnhammer *et al.*, 1998). The prediction for RRV 6K is not as definitive, but nevertheless indicates a single TM domain. If the protein were to cross the membrane

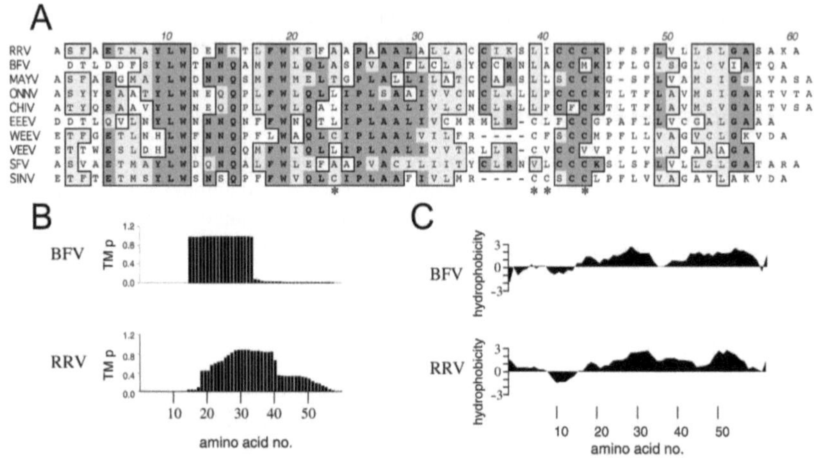

Figure 15.6. A: Homology sequence alignment of alphavirus 6K proteins. Identical residues are shown in gray boxes, homologous residues are shown in plain boxes. Cysteine residues that become palmitoylated in SINV 6K protein are indicated with asterisks (*). B: Transmembrane prediction for BFV and RRV 6K proteins. The probability of a residue being located in a transmembrane helix (TM p) is plotted against amino acid sequence no. using the TMHMM algorithm (Sonnhammer *et al.*, 1998). C: Hydrophobicity plot for BFV and RRV 6K proteins. Hydrophobic regions are above the baseline, hydrophilic regions are below.

twice, the second domain would be squeezed in very tightly, or would have to make an intra-membranous loop.

3.3. The 6K Protein and Virus Budding

To investigate the role of the 6K protein in virus replication, a range of site-directed mutations have been made in SFV and SINV. Removal of the SFV 6K gene affects the efficiency of glycoprotein processing, and the cleavage of p62 and its subsequent transport to the plasma membrane (Schlesinger *et al.*, 1993; Sanz and Carrasco, 2001). The efficiency of SFV budding from BHK cells is dramatically decreased when the 6K gene is deleted (Loewy *et al.*, 1995). It is notable that the relative importance of the 6K protein for budding is highly dependent on the cell type examined. For example, SFV Δ6K virus yield ranged from 3–50% of wild-type SFV (Loewy *et al.*, 1995). This is suggestive of some interaction of 6K protein with plasma membrane components. SINV with residues 35 and 36 of the 6K protein mutated from cysteine to alanine and serine make virions of a multi-cored appearance (Gaedigk-Nitschko *et al.*, 1990). The altered appearance of virions suggests some involvement of 6K protein with budding or membrane fusion. How the 6K protein exerts its effects on budding is not known. It has been reported to be present at low levels in virions and yet it is not contained in any virion reconstructions. Nor does it seem to have any role in docking/internalization processes, as 6K gene deletion has no effect on the infectivity of virions (Liljestrom *et al.*, 1991).

3.4. Ion Channels Formed by BFV and RRV 6K Protein

In order to test the hypothesis that 6K proteins can form ion channels, we expressed the 6K protein gene sequences of RRV and BFV in *E. coli*, purified the 6K protein and examined whether it formed ion channels in planar lipid bilayers (Melton *et al.*, 2002). Examples of typical current traces caused by BFV 6K protein are shown in Figure 15.7A and the currents can be seen to reverse at +40 mV in Figure 15.7B.

Solutions used in these experiments were 510 mM NaCl in the *cis* chamber and 60 mM NaCl in the *trans* chamber. The currents can be seen to reverse between +30 and +60 mV, indicating a preference for Na^+ over Cl^-. The average reversal potential in the presence of a 510 mM *cis*/60 mM *trans* NaCl gradient was 49.1 +/− 0.7 mV. When no protein was added to the *cis* solution, no activity was seen (n = 20, average waiting time = 7 min).

The open probability of RRV 6K channels was found to be voltage-dependent. Channels were much more active at −100 mV than at 100 mV. A channel that was closed at positive holding potentials could often be reactivated by switching to negative holding potentials.

The permeability of RRV 6K channels to cations other than Na^+ was tested by changing the solution in the *cis* bath to either KCl or $CaCl_2$ while maintaining the holding

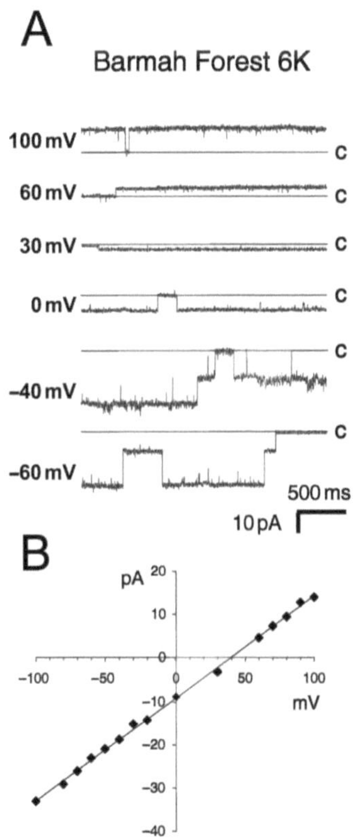

Figure 15.7. BFV 6K ion channels. A: Channel activity is shown for a range of potentials. The closed state is shown as a solid line. Openings are deviations from this line. Scale bars are 500 ms and 10 pA. Solutions contained *cis*: 510 mM NaCl, 10 mM TES (N-tris[Hydroxymethyl] methyl-2-aminoethanesulfonic acid), pH 7.0; *trans*: 60 mM NaCl, 10 mM TES (N-tris[Hydroxymethyl] methyl-2-aminoethanesulfonic acid), pH 7.0. B: Current–voltage relationship for bacterially expressed BFV 6K. Largest single opening events of a single channel are plotted for each holding potential. *Cis* and *trans* solutions were identical to those in part A.

potential at −100 mV. When the *cis* chamber contained potassium or calcium ions, ionic currents were observed. Thus, the channels were also permeable to the cations K^+ and Ca^{2+}.

3.5. Antibody Inhibition of RRV 6K Channels

It is not possible to obtain proteins at 100% purity from expression systems or by synthesis with a peptide synthesizer. Because only a few protein molecules are needed to form an ion channel, it is important to be able to demonstrate that it is indeed the predominant protein of interest, not some contaminant that is forming the ion channel. One way of doing this is to demonstrate effects of specific antibodies to the protein of interest on the ion channel activity. Polyclonal antibodies raised in rabbits immunized with synthetic peptides corresponding to the N- or C-termini of RRV 6K were used to confirm that the 6K protein was indeed the channel-forming molecule in preparations of the purified recombinant protein. When antibody recognizing the N-terminal 20 amino acids of RRV 6K (α-R6N) was added to the *cis* chamber, a reduction in current to baseline levels occurred. An example of this is shown in Figure 15.8A and was seen in eight bilayers. The α-R6N antibody had no effect when added to the *trans* chamber ($n = 5$, data not shown).

Conversely, antibody against the C-terminal of RRV 6K (α-R6C) inhibited channel openings when added to the *trans* bath ($n = 6$, Figure 15.8B), but not the *cis* bath ($n = 7$, not shown). All points histograms of currents recorded before and after addition of antibody are shown in Figures 15.8C and D.

The chamber-dependent effect of antibodies demonstrates that channel inhibition was specific to the particular epitope recognized by the antibody.

The antibody inhibition experiments also indicate that the RRV 6K protein is stably oriented in bilayers. This is a corollary of the specific topological requirements of vesicle fusion with planar lipid bilayers, that is, proteoliposomes placed in the *cis* solution, will fuse with a bilayer so that the intra-vesicular domains of TM proteins will be exposed to the *trans* solution.

The use of affinity-purified antibodies to specifically inhibit channel currents from both sides of the bilayer (Figure 15.8) supports the conclusion that the 6K protein forms an ion channel. The chamber-specific effect of antibody inhibition suggests further that 6K proteins are oriented in bilayers with the N-terminal facing the *cis* bath, and the C-terminal facing the *trans* bath. Given the length of the 6K polypeptide chain, location of N- and C-termini on opposite sides of the membrane suggests that the hydrophobic domain consists of a single TM α-helix. Earlier reports have suggested that 6K crosses the ER membrane twice, with both termini in the lumen. However, these data do not exclude the possibility that the C-terminus of 6K is only transiently located in the ER lumen. The C-terminus of the E2 protein of SINV has been shown to retract through the ER membrane following cleavage by the signalase enzyme. A similar retraction of the C-terminal of the RRV and BFV 6K proteins may occur following cleavage by signalase. The structures of most virus-encoded ion channels discovered to date consist of a single TM domain. Thus, it seems likely that the biologically active form of 6K protein has a single TM domain.

Following the onset of viral RNA translation in alphavirus-infected cells, the plasma membrane becomes more permeable to monovalent cations (Ulug *et al.*, 1984). This is followed by an increase in permeability to larger molecules, such as translation inhibitors (Munoz *et al.*, 1985). It is possible that the ion channels formed by 6K proteins are responsible for both of these changes. This hypothesis is supported by previous experiments on the

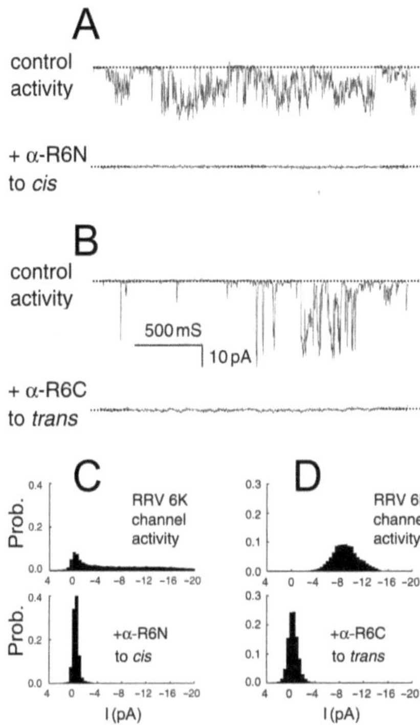

Figure 15.8. Antibody inhibition of RRV 6K channels. A: Channel activity is shown before (control), and after (+α-R6N) addition of anti-RRV 6K N-terminus antibody to the *cis* chamber. B: Channel activity is shown before (control), and after (+α-R6C) addition of anti-RRV 6K C-terminus antibody to the *trans* chamber. A dashed line represents the closed (baseline) state. Channel openings are downward deflections from the baseline. C,D: All points histograms showing the effect of antibodies on RRV 6K channels. Paired histograms are shown before (above, "RRV 6K channel activity"), and after addition of antibody to the stated chamber (below). C: Anti-RRV 6K N-terminus (α-R6N) was added to the *cis* chamber. D: Anti-RRV 6K C-terminus (α-R6C) was added to the *trans* chamber. Current amplitude probability histograms were generated from 30 s of record. NB: baseline current = 0 pA.

fate of 6K protein in infected cells (Lusa *et al.*, 1991). During alphavirus infection, the cell's secretory pathway is used to export the structural proteins of the virus (excepting capsid protein) to the plasma membrane. The topology of the membrane-bound proteins is such that the N-terminal of the 6K protein becomes exterior to the cell: fusion of ER-derived vesicles with the plasma membrane results in a cell-exterior location for residues in the ER lumen. The bilayer experiments with 6K proteins at a holding potential of −100 mV approximate the voltage and orientation conditions for 6K proteins at the plasma membrane of infected cells. Thus, the dramatic activation of 6K ion channels by negative voltages (see Figure 15.6) would also occur at the plasma membrane of infected cells.

If the 6K protein is indeed located at the cell surface then it is also possible that the increased permeability of infected cells to monovalent cations (Ulug *et al.*, 1984) is due to the preference of 6K channels for monovalent cations over divalent cations. The wide range of conductance values obtained for 6K channels may reflect the protein's ability (as infection proceeds) to form channels with larger pores and conductances.

4. NB of Influenza B Virus

In 1984 NB was first described (Shaw and Choppin, 1984) and is unique to the influenza B virus. It was thought to be the homolog of the influenza A virus M2 protein (Shaw and Choppin, 1984; Williams and Lamb, 1988). NB is abundantly expressed in virus-infected cells (Shaw and Choppin, 1984) and found in low copy numbers in the influenza B virus (Betakova *et al.*, 1996; Brassard *et al.*, 1996).

Using [^{35}S]cysteine and [^{3}H]isoleucine, Shaw (Shaw and Choppin, 1984) detected the presence of NB in cells infected with three different strains of influenza B virus. The presence of NB in all three strains of influenza B indicated that this protein was not unique to any one influenza B strain. Shaw and Choppin (1984) used pulse-chase experiments to study the kinetics, synthesis, and stability of NB in virus infected cells. Their observations showed that both NA and NB accumulated in approximately equal quantities in the cells 4 hr after infection. It was observed that after 8–10 hr of infection, NB was the major protein and that there was less NA than NB during this stage of infection. They reported that NB disappeared from the cells 4 hr after it was synthesized. This result is contrary to other findings (Betakova *et al.*, 1996) showing that the protein is stable for up to 12 hr after infection.

Shaw and Choppin (1984) used hyper-immune mouse serum to demonstrate the production of NB during a respiratory infection of influenza B in mice. They showed that mice infected with the influenza B virus develop antibodies to the NB protein. The B/Lee/40 strain seems to have a stronger anti-NB response than was detected with the NA or M1 proteins. Hyper-immune mouse serum was used to detect the presence of NB in cells infected with different strains of influenza B. The serum reacted with all the strains that were tested suggesting that the NB had retained its antigenic determinants sites over a period of 43 years separating the oldest and newest isolates studied.

4.1. Structure of NB

It has been found that NB is an integral membrane protein associated with the same membrane fractions as hemagglutinin (HA) and neuraminidase (NA). Like other viral and cellular membrane proteins 2% TritonX-100 and 0.5M KCl are required to solubilize NB (Williams and Lamb, 1986). NB is a 100 amino acid protein and has a molecular weight of 12,000 Da whereas the glycosylated protein has a molecular weight of 18,000 Da. The nucleotide sequence of NB contains 7 cysteine residues, 18 isoleucine residues, and 4 potential glycosylation sites (Asn-X-Ser/Thr), of which two are found on either side of the TM region. Asn at residue 3 and Asn at position 7 have been identified as the precise glycosylation sites. NB contains carbohydrate side chains of the high-mannose type that are processed to complex sugars in either MDCK or CV1 cells.

NB is a Class III integral membrane protein. The N-terminal domain of 17 amino acids is extracellular and the large C-terminus of 64 amino acids is intracellular (Williams and Lamb, 1988). The amino acid sequence of NB shows that it has a region of uncharged hydrophobic residues (19–40) that could span a membrane. This region has a hydropathic index greater than two indicating that the protein interacts with membranes.

The NB protein has two stretches of hydrophobic amino acids the first one at the N-terminus, residues 19–40, and at the C-terminus, residues 84–95. NB is anchored in the membrane via its N-terminal hydrophobic domain. William and Lamb (1986) used site-directed mutagenesis involving the deletion of the N-terminal region to show that the N-terminus of 18 amino acids is exposed at the cell surface.

The sequence of the NB protein is shown below.

A

M	N	N	A	T	F	N	C	T	N	I	N	P	I	T	15
H	I	R	*G*	*S*	*I*	*I*	*I*	*T*	*I*	*C*	*V*	*S*	*L*	*I*	30
V	*I*	*L*	*I*	*V*	*F*	*G*	*C*	*I*	*A*	K	I	F	I	N	45
K	N	N	C	T	N	N	V	I	R	V	H	K	R	I	60
K	C	P	D	C	E	P	F	C	N	K	R	D	D	I	75
S	T	P	R	A	G	V	D	I	P	S	F	I	L	P	90
G	L	N	L	S	E	G	T	P	N						

4.2. Similarities Between M2 and NB

It was suggested that NB is the influenza B homolog of the influenza A M2 protein (Shaw *et al.*, 1983; Williams and Lamb, 1986). Recently there have been suggestions that the BM2 protein of the influenza B is the counterpart of the M2 protein of influenza A and that BM2 forms ion channels that conduct protons (Mould *et al.*, 2000). Nevertheless, the NB protein of influenza B virus is analogous to the M2 protein of influenza A virus in many respects. There is no amino acid sequence homology between the two proteins but they share a number of common characteristics. Both are small proteins, NB 100 amino acids and M2 96 amino acids. Both are class III integral membrane proteins (von Heijne, 1988) with a single highly hydrophobic TM region. This region in both proteins has a hydropathic index greater than 2, indicating that the proteins interact with the membrane. The proteins are anchored in the membrane via their hydrophobic regions and are oriented with a short extracellular N-terminal region (M2 = 24, NB = 18) and a long cytoplasmic tail (M2 53 amino acids and NB 60). The oligomeric structure of both proteins is similar; like M2, NB runs as a dimer or tetramer on nonreducing gels. The level of expression of both proteins in virus-infected cells is the same. Both proteins are abundantly expressed on virus-infected cells and are found associated with the same membrane fractions as HA and NA. On the other hand both proteins are found in low copy numbers in the viral envelope. The number of molecules per virion is 14 to 68 molecules of M2, 15 to 100 molecules of NB.

The function of NB in the virus still remains uncertain. As NB is located at the host cell surface it has been suggested that it may be involved in transcription or replication of RNA, or that it may be involved in organizing proteins on the cell surface during budding. The function of NB is probably important to the viral life cycle, otherwise the overlapping reading frame would have been lost by natural mutation. A single mutation in the initiation codon of NB can lead to elimination of the NB reading frame. This mutation would not affect the coding of NA but this has not occurred and NB is found both in virus-infected cells and the virus particle.

There is direct evidence that M2 forms a proton channel and since NB is similar in so many respects to M2, it is possible that NB may also function as an ion channel. To test the hypothesis that NB forms an ion channel like the M2, we cloned the NB protein from the strain influenza/B/Lee/40 into bacterial expression vectors, and then purified and incorporated NB into artificial lipid bilayers to determine whether it functions as an ion channel.

4.3. Channel Activity of the NB Protein

We cloned the cDNA for NB into two expression systems (Sunstrom *et al.*, 1996). In the first the cDNA was cloned into the plasmid vector pQE-70 (Qiagen) so that the C-terminus of the protein had a tag of six histidines at the end of the open reading frame. In the second expression system the NB cDNA was cloned into the BamH1 site of the plasmid vector pGEX-2T. In this plasmid the NB open reading frame is fused to the C-terminus of GST (Glutathione-*S*-transferase). The NB produced by the GST-fusion protein was truncated at the C-terminus. The protein failed to react with the C-terminal polyclonal antibody. The protein was, however, detected on silver-stained gels. Both the polyHis and GST purified NB showed channel activity.

4.4. Currents at pH 6.0

The planar lipid bilayer technique was used to see whether the NB protein formed ion channels. The purified protein was incorporated into artificial bilayers by adding 10 to 50 μl of the fractions containing the protein (as detected from Western blots) to the *cis* chamber. The contents of the chamber were stirred until channel activity was seen. Channel activity was observed between 1 and 30 min after addition of the protein at a potential of 0 mV. The *cis* and *trans* chambers contained 150–500 mM NaCl and 50 mM NaCl solutions, respectively, at pH 6.0. Typical currents recorded are shown in Figure 15.9A.

The average reversal potential obtained with 150/50NaCl mM in three experiments was found to be $+20 \pm 2.1$ mV. With 500/50 NaCl mM in four experiments the reversal potential was found to be $+36.8 \pm 1.85$ mV. The sodium equilibrium potentials for solutions containing 500 mM NaCl in the *cis* chamber and 50 mM in the *trans* chamber would be +53 mV and that for 150 mM NaCl in the *cis* chamber and 50 mM in the *trans* chamber

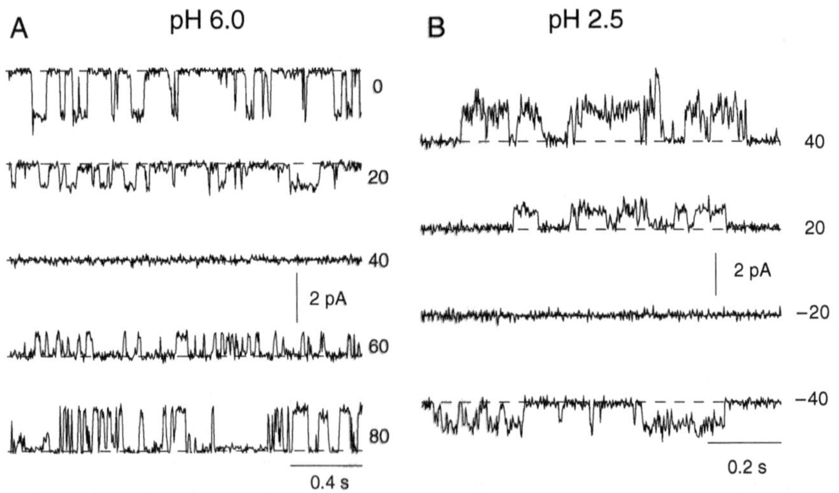

Figure 15.9. Channels formed by NB. Current traces recorded after the addition of NB to the *cis* chamber containing 500 mM NaCl and the *trans* chamber contained 50 mM NaCl. The broken lines indicate the zero current level. A. Currents recorded at pH 6. B. Currents recorded at pH 2.5.

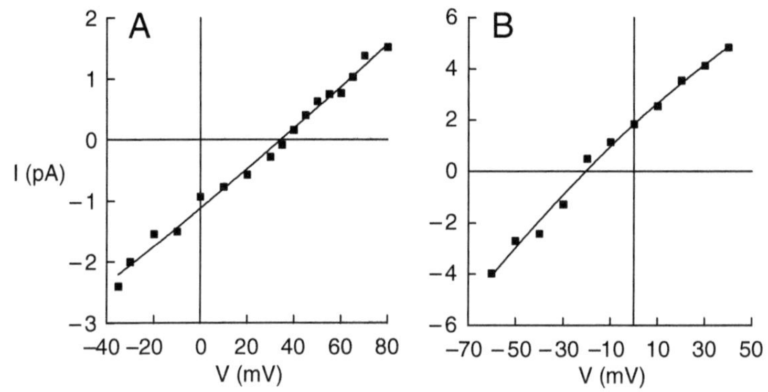

Figure 15.10. The relationship between average current and voltage recorded after the addition of NB to the *cis* chamber containing 500 mM NaCl and the *trans* chamber contained 50 mM NaCl at pH 6.0. A. Currents recorded at pH 6. B. Currents recorded at pH 2.5. The lines through the data points are second order polynomials.

would be +27.75 mV. Currents recorded using the GST-purified NB (the *cis* chamber contained 500 mM NaCl and the *trans* 50 mM NaCl) were downward, that is *cis* to *trans* at 0 mV and +20 mV and upward (*trans* to *cis*) at +60 mV and +80mV (Figure 15.9A). The average current amplitudes plotted against potential in another experiment with the same solutions gave a linear relationship between current and voltage (Figure 15.10A) and a reversal potential of +37 mV.

This type of channel activity was seen in 52 experiments with NaCl solutions at pH 6.0. It was concluded that the channels formed by NB were more permeable to Na^+ than to Cl^- ions. Assuming there is no proton current at this pH the P_{Na}/P_{Cl} ratio is about 9 for both 500 and 150 mM NaCl in the *cis* chamber, indicating that the NB channels are more permeable to Na^+ than Cl^- ions.

In some experiments the purified NB was incorporated into liposomes at relatively low concentrations of proteins. The liposomes were then added to the *cis* chamber to test for channel activity. Channels with conductances as low as 10 pS were recorded, increasing the concentration of NB in the liposomes increased the conductances to several hundred picosiemens.

A synthetic peptide corresponding to the TM domain of the NB protein was also found to form channels in artificial lipid bilayers (Fischer *et al.*, 2000b). A 28 amino acid synthetic peptide mimicking amino acids 16 to 44 of the NB protein from the strain influenza/B/Lee/40 was used to test for channel activity in artificial lipid bilayers. The solution used was symmetrical 0.5 M KCl. The channels showed a range of conductances from 20 pS to 107 pS.

4.5. Currents at pH 2.5

A channel that is permeable to Na^+ could also be permeable to protons. However, the relative concentration of protons at pH 6 in the above experiments would have been very low and a proton current would be hidden by a sodium current if the channel did not have a very high H^+/Na^+ permeability ratio. In an attempt to test whether the NB protein formed a proton channel, we used 50 mM glycine–HCl solutions at pH 2.5 in both the chambers

(Sunstrom *et al.*, 1996). It was assumed that at this pH the only cations that pass through the channels would be protons. The concentration of proton might be sufficiently high to detect proton currents. Currents were seen in 22 experiments with a reversal potential of 0 mV. In symmetrical glycine–HCl solutions, this current could have been carried by protons, glycine, or chloride ions. To confirm the currents were formed by protons symmetrical glycine–H_2SO_4 solutions at pH 2.5 were tried but no channel activity was seen suggesting the currents in glycine–HCl were chloride currents. To support this, high concentrations of NaCl solutions containing 150–500 mM in the *cis* chamber and 50 mM NaCl in the *trans* chamber at pH 2.5 were used. Currents were upward, *trans* to *cis* at 0 mV indicating that the predominant current could not be carried by Na ions. In a typical experiment (Figure 15.9B) in which there was 150 mM NaCl in the *cis* chamber and 50 mM in the *trans*, the currents were upward at +40 mV and +20 mV, downward at about −40 mV and there was no current at about −20 mV. Average current amplitude plotted against potential in another experiment is shown in Figure 15.10B. The *cis* chamber contained 500 mM NaCl and the *trans* chamber contained 50 mM NaCl. There was a linear relationship between current and voltage with a reversal potential at about −22.5 mV. In five experiments in which the solutions used were 150 and 50 mM NaCl, the reversal potential was −8 ± 2.9 mV giving a P_{Na}/P_{Cl} ratio of 0.5. When the solutions used were 500 and 50 mM NaCl, in three experiments the reversal potential was −25 ± 3.1 mV giving a P_{Na}/P_{Cl} ratio of 0.26. These reversal potentials indicate that the channels were more permeable to chloride ions than sodium ions at pH 2.5. It appears from the results that the NB channel changes from a cation-selective channel at physiological pHs to an anion-selective channel at acidic pH.

Solutions						
cis	*trans*	pH	E_{Na}	E_{Cl}	E_0	P_{Na}/P_{Cl}
150	50	6	+25.9	−25.96	+20 ± 2.1	8.8
500	50	6	+53.8	−53.8	+36 ± 1.9	8
150	50	2.5	25.96	−25.96	−8± 2.9	0.5
500	50	2.5	+53.8	−53.8	−25 ± 3.1	0.26

4.6. Effect of C-Terminal Antibody

10–20 μl of polyclonal antibody raised to a synthetic peptide corresponding to the last 24 amino acid residues of NB was added to both the *cis* and *trans* chambers of bilayers showing NB channel activity at pH 6. The high level of channel activity seen with the NB protein was completely abolished within 5 min of addition of the antibody. This effect of the antibody is illustrated in Figure 15.11.

As a control for the antibody experiments, pre-immune serum was used in both chambers in the same manner as the antibody. The pre-immune serum did not affect the NB channel activity, indicating that the effect of the C-terminal NB antibody was specific.

4.7. Effect of Amantadine

High concentrations (2–3 mM) of the anti-influenza-A drug amantadine (which inhibits the M2 ion channel) reduced NB channel activity: lower concentrations had no effect

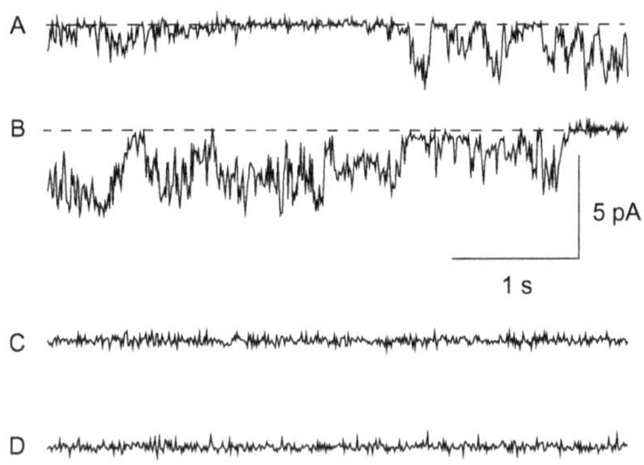

Figure 15.11. An antibody to the C-terminal of NB (NBC) blocks currents generated by the NB protein. The *cis* chamber contained 150 mM NaCl, and the *trans* chamber 50 mM NaCl. The pH in both chambers was 6.0, and the potential across the bilayer was 0 mV. A and B show currents recorded before the addition of the antibody. C and D are traces recorded after the addition of antibody NBC to both the *cis* and *trans* chambers. The broken lines show the zero current level.

(Sunstrom *et al.*, 1996). This concentration is 100 times more than the concentration required to inhibit influenza A virus M2 ion channels (Duff and Ashley, 1992; Pinto *et al.*, 1992; Tosteson *et al.*, 1994). However this high concentration of the drug has been reported to prevent replication of the influenza B virus and some other viruses that enter the host through endosomes and that require low pH for uncoating of the virus (Davies *et al.*, 1964).

Amantadine has been reported to block ion channels formed by a 28 amino acid NB-peptide in a voltage-dependent fashion (Fischer *et al.*, 2001). 0.1 of mM amantadine completely blocked NB peptide channels. The optimal concentration of amantadine was between 0.1 and 2.1 mM. Application of a potential of +100 mV potential to a channel blocked by 2.1 mM amantadine restored the channel activity after several seconds.

4.8. Proton Permeability of NB Ion Channels

In a further attempt to test whether the NB channels were permeable to protons, we (Ewart *et al.*, 1996) used an NADH-dependent quinacrine fluorescence quenching assay. This assay determines the ability of *E. coli* membranes to maintain a proton gradient during oxidative phosphorylation. The molecule of the quinacrine dye, atebrin, has two protonable nitrogen atoms. In the unprotonated state, atebrin is fluorescent, electrically neutral, and can equilibrate across membranes. In the protonated state it loses its fluorescence and the ability to traverse membranes. In inside-out vesicles in the presence of NADH and oxygen, the internal concentration of protons increases due to respiratory chain activity. This leads to protonation of atebrin inside the vesicles resulting in the quenching of fluorescence. Vesicles that are leaky to protons are unable to maintain the proton gradient (H^+_{in}/H^+_{out}), resulting in reduced fluorescence quenching. The uncoupler CCCP (carbonyl cyanide *m*-chlorophenyl-hydrazone) is added to return the fluorescence back to its maximum level. The amount of active inverted membranes added was calibrated by varying the amount of membranes. The time taken for the fluorescence to return (due to a lack of oxygen in the system) was kept within 5 s of 1 min after the addition of NADH.

Membrane vesicles prepared from *E. coli* cells expressing the NB protein from the polyHis plasmid were used in the fluorescence quenching assay. Vesicles containing NB showed a clear reduction in quenching compared to the control vesicles. The control vesicles were prepared from the same *E. coli* strain, M15, carrying the plasmid pQE-70 without any insert. The control vesicles did not show proton leakiness in four out of four experiments (Average quench was $86 \pm 0.89\%$). In all four experiments the vesicles expressing the NB protein showed an average proton permeability of $55.6 \pm 6.4\%$, indicating that the NB channel does conduct protons in membranes.

In summary, purified NB, when incorporated into planar lipid bilayers, formed cation-selective ion channels at physiological pHs. These channels were specifically inhibited by the polyclonal antibody to the C-terminus of the protein but not by pre-immune serum. This is strong evidence that the channel activity is indeed caused by the NB protein and not by any other contaminating protein. Attempts to test proton permeability of the channels by reducing the pH of solutions to 2.5 changed the ion selectivity of the channel so that it became predominantly chloride-selective. This could be due to the protonation of acidic residues in the selectivity filter of the channel. The NB TM region most likely forms an α-helix. The TM region of NB, from amino acid 19 to 40, contains three amino acids with hydroxyl side chains: 20-serine, 24-threonine, and 28-serine. Modeling of this region as an α-helix places these polar residues along the same face of the α-helix (Fischer *et al.*, 2000b). Because the high resolution structure of NB has not been determined, it is not known exactly how many α-helices make up the bundle that form the NB channel pore. Using restrained MD simulation, computational modeling of NB suggests that it probably exists as a tetrameric, pentameric, or hexameric channel. In 500 mM KCl the predicted conductance levels of the tetrameric, pentameric, and hexameric channels are 13, 34, and 53 pS, respectively which are very similar to experimental conductances of 20, 61, and 107 pS, respectively for each of the multimers studied (Fischer *et al.*, 2000b).

Replication of the influenza A virus is very similar to that of the influenza B virus. A low pH step similar to the one seen in influenza A virus is required for the matrix protein M1 to dissociate from the RNPs in the influenza B virus (Zhirnov, 1990, 1992). It is possible that NB functions as a proton channel like M2 during this stage in the viral life cycle. A blocker of the channel during the uncoating stage in the viral life cycle could lead to the arrest of infection and viral replication and possibly a cure for flu.

A reverse genetics system has been used to generate influenza B virus entirely from cloned cDNAs (Hatta and Kawaoka, 2003). Using this strategy, it was shown that the NB protein is not essential for replication of the influenza B virus in cell cultures. They generated a mutant virus that did not express the NB protein and showed that the mutant virus replicated as efficiently as the wild-type virus in multiple cycles of replication. However, when tested in mice, the NB-knockout viruses showed restricted growth. The titer of viruses was more than one log less than that of wild-type viruses, suggesting that the NB protein is required for efficient replication of the virus in mice. It was suggested (Hatta and Kawaoka, 2003) that channel activity is not the only function of NB and that it probably has other functions that are important to viral replication in nature.

References

Betakova, T., Nermut, M.V., and Hay, A.J. (1996). The NB protein is an integral component of the membrane of influenza B virus. *J. Gen. Virol.* **77**(Pt 11), 2689–2694.

Brassard, D.L., Leser, G.P., and Lamb, R.A. (1996). Influenza B virus NB glycoprotein is a component of the virion. *Virology* **220**, 350–360.

Carrasco, L. (1977). The inhibition of cell functions after viral infection. A proposed general mechanism. *FEBS Lett.* **76**, 11–15.

Carrasco, L. (1995). Modification of membrane permeability by animal viruses. *Adv. Virus Res.* **45**, 61–112.

Carrasco, L., Perez, L., Irurzun, A., Lama, J., Martinez-Abarca, F., Rodrigez, P. *et al.* (1993). Modification of membrane permeability by animal viruses. L. Carrasco, N. Sonenberg, and Wimmer, E. (eds). *Regulation of Gene Expression in Animal Viruses*. Plenum, New York, pp. 283–305.

Carrere-Kremer, S., Montpellier-Pala, C., Cocquerel, L., Wychowski, C., Penin, F., and Dubuisson, J. (2002). Subcellular localization and topology of the p7 polypeptide of hepatitis C virus. *J. Virol.* **76**, 3720–3730.

Davies, W.L., Grunert, R.R., Haft, R.F., McGahen, J.W., Neumayer, E.M., Paulshock, M. *et al.* (1964). Antiviral activity of 1-adamantanamine (amantadine). *Science* **144**, 862–863.

Du, B., Wolf, A., Lee, S., and Terwilliger, E. (1993). Changes in the host range and growth potential of an hiv-1 clone are conferred by the vpu gene. *Virology* **195**, 260–264.

Duff, K.C. and Ashley, R.H. (1992). The transmembrane domain of influenza A M2 protein forms amantadine-sensitive proton channels in planar lipid bilayers. *Virology* **190**, 485–489.

Emerman, M. and Malim, M.H. (1998). HIV-1 regulatory/accessory genes: Keys to unraveling viral and host cell biology. *Science* **280**, 1880–1884.

Ewart, G.D., Mills, K., Cox, G.B., and Gage, P.W. (2002). Amiloride derivatives block ion channel activity and enhancement of virus-like particle budding caused by HIV-1 protein Vpu. *Eur. Biophys. J.* **31**, 26–35.

Ewart, G.D., Nasr, N., Naif, H., Cox, G.B., Cunningham, A.L. and Gage, P.W. (2004). Potential new anti-human immunodeficiency virus type 1 compounds depress virus replication in cultured human macrophages. *Antimicrobial Agents and Chemotherapy* **48**, 2325-2330.

Ewart, G.D., Sutherland, T., Gage, P.W., and Cox, G.B. (1996). The Vpu protein of human immunodeficiency virus type 1 forms cation-selective ion channels. *J. Virol.* **70**, 7108–7115.

Fischer, W.B., Forrest, L.R., Smith, G.R., and Sansom, M.S. (2000a). Transmembrane domains of viral ion channel proteins: A molecular dynamics simulation study. *Biopolymers* **53**, 529–538.

Fischer, W.B., Pitkeathly, M., and Sansom, M.S. (2001). Amantadine blocks channel activity of the transmembrane segment of the NB protein from influenza B. *Eur. Biophys. J.* **30**, 416–420.

Fischer, W.B., Pitkeathly, M., Wallace, B.A., Forrest, L.R., Smith, G.R., and Sansom, M.S. (2000b). Transmembrane peptide NB of influenza B: A simulation, structure, and conductance study. *Biochemistry* **39**, 12708–12716.

Fischer, W.B., and Sansom, M.S. (2002). Viral ion channels: Structure and function. *Biochim. Biophys. Acta* **1561**, 27–45.

Gaedigk-Nitschko, K., Ding, M.X., Levy, M.A., and Schlesinger, M.J. (1990). Site-directed mutations in the Sindbis virus 6K protein reveal sites for fatty acylation and the underacylated protein affects virus release and virion structure. *Virology* **175**, 282–291.

Garoff, H., Hewson, R., and Opstelten, D.J. (1998). Virus maturation by budding. pdf, *hc review*. **62**, 1171–1190.

Garoff, H. and Simons, K. (1974). Location of the spike glycoproteins in the Semliki Forest virus membrane. *Proc. Natl. Acad. Sci. USA* **71**, 3988–3992.

Gonzalez, M.E., and Carrasco, L. (1998). The human immunodeficiency virus type 1 Vpu protein enhances membrane permeability. *Biochemistry* **37**, 13710–13719.

Grice, A.L., Kerr, I.D., and Sansom, M.S. (1997). Ion channels formed by HIV-1 Vpu: A modelling and simulation study. *FEBS Lett.* **405**, 299–304.

Griffin, S.D., Beales, L.P., Clarke, D.S., Worsfold, O., Evans, S.D., Jaeger, J. *et al.* (2003). The p7 protein of hepatitis C virus forms an ion channel that is blocked by the antiviral drug, Amantadine. *FEBS Lett.* **535**, 34–38.

Hatta, M. and Kawaoka, Y. (2003). The NB protein of influenza B virus is not necessary for virus replication in vitro. *J. Virol.* **77**, 6050–6054.

Kerkau, T., Bacik, I., Bennink, J.R., Yewdell, J.W., Hunig, T., Schimpl, A. *et al.* (1997). The human immunodeficiency virus type 1 (HIV-1) Vpu protein interferes with an early step in the biosynthesis of major histocompatibility complex (MHC) class I molecules. *J. Exp. Med.* **185**, 1295–1305.

Lama, J. and Carrasco, L. (1992). Expression of poliovirus nonstructural proteins in *Escherichia coli* cells. Modification of membrane permeability induced by 2B and 3A. *J. Biol. Chem.* **267**, 15932–15937.

Liljestrom, P., Lusa, S., Huylebroeck, D., and Garoff, H. (1991). In vitro mutagenesis of a full-length cDNA clone of Semliki Forest virus: The small 6,000-molecular-weight membrane protein modulates virus release. *J. Virol.* **65**, 4107–4113.

Loewy, A., Smyth, J., von Bonsdorff, C.H., Liljestrom, P., and Schlesinger, M.J. (1995). The 6-kilodalton membrane protein of Semliki Forest virus is involved in the budding process. *J. Virol.* **69**, 469–475.

Lusa, S., Garoff, H., and Liljestrom, P. (1991). Fate of the 6K membrane protein of Semliki Forest virus during virus assembly. *Virology* **185**, 843–846.

Melton, J.V., Ewart, G.D., Weir, R.C., Board, P.G., Lee, E., and Gage, P.W. (2002). Alphavirus 6K proteins form ion channels. *J. Biol. Chem.* **277**, 46923–46931.

Miller, C. (1986). *Ion Channel Reconstitution*, 1st edn. Plenum Press, New York and London.

Montal, M. (2003). Structure–function correlates of Vpu, a membrane protein of HIV-1. *FEBS Lett.* **552**, 47–53.

Moore, P.B., Zhong, Q., Husslein, T., and Klein, M.L. (1998). Simulation of the HIV-1 Vpu Transmembrane domain as a pentameric bundle. *FEBS Lett.* **431**, 143–148.

Mosier, D. and Sieburg, H. (1994). Macrophage-tropic HIV: Critical for AIDS pathogenesis? [see comments]. *Immunol. Today.* **15**, 332–339.

Mould, J.A., Drury, J.E., Frings, S.M., Kaupp, U.B., Pekosz, A., Lamb, R.A. *et al.* (2000). Permeation and activation of the M2 ion channel of influenza A virus. *J. Biol. Chem.* **275**, 31038–31050.

Munoz, A., Castrillo, J.L., and Carrasco, L. (1985). Modification of membrane permeability during Semliki Forest virus infection. *Virology* **146**, 203–212.

Pavlovic, D., Neville, D.C., Argaud, O., Blumberg, B., Dwek, R.A., Fischer, W.B. *et al.* (2003). The hepatitis C virus p7 protein forms an ion channel that is inhibited by long-alkyl-chain iminosugar derivatives. *Proc. Natl. Acad. Sci. USA* **100**, 6104–6108.

Pekosz, A. and Lamb, R.A. (1997). The CM2 protein of influenza C virus is an oligomeric integral membrane glycoprotein structurally analogous to influenza A virus M2 and influenza B virus NB proteins. *Virology* **237**, 439–451.

Piller, S.C., Ewart, G.D., Premkumar, A., Cox, G.B., and Gage, P.W. (1996). Vpr protein of human immunodeficiency virus type 1 forms cation-selective channels in planar lipid bilayers. *Proc. Natl. Acad. Sci. USA* **93**, 111–115.

Pinto, L.H., Holsinger, L.J., and Lamb, R.A. (1992). Influenza virus M2 protein has ion channel activity. *Cell* **69**, 517–528.

Plugge, B., Gazzarrini, S., Nelson, M., Cerana, R., Van Etten, J.L., Derst, C. *et al.* (2000). A potassium channel protein encoded by chlorella virus PBCV-1. *Science* **287**, 1641–1644.

Porter, A.G. (1993). Picornavirus nonstructural proteins: Emerging roles in virus replication and inhibition of host cell functions. *J. Virol.* **67**, 6917–6921.

Sanz, M.A. and Carrasco, L. (2001). Sindbis virus variant with a deletion in the 6K gene shows defects in glycoprotein processing and trafficking: Lack of complementation by a wild-type 6K gene in trans. *J. Virol.* **75**, 7778–7784.

Schlesinger, M.J., London, S.D., and Ryan, C. (1993). An in-frame insertion into the Sindbis virus 6K gene leads to defective proteolytic processing of the virus glycoproteins, a trans-dominant negative inhibition of normal virus formation, and interference in virus shut off of host-cell protein synthesis. *Virology* **193**, 424–432.

Schubert, U., Anton, L.C., Bacik, I., Cox, J.H., Bour, S., Bennink, J.R. *et al.* (1998). CD4 glycoprotein degradation induced by human immunodeficiency virus type 1 Vpu protein requires the function of proteasomes and the ubiquitin-conjugating pathway. *J. Virol.* **72**, 2280–2288.

Schubert, U., Bour, S., Ferrer-Montiel, A.V., Montal, M., Maldarell, F., and Strebel, K. (1996a). The two biological activities of human immunodeficiency virus type 1 Vpu protein involve two separable structural domains. *J. Virol.* **70**, 809–819.

Schubert, U., Bour, S., Willey, R.L., and Strebel, K. (1999). Regulation of virus release by the macrophage-tropic human immunodeficiency virus type 1 AD8 isolate is redundant and can be controlled by either Vpu or Env. *J. Virol.* **73**, 887–896.

Schubert, U., Ferrer-Montiel, A.V., Oblatt-Montal, M., Henklein, P., Strebel, K., and Montal, M. (1996b). Identification of an ion channel activity of the Vpu transmembrane domain and its involvement in the regulation of virus release from HIV-1-infected cells. *FEBS Lett.* **398**, 12–18.

Schuitemaker, H. (1994). Macrophage-tropic HIV-1 variants: Initiators of infection and AIDS pathogenesis? *J. Leukoc. Biol.* **56**, 218–224.

Shaw, M.W. and Choppin, P. W. (1984). Studies on the synthesis of the influenza V virus NB glycoprotein. *Virology* **139**, 178–184.

Shaw, M.W., Choppin, P.W., and Lamb, R.A. (1983). A previously unrecognized influenza B virus glycoprotein from a bicistronic mRNA that also encodes the viral neuraminidase. *Proc. Natl. Acad. Sci. USA* **80**, 4879–4883.

Sonnhammer, E.L., von Heijne, G., and Krogh, A. (1998). A hidden Markov model for predicting transmembrane helices in protein sequences. *Proc. Int. Conf. Intell. Syst. Mol. Biol.* **6**, 175–182.

Sunstrom, N.A., Premkumar, L.S., Premkumar, A., Ewart, G., Cox, G.B., and Gage, P.W. (1996). Ion channels formed by NB, an influenza B virus protein. *J. Membr. Biol.* **150**, 127–132.

Tosteson, M.T., Pinto, L.H., Holsinger, L.J., and Lamb, R.A. (1994). Reconstitution of the influenza virus M_2 ion channel in lipid bilayers. *J. Membr. Biol.* **142**, 117–126.

Ulug, E.T., Garry, R.F., Waite, M.R., and Bose, H.R., Jr. (1984). Alterations in monovalent cation transport in Sindbis virus-infected chick cells. *Virology* **132**, 118–130.

van Kuppeveld, F.J., Hoenderop, J.G., Smeets, R.L., Willems, P.H., Dijkman, H.B., Galama, J.M. *et al.* (1997). Coxsackievirus protein 2B modifies endoplasmic reticulum membrane and plasma membrane permeability and facilitates virus release. *EMBO J.* **16**, 3519–3532.

von Heijne, G. (1988). Transcending the impenetrable: How proteins come to terms with membranes. *Biochim. Biophys. Acta* **947**, 307–333.

Willbold, D., Hoffmann, S., and Rosch, P. (1997). Secondary structure and tertiary fold of the human immunodeficiency virus protein U (Vpu) cytoplasmic domain in solution. *Eur. J. Biochem.* **245**, 581–588.

Williams, M.A. and Lamb, R.A. (1986). Determination of the orientation of an integral membrane protein and sites of glycosylation by oligonucleotide-directed mutagenesis: Influenza B virus NB glycoprotein lacks a cleavable signal sequence and has an extracellular NH_2-terminal region. *Mol. Cell. Biol.* **6**, 4317–4328.

Williams, M.A. and Lamb, R.A. (1988). Polylactosaminoglycan modification of a small integral membrane glycoprotein, influenza B virus NB. *Mol. Cell. Biol.* **8**, 1186–1196.

Wray, V., Kinder, R., Federau, T., Henklein, P., Bechinger, B., and Schubert, U. (1999). Solution structure and orientation of the transmembrane anchor domain of the HIV-1-encoded virus protein U by high-resolution and solid-state NMR spectroscopy. *Biochemistry* **38**, 5272–5282.

Zhirnov, O.P. (1990). Solubilization of matrix protein M1/M from virions occurs at different pH for orthomyxo- and paramyxoviruses. *Virology* **176**, 274–279.

Zhirnov, O.P. (1992). Isolation of matrix protein M1 from influenza viruses by acid-dependent extraction with nonionic detergent. *Virology* **186**, 324–330.

[illegible]

16

The Alphavirus 6K Protein

M.A. Sanz, V. Madan, J.L. Nieva, and Luis Carrasco

1. Introduction

Alphaviruses contain a single-stranded RNA molecule as genome of positive polarity. This genome encodes two polyproteins, one of which is expressed early on in infection, while the other is synthesized during the late phase of the virus life cycle (for reviews see Strauss and Strauss, 1994; Schlesinger and Schlesinger, 1996; Garoff *et al.*, 1998). Therefore, the replication of alphaviruses is clearly divided in two very distinct phases. Early on in infection, several nonstructural proteins and their precursors are generated. These proteins are necessary for viral RNA replication. Synthesis of early proteins is directed by the translation of the genomic RNA that functions as the early viral mRNA. Late on in infection, an internal RNA promoter is recognized on the minus-stranded RNA molecule, generating the subgenomic 26S mRNA. This late subgenomic RNA is translated very efficiently and leads to the production of viral structural proteins. The profound shut-off of host translation is such that exclusive synthesis of virion proteins takes place during the late phase of infection.

Alphaviruses modify membrane permeability at two well-defined times during their lytic cycle (Carrasco, 1981, 1995; Muñoz *et al.*, 1985). Early modifications of the membrane are provoked by the entry of the virus particles in a process that does not require viral gene expression. The virion glycoproteins are responsible for early membrane leakiness. Small molecules, as well as proteins, efficiently co-enter with virus particles during entry of alphaviruses (Fernández Puentes and Carrasco, 1980; Otero and Carrasco, 1987). Late on during infection, ions are redistributed between the outside medium and the cytoplasm, leading to a progressive drop in membrane potential. A number of small molecules, but not macromolecules, are able to diffuse through the plasma membrane of alphavirus-infected cells, starting about 4 hr post-infection (Figure 16.1) (Carrasco, 1978; Garry *et al.*, 1979).

M.A. Sanz, V. Madan, and Luis Carrasco • Centro de Biología Molecular Severo Ochoa (CSIC-UAM) Facultad de Ciencias, Universidad Autónoma, Cantoblanco, 28049 Madrid, Spain. **J.L. Nieva** • Unidad de Biofísica (CSIC-UPV/EHU), Departamento de Bioquímica, Universidad del País Vasco, Aptdo. 644, 48080 Bilbao, Spain.

Viral Membrane Proteins: Structure, Function, and Drug Design, edited by Wolfgang Fischer.
Kluwer Academic / Plenum Publishers, New York, 2005.

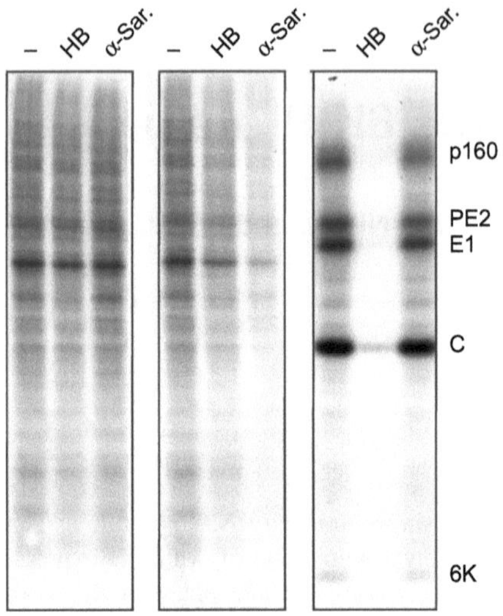

Figure 16.1. Permeabilization of BHK cells to α-sarcin or hygromycin B by SV. Translation inhibition was assayed during virus entry (early) and at 6 hr post-infection (late). No effect is observed on translation in control BHK cells (left panel). Co-entry of virus with the inhibitors, specially α-sarcin, leads to blockade of protein synthesis at early times (central panel). At late time hygromycin B but not α-sarcin entered into cells (right panel). The inhibitors (hygromycin B 1 mM or α-sarcin 100 μM) were added 15 min before and during the 30 min labeling with radioactive cys/met. The proteins were separated by SDS-PAGE as described (Sanz *et al.*, 2003).

2. Methods to Assess Whether 6K is a Membrane-Active Protein

2.1. Hydrophobicity Tests

Prediction of membrane protein structure and stability has recently been implemented with the introduction of the WW hydrophobicity scales (Wimley and White, 1996). The hydrophobicity-at-interface is a new whole-residue hydrophobicity scale (i.e., it includes contributions arising from peptide bonds as well as from side chains), based on the free energy of partitioning of each amino acid into membrane bilayers. The interfacial hydrophobicity (WW-1) of an arbitrary sequence reflects the tendency to promote protein–membrane interactions leading to integration. Predilection of a given sequence for membrane-interfaces as compared to bulk apolar or polar phases appears to be dictated by the presence of aromatic residues. WW-1 analysis detects within the N-terminal region of 6K protein ectodomain the presence of two stretches that have a strong tendency to partition into membrane interfaces and that are segregated from the transmembrane domain (TMD) detected by the Kyte–Doolittle algorithm (Figure 16.2A). Mutagenesis studies indicated that the preservation of 6K interfacial sequences is crucial for enhanced membrane permeability (Sanz *et al.*, 2003). Thus, the integration of 6K into the membrane is not sufficient for membrane destabilization, but this 6K N-terminal region is nevertheless necessary for permeabilizing membranes and for efficient virus budding. Clusters of aromatic residues containing invariant Trp residues at membrane proximal regions are known to occur in viral proteins inducing fusion and fission (Suárez *et al.*, 2000; Jeetendra *et al.*, 2002; Garry and Dashb, 2003). These "pre-TM" or "aroma" domains (Figure 16.2B) may be of functional importance for viral proteins that induce membrane perturbations necessary for an efficient viral cycle.

Wimley and White also measured free energies of partitioning from water to *n*-octanol for each amino acid, the latter representing a bulk apolar phase analogous to the aliphatic core of the bilayer. The octanol scale (WW-2) provides a reasonable estimate of the free energy of inserting α-helical TMDs into bilayers (White *et al.*, 2001). The 6K transmembrane region is thought to allow the early formation of a membrane-embedded hairpin structure that positions the polypeptide amino and carboxyl termini in the endoplasmic reticulum (ER) lumen (Liljeström and Garoff, 1991). WW-2 analysis indicates that whereas 17FFWVQLCI-PLAAFIVLMRCCS37 (o-i)-TMD insertion into membranes would be energetically

A

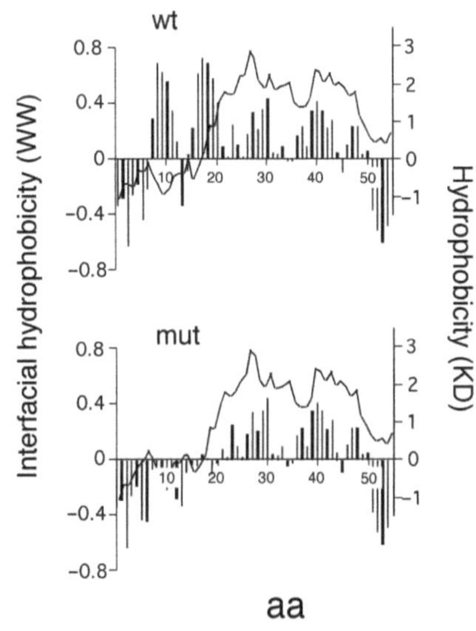

B

"aroma"

```
rrvt     -ASFAETMAYLWDENKTLFWM...
sfv      -ASVAETMAYLWDQNQALFWL...
onnvg    -ASYYEAATYLWNEQQPLFWL...
weev     -ETFGETLNHLWFNNQPFLWA...
sindv    -ETFTETMSYLWSNSQPFFWV...
eevv3    -ETTWESLDHLWNNNQQMFWI...
eeev     ADDTLQVLNYLWNNNQNFFWM...
```

Figure 16.2. (A) Hydropathy plots corresponding to 6K (wt) and 9YLW11xAAA/18FWV20xAAA mutant (mut) sequences. Plots were produced using the Wimley–White interfacial hydrophobicity (bars) and Kyte–Doolittle hydropathy index (line) scales for individual residues. Mean values for windows of 5 and 11 amino acids were used, respectively. (B) Alignment of N-terminal sequences (first 20 aa) derived from divergent alphavirus 6K proteins. Aromatic residues are in bold characters. "Aroma" domain is designated by the square. Protein sequences: rrvt, Ross River; sfv, Semliki Forest; onnvg, O'Nyong-Nyong; weev, Western Equine Encephalitis; sindv, Sindbis; eevv3, Venezuelan Equine Encephalitis; eeev, Eastern Equine Encephalitis.

favorable ($\Delta G \approx -11$ kcal mol^{-1}) the 38CCLPFLVVAGAYLAKVDA55 (i-o)-TMD would not spontaneously remain inserted at the membrane($\Delta G \approx 2$ kcal mol^{-1}). Exposure to increasing membrane thickness during protein trafficking might affect 6K membrane topology. In particular we proposed that the interfacial "aroma" region helix, whose main axis is approximately parallel to the bilayer plane, followed by a single TMD, would be a more favorable topology adopted by the secreted versions of 6K (Sanz *et al.*, 2003). Indeed, WW-2 suggests that a single 22LCIPLAAFIVLMRCCSCCLPFLVVAGAYLA51 (o-i)-TMD would be energetically favored ($\Delta G \approx -10$ kcal mol^{-1}).

2.2. Inducible Synthesis of 6K in *E. coli*

The system developed by Studier's group (Studier and Moffat, 1986; Rosemberg *et al.*, 1987; Studier *et al.*, 1990) has been used successfully for the expression of toxic proteins in *E. coli*. This system employs an *E. coli* strain, BL21(DE3), bearing a prophage (DE3) integrated in its genome that includes the sequence codifying for T7 RNA polymerase under the control of a lac promoter. The gene encoding the toxic protein to be induced is cloned into pET plasmids under the control of a T7 RNA polymerase promoter. A variant of this system, which is even more repressed, makes use of the plasmid pLys, which encodes lysozyme, an inhibitor of T7 RNA polymerase. Besides, rifampicin selectively blocks transcription driven by bacterial polymerase, but has no effect on the T7 RNA polymerase itself. Under these conditions, only recombinant proteins are synthesized.

Inducible expression of the 6K gene is achieved in both systems, BL21(DE3) and BL21(DE3)pLys *E. coli* cells. Significant synthesis of the highly toxic 6K protein is found when 6K is cloned in pET11 and expressed in BL21(DE3)pLys cells in the presence of IPTG and rifampicin. 6K synthesis cannot be detected in cultures of BL21(DE3) and plasmid pET3-6K that are less repressed. The toxicity of 6K induces a positive selection of cells that do not synthesize this protein. This selection results from the loss of the prophage. The *E. coli* cells that have lost the prophage do not contain T7 RNA polymerase and consequently do not express 6K. All of the colonies obtained after transfection of *E. coli* BL21(DE3) with pET3-6K plasmid are unable to produce 6K. Even the expression of 6K under the more repressed conditions may frequently give rise to a mixture of cells that are capable and incapable of expressing 6K. This is an important point to consider if results are to be correctly interpreted.

2.3. Synthesis of 6K in Mammalian Cells

The expression of 6K in mammalian cells has been achieved using Sindbis Virus (SV) replicons derived from an infective SV cDNA clone (Hahn *et al.*, 1992; Sanz *et al.*, 2001). These replicons retain the sequence for the nonstructural proteins and the noncodifying sequences at the 3′ and 5′ ends of the genomic RNA (Figure 16.3). The genes encoding for the structural proteins can be deleted without affecting its replication ability. Moreover, the presence of the capsid sequence enhances the translation of the subgenomic mRNAs (Frolov and Schlesinger, 1994, 1996). The SV sequences in the form of cDNA are located after a T7 promoter. Therefore, they can be *in-vitro* transcribed from the corresponding plasmid using the T7 phage RNA polymerase. These transcribed RNAs from SV cDNA are then electroporated into BHK cells (Liljeström *et al.*, 1991; Soumalainen and Garoff, 1994).

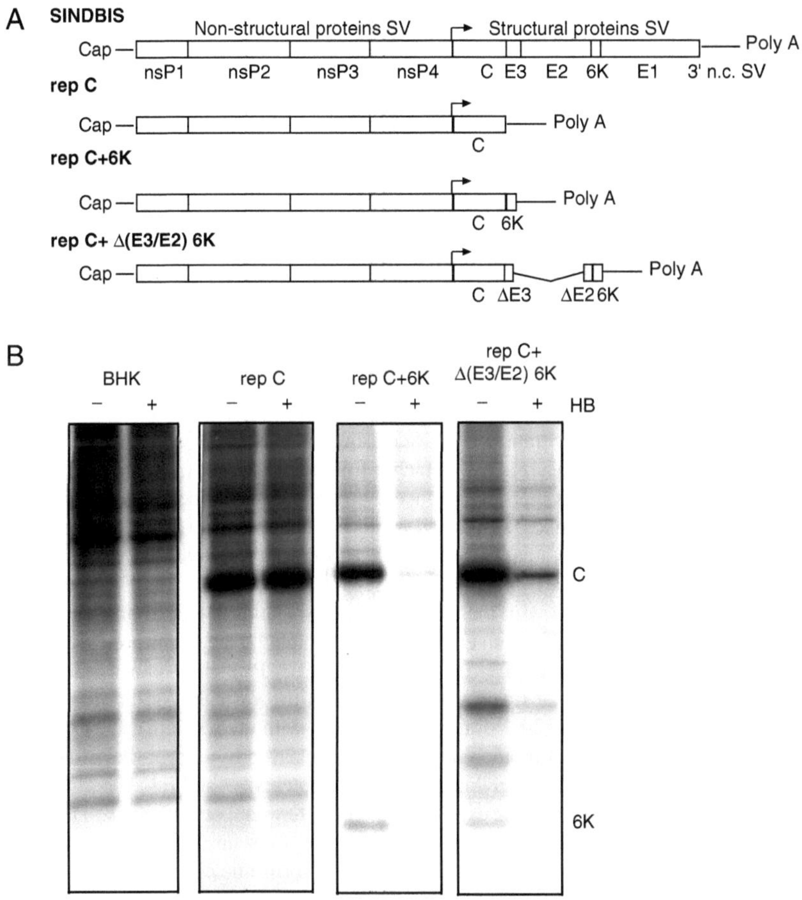

Figure 16.3. Permeabilization of BHK cells to hygromycin B by the expression of SV 6K. (A) Schematic representation of the SV genome, the replicon that only encodes C protein, and the two constructs that express SV 6K. (B) Permeabilization to hygromycin. BHK cells were transfected with the indicated RNAs. Protein synthesis was estimated at 16 hr post-transfection, as indicated in Figure 16.1.

Systems to express the alphavirus 6K gene, based on SV replicons that synthesize the 6K protein in large amounts, have been developed. Two different constructs have been employed (Figure 16.3). One of them uses a construct containing the 6K sequence adjacent to the capsid gene. In this replicon, the 6K product released after C autoproteolysis contains four additional amino acids in its N-terminal end, two from the E3 sequence and two from the Nde I site introduced in the cloning process. Upon translation of the subgenomic mRNA, the Capsid (C) protein is liberated by its autocatalytic activity and the rest of the protein is made at equimolar levels as compared with C. As 6K is an integral membrane protein and its insertion into the membrane is necessary to position the polyprotein precursor correctly it is possible that this 6K may not acquire a genuine topology in the membrane. To circumvent this problem, another construct was engineered, which retains the ER translocation sequence from E3 and the end of the E2 sequence. The two sites of proteolytic cleavage, those between C–E3 and E2–6K, have been maintained in this construct in order to synthesize a genuine 6K protein with no extra amino acids (Figure 16.3). The cleavage between E2–6K which is made by a signalase present at the ER demonstrates that 6K has reached this cellular compartment. Large amounts of 6K are synthesized from both replicons and the permeabilizing ability of 6K in both cases is similar (Figure 16.3).

3. Synthesis of 6K During Virus Infection

The schematic representation of the alphavirus lytic cycle is depicted in Figure 16.4. After virus entry, the genomic single-stranded RNA is translated and subsequently transcribed to generate the late subgenomic 26S mRNA. The alphavirus structural proteins are synthesized from this subgenomic mRNA, which encodes a proteolytically processed polyprotein (for reviews see Strauss and Strauss, 1994; Schlesinger and Schlesinger, 1996; Garoff *et al.*, 1998). C protein is first synthesized and detaches from the rest of the polyprotein by autocatalytic proteolysis at its carboxy terminus. Once C protein has been liberated to the cytoplasm, polyprotein synthesis continues associated with ER membranes. After C cleavage, the exposed amino terminus contains a signal sequence that interacts with ER membranes and directs the glycoprotein precursor (E3-E2-6K-E1) into the lumen of the ER vesicles. This precursor becomes associated with the ER membrane, spanning the lipid bilayer five times. This precursor is then cleaved at both ends of the 6K protein by a cellular protease present in the ER, generating the products PE2 (E3+E2), 6K and E1. PE2 and E1 then associate with each other to form dimers that travel with 6K through the vesicular system to the plasma membrane. In addition, the PE2-E1 dimers oligomerize, with the result that the dimers adopt a trimeric structure. As a final proteolytic step, PE2 is cleaved by a furin-like protease present in a post-Golgi compartment, giving rise to glycoproteins E3 and E2. The glycoproteins transported to the plasma membrane expose their amino-terminal ectodomains to the external medium, while the carboxy domains remain facing the cytoplasm. The virus genomes replicated in the cytoplasm interact with the C protein to form nucleocapsids. The assembled nucleocapsids subsequently interact with the carboxy domain of E2. This interaction provokes the wrapping of the capsid by the lipid envelope, concomitant with the budding of virus particles.

4. Cell Membrane Permeabilization by 6K

Alphavirus 6K is a very hydrophobic protein of about 60 amino acid residues that has been classified as a viroporin (Carrasco, 1995). 6K is a heavily acylated integral membrane protein (Gaedigk-Nitscho and Schlesinger, 1990; Lusa *et al.*, 1991). There are a number of methods for testing the permeabilization ability of a protein based on the entry or release of compounds that usually do not cross the cellular plasma membrane (Lama and Carrasco, 1992; Sanz *et al.*, 1994; Carrasco, 1995). The hygromycin B test has been widely employed to analyze permeability changes produced by the isolated expression of alphavirus 6K in bacterial and mammalian cells. The entry of this aminoglycoside antibiotic into cells inhibits translation that can be easily detected by a radioactive protein labeling assay.

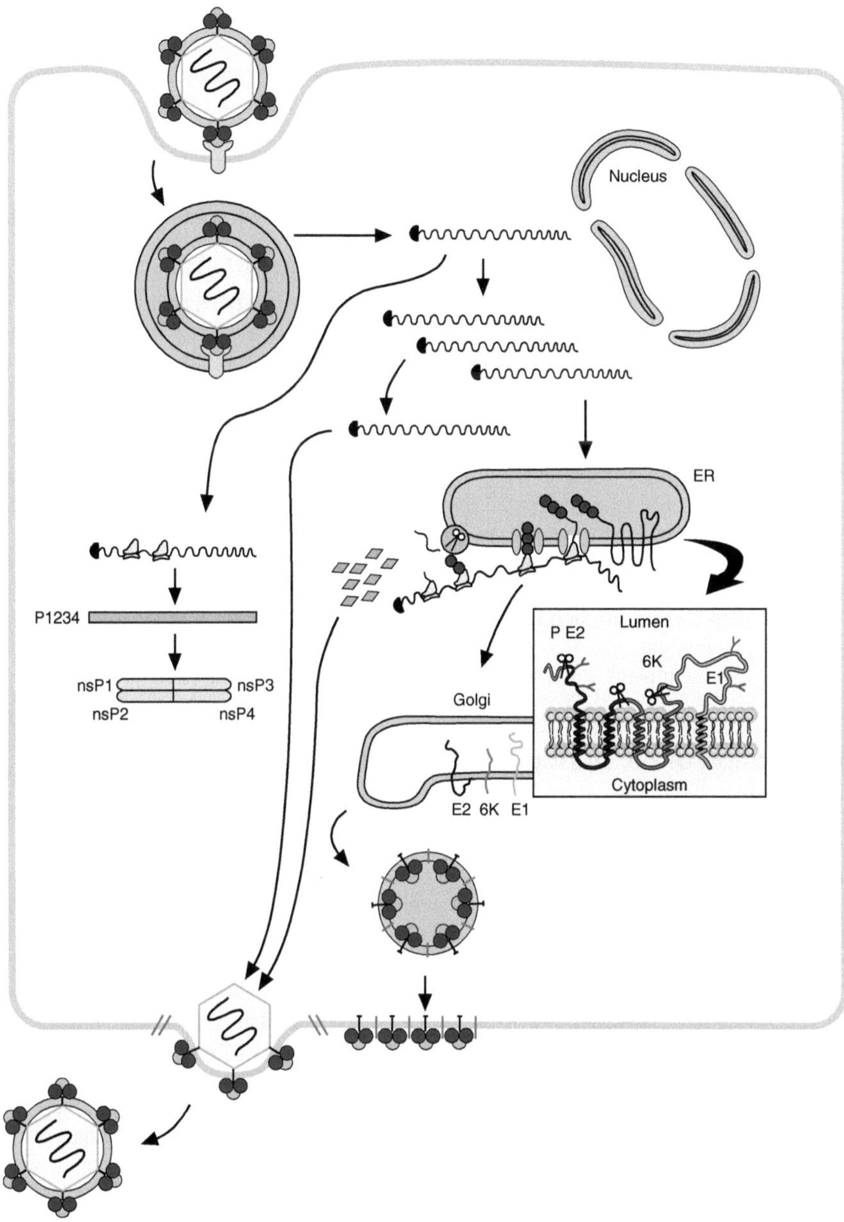

Figure 16.4. Schematic representation of the alphavirus lytic cycle. For details, see text.

Synthesis of 6K is very toxic for bacterial cells as judged by the rapid lysis observed after its induction (Sanz *et al.*, 1994). In addition to the entry of hygromycin B, the efflux of ^{3}H-coline from preloaded *E. coli* cells can also be detected after inducible expression of 6K. However, enhanced passive diffusion of compounds with a greater MW, such as α-sarcin, does not take place (Figure 16.1). The function of alphavirus 6K in cell membrane permeabilization is more probably related to its pore-formation ability. The simple insertion of 6K variants into the bacterial membrane is not sufficient to promote permeabilization, since 6K variants with mutations in the aromatic residues located at the N-terminal region outside the TMD are defective for this function (Sanz *et al.*, 2003).

As described above, a system has been designed, based on SV replicons that synthesize large amounts of 6K protein in BHK cells. HB readily entered into BHK cells transfected with full-length RNA from wt SV, leading to a profound inhibition of viral translation. The expression of wt 6K enhanced membrane permeability to HB, whereas this effect was not as strong with the 6K variants in the interface region located at the amino terminus (Sanz *et al.*, 2003). These findings are consistent with the hypothesis that the integrity of the interfacial sequence of 6K is crucial for the correct activity of this protein for enhancing membrane permeability. Notably, a 6K expressed just after the C sequence exhibited the same permeabilization ability as the genuine 6K containing the E3 signal sequence (Figure 16.3). These results suggest that the 6K is able by itself to interact with and permeabilize the plasma membrane.

The 6K proteins of Ross River virus and Barmah Forest virus exhibit the ability to form cation-selective ion channels in planar lipid bilayers (Melton *et al.*, 2002). Bacterially expressed 6K proteins were purified and inserted into lipid bilayers with a defined orientation that is, N-terminal *cis*, C-terminal *trans*. Channel conductances varied from 40–800 ps, suggesting that the protein is able to form channels with different oligomerization states.

5. Function of 6K During the Alphavirus Life Cycle

Despite the relative structural simplicity of 6K, this protein plays several roles during the virus life cycle, including those we shall now mention. The sequences of 6K located at the amino and carboxy termini provide cleavage sites during polyprotein processing (Welch and Sefton, 1980). The second hydrophobic region of 6K directs the translocation of glycoprotein E1 into the lumen of the ER (Liljeström and Garoff, 1991). After synthesis and proteolytic processing, 6K may interact with glycoproteins E1 and pE2 to regulate their trafficking to the plasma membrane (Yao *et al.*, 1996). Another well-documented activity of 6K is its membrane permeabilization capacity (Sanz *et al.*, 1994, 2003; Melton *et al.*, 2002). Finally, the major function of 6K is to participate in efficient virus budding from infected cells (Gaedigk-Nitschko *et al.*, 1990; Gaedigk-Nitschko and Schlesinger, 1991; Liljeström *et al.*, 1991; Schlesinger *et al.*, 1993; Ivanova *et al.*, 1995; Yao *et al.*, 1996; Loewy *et al.*, 1995; Sanz and Carrasco, 2001; Sanz *et al.*, 2003). Despite the association of the 6K protein with the plasma membrane and its interaction with E1-E2, very little 6K is incorporated into mature virus particles (Gaedigk-Nitschko *et al.*, 1990; Lusa *et al.*, 1991).

The acquisition and analysis of a number of variants of alphavirus, defective in 6K function, have helped to elucidate its role during the virus life cycle. These studies have yielded evidence indicating that 6K may be considered an accessory protein, since defective 6K virions can be obtained, albeit in small amounts (Liljeström *et al.*, 1991; Loewy *et al.*, 1995; Sanz *et al.*, unpublished results). Although 6K protein provides the cleavage sites in the glycoprotein precursor for signalase activity, a Semliki Forest virus (SFV) variant lacking the entire 6K is processed between E2 and E1 (Liljeström *et al.*, 1991; Loewy *et al.*, 1995). This mutant virus is not defective in synthesis and transport of glycoproteins or in nucleocapsid formation; its major defects concern the budding process. Notably, E1 is properly translocated in the 6K-deleted SFV mutant. Similarly, SV variants with single or multiple amino acid substitutions in the 6K have defects in virion release, leading to the formation of multinucleated virus particles (Gaedigk-Nitschko *et al.*, 1990; Gaedigk-Nitschko and Schlesinger, 1991; Ivanova *et al.*, 1995). Proper proteolytic processing of the virus glycoproteins is hampered in an SV variant bearing an insertion of 15 amino acids in the 6K protein. This variant exhibits a transdominant phenotype, but virus particles display a similar morphology to that observed with wt virus (Schlesinger *et al.*, 1993). An SV variant with deleted 6K 22 amino acids shows defects in glycoprotein proteolytic processing (Sanz *et al.*, 2001). A revertant of this mutant that had corrected these defects was subsequently isolated, but virus release was still impaired. Further, the expression *in trans* of a genuine 6K from an extra subgenomic promoter placed in the same genome did not produce appreciable reversion. In addition, the functions of the 6K protein cannot be rescued by the corresponding counterparts from related virus species. Thus, the substitution of the SV 6K gene by the 6K counterpart from Ross River virus leads to the small plaque phenotype and reduced formation of infectious virus (Yao *et al.*, 1996). This SV variant with the 6K gene from Ross River virus was able to cleave the glycoprotein precursors and to transport them into the plasma membrane, although the budding process was impaired. More recently, a clear defect has been observed with three SV 6K variants in the interface region located at the amino terminus (Sanz *et al.*, 2003). In all three cases, virus particles accumulated at the plasma membrane. Although their morphology showed no anomaly, these particles were unable to detach efficiently from cells. Together, these observations suggest a function for 6K in the release of virions from infected cells.

6. A Model of 6K Function in Virion Budding

A major gap in our understanding of 6K function concerns the link between pore activity and the stimulation of virus budding. Several models have been put forward to explain this connection. One such model proposes that the function of 6K is merely mechanical. The interaction of 6K with membranes would lead to phospholipid bilayer bending (Gaedigk-Nitschko *et al.*, 1990; Loewy *et al.*, 1995). This deformation of the membrane promotes virus budding. The most significant aspect of this model is the actual interaction of 6K with the membrane, but it does not take into consideration the ion-channel activity of 6K.

Another possibility, which we have advanced, is that pore formation by 6K serves to dissipate ion gradients, thereby providing the energy to push the viral particles out of the cell (Sanz *et al.*, 2001, 2003). In addition, the interaction of 6K with membranes may also participate in the bending of the phospholipid bilayer. The initial synthesis of 6K-E2-E1 trimers is followed by its oligomerization into trimers of PE2-6K-E1 trimers. These oligomers then travel to the plasma membrane, concentrating in some selected regions at the cell surface. The interaction of the nucleocapsid with the cytoplasmic tail of E2, may expel 6K from these hetero-oligomers, giving rise to 6K homo-oligomerization and consequent pore formation. Thus, these pores would be generated surrounding the budding virions (Figure 16.5). This leads to the dissipation of membrane potential in the immediate vicinity of these particles. The dissipation energy pushes the virus particles out of the cells. In fact, ionic

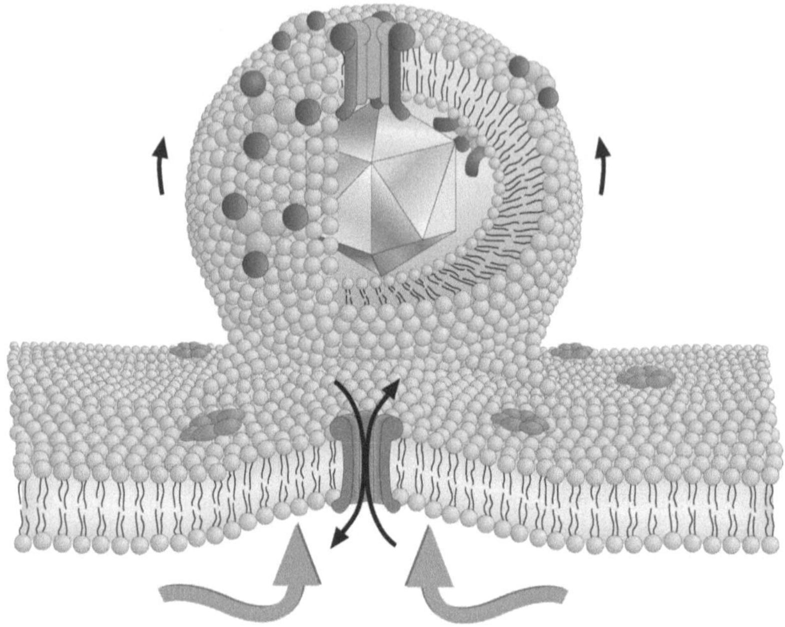

Figure 16.5. Connection between viroporin activity and alphavirus budding. Pore formation by the alphavirus 6K around the budding area leads to dissipation of membrane potential providing the energy to push the viral particle out of the cell.

changes in the culture medium that modifies membrane potential hamper the release of virions into the medium (Waite and Pfefferkorn, 1970; Li and Stollar, 1995).

Acknowledgments

We acknowledge the financial support of the CAM (project number 07B/0010/2002) and DGICYT (project numbers PM99-0002 [MAS, VM, and LC] and BIO2000-0929 [JLN]). Further support to JLN was obtained from the Basque Government (PI-1998-32) and the University of the Basque Country (UPV 042.310-G03/98). CBM was awarded an institutional grant by the Fundación Ramón Areces, Spain.

References

Carrasco, L. (1978). Membrane leakiness after viral infection and a new approach to the development of antiviral agents. *Nature* **272**, 694–699.

Carrasco, L. (1981). Modification of membrane permeability induced by animal viruses early in infection. *Virology* **113**, 623–629.

Carrasco, L. (1995). Modifications of membrane permeability by animal viruses. *Adv. Virus Res.* **45**, 61–112.

Fernández-Puentes, C. and Carrasco, L. (1980). Viral infection permeabilizes mammalian cells to protein toxins. *Cell* **20**, 769–775.

Frolov, I. and Schlesinger, S. (1994). Translation of Sindbis virus mRNA: Effects of sequences downstream of the initiating codon. *J. Virol.* **68**, 8111–8117.

Frolov, I. and Schlesinger, S. (1996). Translation of Sindbis virus mRNA: Analysis of sequences downstream of the initiating AUG codon that enhance translation. *J. Virol.* **70**, 1182–1190.

Gaedigk-Nitschko, K. and Schlesinger, M.J. (1990). The Sindbis virus 6K protein can be detected in virions and is acylated with fatty acids. *Virology* **175**, 274–281.

Gaedigk-Nitschko, K. and Schlesinger, M.J. (1991). Site-directed mutations in Sindbis virus E2 glycoproteins cytoplasmic domain and the 6K protein lead to similar defects in virus assembly and budding. *Virology* **183**, 206–214.

Gaedigk-Nitschko, K., Ding, M.X., Levy, M.A., and Schlesinger, M.J. (1990). Site-directed mutations in the Sindbis virus 6K protein reveal sites for fatty acylation and the underacylated protein affects virus release and virion structure. *Virology* **175**, 282–291.

Garoff, H., Hewson, R., and Opstelfen, D.-J.E. (1998). Virus maturation by budding. *Microbiol. Mol. Biol. Rev.* **62**, 1171–1190.

Garry, R.F., and Dashb, S. (2003). Proteomics computational analyses suggest that hepatitis C virus E1 and pestivirus E2 envelope glycoproteins are truncated class II fusion proteins. *Virology* **307**, 255–265.

Garry, R.F., Bishop, J.M., Parker, J., Westbrook, K., Lewis, G., and White, M.R.F. (1979). Na^+ and K^+ concentrations and the regulation of protein synthesis in Sindbis virus-infected chick cells. *Virology* **96**, 108–120.

Hahn, C.S., Hahn, Y.S., Braciale, T.J., and Rice, C.M. (1992). Infectious Sindbis virus transient expression vectors for studying antigen processing and presentation. *Proc. Natl. Acad. Sci. USA* **89**, 2679–2683.

Ivanova, L., Le, L. and Schlesinger, M.J. (1995). Characterization of revertants of a Sindbis virus 6K gene mutant that affects proteolytic processing and virus assembly. *Virus Res.* **39**, 165–179.

Jeetendra, E., Robison, C.S., Albritton, L.M., and Whitt, M.A. (2002). The membrane-proximal domain of vesicular stomatitis virus G protein functions as a membrane fusion potentiator and can induce hemifusion. *J. Virol.* **76**, 12300–12311.

Lama, J. and Carrasco, L. (1992). Expression of poliovirus nonstructural proteins in Escherichia coli cells. Modification of membrane permeability induced by 2B and 3A. *J. Biol. Chem.* **267**, 15932–15937.

Li, M.C. and Stollar, V. (1995). A mutant of Sindbis virus which is released efficiently from cells maintained in low ionic strength medium. *Virology* **210**, 237–243.

Liljeström, P. and Garoff, H. (1991). Internally located cleavable signal sequences direct the formation of Semliki Forest virus membrane proteins from a polyprotein precursor. *J. Virol.* **65**, 147–154.

Liljeström, P., Lusa, S., Huylebroeck, D., and Garoff, H. (1991). In vitro mutagenesis of a full-length cDNA clone of Semliki Forest virus: The small 6.000-molecular-weight membrane protein modulates virus release. *J. Virol.* **65**, 4107–4113.

Loewy, A., Smyth, J., Von, C.-H., Bonsdorff, Liljeström, P., and Schlesinger, M.J. (1995). The 6-kilodalton membrane protein of Semliki Forest virus is involved in the budding process. *J. Virol.* **69**, 469–475.

Lusa, S., Garoff, H., and Liljeström, P. (1991). Fate of the 6K membrane protein of Semliki Forest virus during virus assembly. *Virology* **185**, 843–846.

Melton, J.V., Ewart, G.D., Weir, R.C., Board, P.G., Lee, E., and Gage, P.W. (2002). Alphavirus 6K proteins form ion channels. *J. Biol. Chem.* **277**, 46923–46931.

Muñoz, A., Castrillo, J.L., and Carrasco, L. (1985). Modification of membrane permeability during Semliki Forest virus infection. *Virology* **146**, 203–212.

Otero, M.J. and Carrasco, L. (1987). Proteins are cointernalized with virion particles during early infection. *Virology* **160**, 75–80.

Rosemberg, A.H., Lade, B.N., Chui, D., Lin, S., Dunn, J.J., and Studier, F.W. (1987). Vectors for selective expression of cloned DNAs by T7 RNA polymerase. *Gene* **56**, 125–35.

Sanz, M.A. and Carrasco, L. (2001). Sindbis virus variant with a deletion in the 6K gene shows defects in glycoprotein processing and trafficking: Lack of complementation by a wild-type 6K gene in trans. *J. Virol.* **75**, 7778–7784.

Sanz, M.A., Madan, V., Carrasco, L., and Nieva, J.L. (2003). Interfacial domains in Sindbis virus 6K protein. Detection and functional characterization. *J. Biol. Chem.* **278**, 2051–2057.

Sanz, M.A., Pérez, L. and Carrasco, L. (1994). Semliki Forest virus 6K protein modifies membrane permeability after inducible expression in *Escherichia coli* cells. *J. Biol. Chem.* **269**, 12106–1211.

Schlesinger, S. and Schlesinger, M.J. (1996). In Fields, B.N. (ed.), *Virology*, Lippincott-Raven, Philadelphia, pp. 825–841.

Schlesinger, M.J., London, S.D. and Ryan, C. (1993). An in-frame insertion into the Sindbis virus 6K gene leads to defective proteolytic processing of the virus glycoproteins, a trans-dominant negative inhibition of normal virus formation, and interference in virus shut off of host-cell protein synthesis. *Virology* **193**, 424–432.

Suomalainen, M. and Garoff, H. (1994). Incorporation of homologous and heterologous proteins into the envelope of Moloney murine leukemia virus. *J. Virol.* **68**, 4879–4889.

Strauss, J.H. and Strauss, E.G. (1994). The alphaviruses: Gene expression, replication, and evolution. *Microbiol. Rev.* **58**, 491–562.

Studier, F.W. and Moffatt, B.A. (1986). Use of bacteriophage T7 RNA polymerase to direct selective high level expression of cloned genes. *J. Mol. Biol.* **198**, 113–130.

Studier, F.W., A.H. Rosenberg, and Dunn, J.J. (1990). Use of T7 RNA polymerase to direct expression of cloned genes. *Methods in Enzymology* **185**, 60–89.

Suárez, T., Gallaher, W.R., Agirre, A., Goñi, F.M., and Nieva, J.L. (2000). Membrane interface-interacting sequences within the ectodomain of the HIV-1 envelope glycoprotein: Putative role during viral fusion. *J. Virol.* **74**, 8038–8047.

Waite, M.R. and Pfefferkorn, E.R. (1970). Inhibition of Sindbis virus production by media of low ionic strength: Intracellular events and requirements for reversal. *J. Virol.* **5**, 60–71.

Welch, W.J. and Sefton, B.M. (1980). Characterization of a small, nonstructural viral polypeptide present late during of BHK cells by Semliki Forest virus. *J. Virol.* **33**, 230–237.

White, S.H., Ladokhin, A.S., Jayasinghe, S., and Hristova, K. (2001). How membranes shape protein structure. *J. Biol. Chem.* **276**, 32395–32398.

Wimley, W. and White, S.H. (1996). Experimentally determined hydrophobicity scale for proteins at membrane interfaces. *Nature Struct. Biol.* **3**, 842–848.

Yao, J.S., Strauss, E.G., and Strauss, J.H. (1996). Interactions between PE2, E1, and 6K required for assembly of alphaviruses studied with chimeric viruses. *J. Virol.* **70**, 7910–7920.

Part IV

Membrane-Spanning/Membrane Associated

Part IV

17

The Structure, Function, and Inhibition of Influenza Virus Neuraminidase

Elspeth Garman and Graeme Laver

1. Introduction

The neuraminidase story started in the early 1940s, almost a decade after the first human influenza virus was isolated.

George Hirst, working in the Rockefeller Institute in New York City, found that when allantoic fluid from embryonated chicken eggs, which had been infected with influenza virus, was mixed with red blood cells at 0°C, the cells were very heavily agglutinated (Hirst, 1941). He then found that when the agglutinated cells were warmed to 37°C they dispersed as the virus eluted, and the cells could not be re-agglutinated when they were mixed with fresh infected allantoic fluid at 0°C. The eluted virus, on the other hand, could agglutinate fresh red cells in the cold.

Hirst's interpretation of this finding was that the virus had an enzyme which removed receptors for the virus from the agglutinated red cells when they were warmed to 37°C where the enzyme was more active. The enzyme therefore became known as "receptor-destroying enzyme" or RDE.

Alfred Gottschalk, at the Walter and Eliza Hall Institute in Melbourne, Australia, reasoned that the action of RDE on its substrate would probably yield a "split product." This split product was eventually isolated and characterized as sialic acid, or *N*-acetyl neuraminic acid and the RDE of influenza virus became known as sialidase or neuraminidase (Gottschalk, 1957). Subsequently it was discovered that sialidases are quite widespread in nature. Other viruses, bacteria, mammalian cells, and some parasites all have their own sialidase enzymes.

In 1948, MacFarlane Burnet realized that specific inhibitors of flu neuraminidase might be effective antiviral agents. "An effective competitive poison for the virus enzyme might be administered which, when deposited on the mucous film lining the respiratory tract, would render this an effective barrier against infection, both initial infection from without and the spreading surface infection of the mucosa which follows the initiation of infection" (Burnet, 1948).

Elspeth Garman • Laboratory of Molecular Biophysics, Department of Biochemistry, University of Oxford, South Parks Road, Oxford OX1 3QU, United Kingdom. **Graeme Laver** • Barton Highway, Murrumbateman, NSW 2582, Australia.

Viral Membrane Proteins: Structure, Function, and Drug Design, edited by Wolfgang Fischer.
Kluwer Academic / Plenum Publishers, New York, 2005.

Burnet's comment was that this approach did not seem even remotely possible. Now, more than 50 years later, although we know that "poisoning" the viral neuraminidase does not stop the virus infecting cells, the subsequent "spreading surface infection" is effectively quelled.

So far, four different potent and specific "competitive poisons" for flu neuraminidase have been developed, two of which are now being used worldwide to control influenza infections in people.

1.1. Structure of Influenza Viruses

Two serologically different types of influenza virus exist; Type A and Type B. Type A influenza infects a wide variety of animals; pigs, horses, seals, whales, and many different kinds of birds. Type B influenza seems to be confined to the human population, though one isolation of type B flu from harbor seals has been reported (Osterhaus *et al.*, 2000).

Influenza virus particles are pleomorphic, consisting of misshapen spherical objects or long spaghetti like filaments (Figure 17.1).

The genome of influenza A and B viruses consists of single stranded RNA of negative sense. The RNA exists in eight separate pieces, each of which codes one of the virus proteins (in some cases two, using overlapping reading frames). The eight RNA pieces are packaged in an orderly fashion within the virus particle (Fujii *et al.*, 2003).

The RNA is associated with a nucleoprotein and with three proteins, PB1, PB2, and PB3, involved in RNA replication and transcription. This replication complex is enclosed within a membrane composed of a matrix protein associated with a lipid bilayer. Embedded in the lipid bilayer are the two surface glycoprotein spikes, one of which is the hemagglutinin and the other the neuraminidase, described below.

(a)

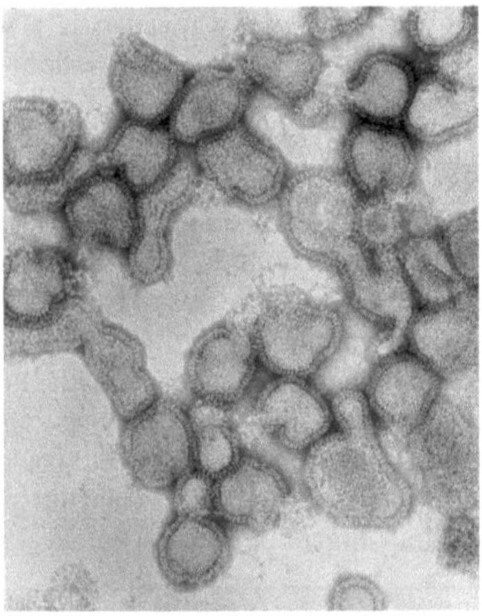

(b)

Figure 17.1. (a) Electron micrographs of negatively stained particles of influenza virus showing their pleomorphic nature and the surface layer of "spikes" which have been identified as the hemagglutinin and neuraminidase antigens. The particles are approximately 80–120 nm in diameter. (b) Diagram of the influenza virus showing the eight segments of negative sense ssRNA, the M2 ion channel spanning the membrane, and the two surface glycoproteins, neuraminidase (boxes on stalks) and hemagglutinin (rods).

Also spanning the lipid bilayer of Type A influenza virus are a small number of M2 protein molecules, which function as ion channels. Until recently the only antiviral drugs available for treating influenza A infections were the ion channel blockers, amantadine and rimantadine. These, however, have no effect on influenza Type B (which does not possess the M2 ion channel), they have undesirable side effects and resistance to these drugs develops very rapidly.

It is perhaps amazing that despite the widespread occurrence of influenza in the world, the huge number of deaths each year, the misery of flu sufferers, and the enormous economic cost of influenza, amantadine and rimantadine have been the only compounds found, until recently, to be effective in treating influenza. This is despite a vast research effort which included the random screening of many thousands of compounds by pharmaceutical companies, none of which was found to be an effective antiviral drug.

The two safe and effective drugs, Relenza and Tamiflu, now being used to treat influenza Type A and Type B, were rationally designed from a knowledge of the three-dimensional structure of flu neuraminidase. The development of these and other neuraminidase inhibitors will now be described.

2. Structure of Influenza Virus Neuraminidase

For some time after flu virus neuraminidase was discovered it was assumed that the agglutination of red cells by influenza virus particles was due to the neuraminidase on the virus binding to its substrate, sialic acid, on the surface of the red cells so linking them together in large clumps. It is now known that this idea is incorrect. The first indication that the neuraminidase enzyme was not responsible for aggutinating red cells came from the finding that when some strains of influenza virus were heated to 55°C, the neuraminidase was inactivated while the hemagglutinin was still fully active (Stone, 1949). Then, in 1961, further doubts began to appear. Mayron and colleagues found that a soluble sialidase could be separated from the PR8 strain of Type A influenza virus, and that this soluble enzyme did not adsorb to red cells (Mayron *et al.*, 1961). Hans Noll then discovered that when influenza B virus particles were treated with trypsin, almost 100% of the neuraminidase was liberated as a soluble molecule with a sedimentation coefficient of 9S (equivalent to about 200,000 molecular weight), leaving all of the hemagglutinin activity still associated with the virus particles (Noll *et al.*, 1962).

Experiments were then done in which influenza virus particles were disrupted with detergents, and the disrupted virus particles subjected to electrophoresis on cellulose acetate strips. This resulted in a clear separation of hemagglutinin and neuraminidase activities and, since the procedure used did not cleave any covalent bonds, it proved that the hemagglutinin and neuraminidase activities resided in separate protein molecules on the surface of the virus particle (Laver, 1964).

At about this time the first electron microscope images of negatively stained influenza virus particles were obtained. These showed pleomorphic objects completely covered with a densely packed layer of surface projections or "spikes" (Figure 17.1). These were the two surface antigens, the hemagglutinin and the neuraminidase.

Electron micrographs of pure preparations of influenza virus neuraminidase molecules, separated from virus particles which had been disrupted with detergents, showed that the neuraminidase consisted of a square, box-shaped head atop a long thin stalk with a small hydrophobic knob at the end. This served to attach the neuraminidase to the lipid membrane of the virus (Laver and Valentine, 1969) (Figure 17.2) and also caused the isolated neuraminidase to form rosettes in the absence of detergents. It was estimated that each virus particle possesses about 500 neuraminidase "spikes" which account for about 5% of the protein in the virus particle. These numbers are approximations only, as they vary from strain to strain.

Further electron micrographs of isolated neuraminidase "heads" by Nick Wrigley showed these to be tetramers (Wrigley *et al.*, 1973). The stalk of the neuraminidase, which serves to attach the molecule to the lipid bilayer of the virus can vary in length (Els *et al.*, 1985; Mitnaul *et al.*, 1996). This is shown in Figure 17.3 where neuraminidase molecules with shortened stalks ("stubbies") can be seen in electron micrographs.

The neuraminidase tetramer is composed of four identical monomers, each of which contains a single polypeptide chain coded by RNA segment number 6. The neuraminidase is anchored in the lipid bilayer of the viral membrane by a series of hydrophobic amino

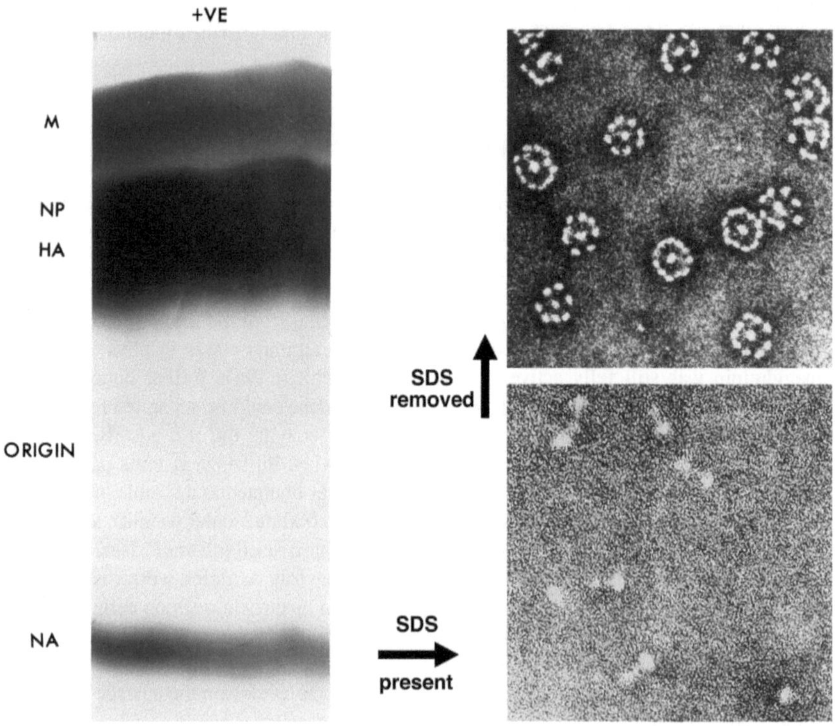

Figure 17.2. Pure preparations of intact, biologically active neuraminidase "spikes" from influenza virus particles. The virus particles were disrupted with sodium dodecyl sulfate (SDS) at room temperature and electrophoresed on cellulose acetate strips at pH 9.0. Strips stained with Coomassie Blue are shown on the left-hand side. Following electrophoresis, the neuraminidase, which was completely separated from all of the other virus proteins, was eluted from the (unstained) strips with water. The eluted neuraminidase (in the presence of SDS) existed as single molecules (bottom right), consisting of a box-like head atop a long thin stalk (approx 12 nm long). Following removal of the detergent, the neuraminidase molecules all aggregated by their hydrophobic membrane attachment sequences at the ends of the stalks to form the rosettes shown (top right, which is ~2.5 lower magnification than bottom right). Stalk length varied between strains but was usually approximately 10 nm, giving rosettes of approximately 24 nm diameter.

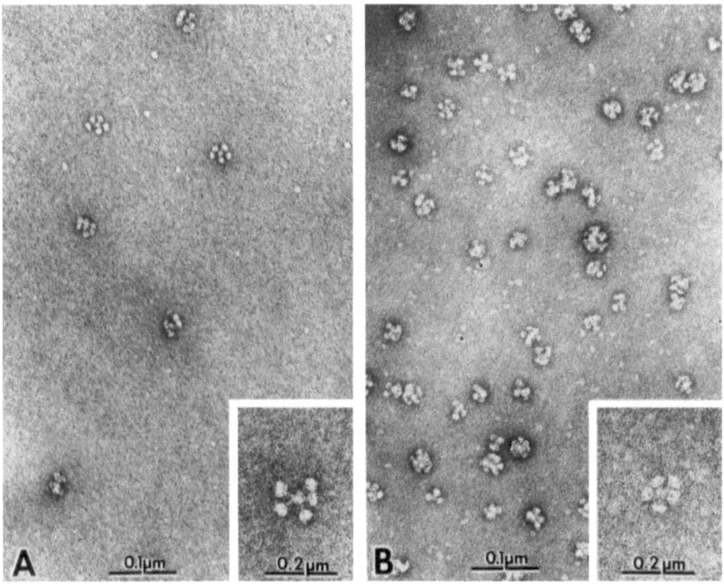

Figure 17.3. Rosettes of intact neuraminidase molecules isolated from SDS disrupted wild-type influenza virus and from a virus with an 18-amino acid residue deletion in the stalk ("stubby").

acids near the N-terminal end of the polypeptide (Figure 17.4). This contrasts with the hemagglutinin which is anchored by a hydrophobic sequence near the C-terminal end of the hemagglutinin polypeptide.

No post-translational cleavage of the neuraminidase polypeptide occurs, no signal peptide is split off and even the initiating methionine is retained. Nor is there any processing at the C-terminus; the sequence Met-Pro-Ile predicted from the gene sequence of N2 neuraminidase is found in intact neuraminidase molecules isolated from the virus. A sequence of six polar amino acids at the N-terminus of the neuraminidase polypeptide, which are totally conserved in all nine flu Type A neuraminidase subtypes (but not in flu Type B), is followed by a sequence of hydrophobic amino acids that must represent the TM region of the neuraminidase polypeptide. This sequence is not conserved at all among subtypes (apart from conservation of hydrophobicity).

Intact neuraminidase molecules can be isolated after disruption of influenza virus particles with detergents. Remarkably, the neuraminidase from a number of strains of flu virus is 100% active after disruption of the virus with the powerful detergent, sodium dodecyl sulfate (SDS). Even more remarkably, when virus particles from these strains were disrupted with SDS and electrophoresed on cellulose acetate strips, the intact, active neuraminidase molecules migrated in one direction completely free of any of the other virus proteins, all of which migrated in the opposite direction (Figure 17.2) (Laver, 1964).

Neuraminidase molecules eluted from the strips following such electrophoresis existed as single molecules. When the detergent was removed, for example, by cold ethanol precipitation of the protein, the single neuraminidase molecules aggregated by the hydrophobic tips of their tails, forming the rosettes seen in electron micrographs (Figure 17.2).

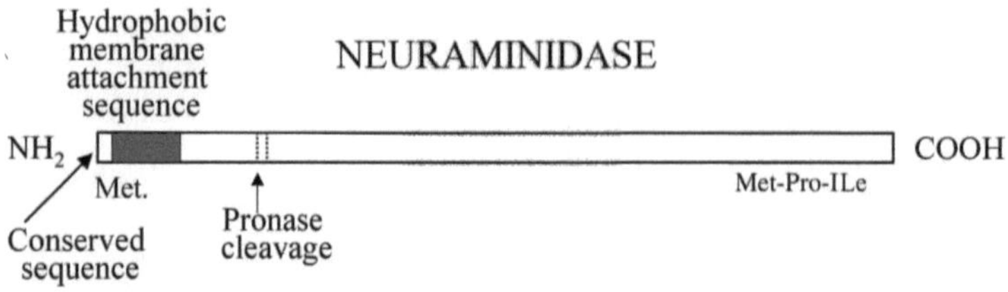

Figure 17.4. Diagram showing certain features of the neuraminidase polypeptide. The neuraminidase is oriented in the virus membrane in the opposite way to the hemagglutinin. No post-translational cleavage of the neuraminidase polypeptide occurs, no signal peptide is split off and even the initiating methionine is retained. No processing at the C-terminus takes place—the C-terminal sequence, Met-Pro-Ile predicted from the gene sequence is found in intact neuraminidase molecules isolated from virus and in the pronase-released neuraminidase heads. A sequence of six polar amino acids at the N-terminus of the neuraminidase polypeptide, which is totally conserved in at least eight different neuraminidase subtypes, is followed by a sequence of hydrophobic amino acids which probably represents the transmembrane region of the neuraminidase stalk. This sequence is not conserved at all between subtypes (apart from conservation of hydrophobicity). Pronase cleaves the polypeptide at the positions shown, removing the stalk and releasing the enzymatically and antigenically active head of the neuraminidase, which, in some cases, can be crystallized.

Soluble neuraminidase "heads" can be released from some strains of influenza virus by treating the particles with proteolytic enzymes, for example, pronase or trypsin. These proteases cleave the stalk of the neuraminidase at about residue 75 of the neuraminidase polypeptide (Figure 17.4) releasing the box-shaped "head" which carries all of the enzymatic and antigenic activity of flu virus neuraminidase.

2.1. Crystallization of Influenza Virus Neuraminidase

In 1978, neuraminidase "heads" released by pronase from a number of strains of H2N2 and H3N2 influenza virus were crystallized (Laver, 1978). The three-dimensional structure of the subtype N2 enzyme at 2.9 Å resolution was then determined using X-ray crystallography (Varghese *et al.*, 1983). Figure 17.5 shows the structure of subtype N9 neuraminidase.

This showed that each monomer in the tetrameric enzyme is composed of six topologically identical beta sheets arranged in propellor formation. The tetrameric enzyme has circular 4-fold symmetry partially stabilized by metal ions bound on the symmetry axis. Deep pockets occur on the upper corners of the box-shaped tetramer. These pockets were identified as the catalytic sites by soaking substrate (sialic acid) into neuraminidase crystals and solving the structure of the complex (Colman *et al.*, 1983).

Sugar residues are attached to four of the five potential glycosylation sequences in N2 neuraminidase and in one case, the carbohydrate contributes to the interaction between the monomers in the tetramer.

2.2. Structure of the Conserved Catalytic Site

Sequences of neuraminidase from influenza A and B strains can differ by as much as 75%. Nevertheless, scattered along the neuraminidase polypeptide, are charged residues which are totally conserved among all strains. These include Arg 118, Glu 119, Asp 151,

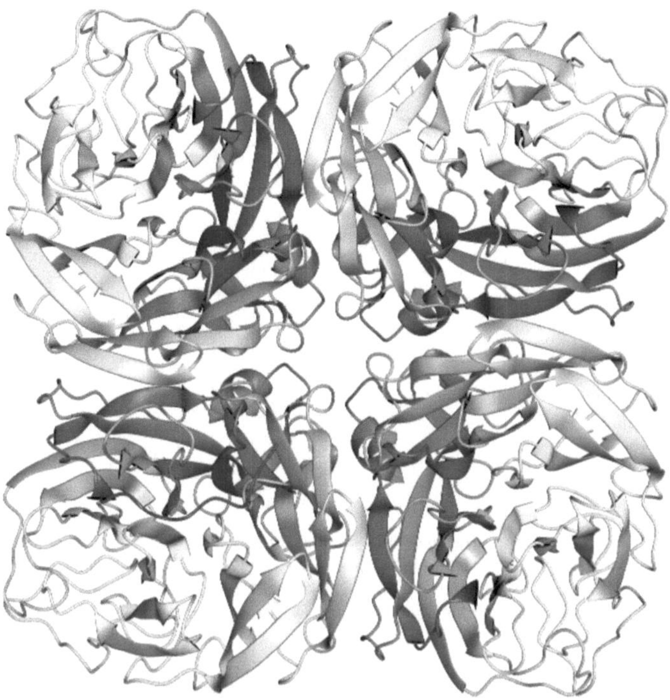

Figure 17.5. Three-dimensional structure of an N9 neuraminidase tetramer, [PDB entry 7NN9]. (Figure drawn with AESOP [Noble, 1995].)

Arg 152, Asp 198, Arg 224, Glu 227, Asp 243, His 274, Glu 276, Glu 277, Arg 292, Asp 330, Lys 350, and Glu 425 (N2 numbering).

When the linear neuraminidase polypeptide folded into its three-dimensional structure, these conserved residues all came together and clustered on the rim and walls of the pocket (Figure 17.6). This suggested that if an inhibitor, a "plug-drug," could be devised which blocked one flu neuraminidase active site, it would also block the sites on all other influenza virus strains, even those which have not yet been found infecting humans.

2.3. Structures of Other Influenza Virus Neuraminidases

All influenza neuraminidases except N4 and N7 have been crystallized (Figure 17.7) though not all crystals were suitable for X-ray structure determination. Structures have been obtained for N2 (Varghese *et al.*, 1983), N9 (Baker *et al.*, 1987), N8 (Taylor *et al.*, 1993), N6 (Garman *et al.*, 1995), and type B (Burmeister *et al.*, 1992) neuraminidases.

N9, N8, N6, and influenza type B neuraminidases have a similar overall topology to N2 neuraminidase despite having up to 75% differences in amino acid sequence of the neuraminidase polypeptide chain.

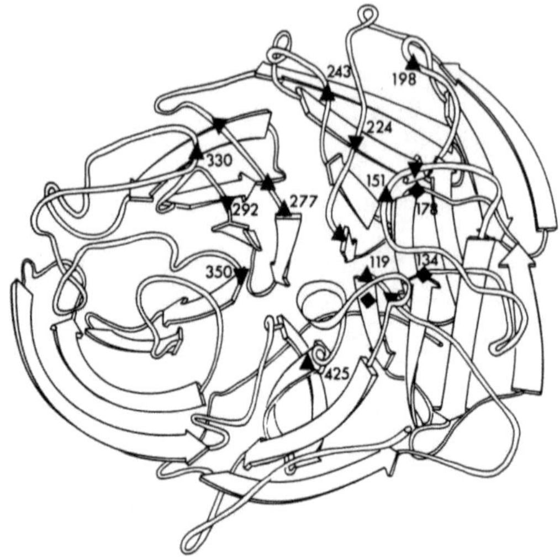

Figure 17.6. Ribbon diagram of a monomer of N2 influenza neuraminidase showing the active-site residues which are conserved across flu strains: ▲ Glu 119, Asp 151, Asp 198, Glu 227, Asp 243, Glu 276, Glu 277, Asp 330, Glu 425 ▼, Arg 118, Arg 152, Arg 224, His 274, Arg 292, Lys 350, ◆ Tyr 121, Leu 134, Trp 178. (Reproduced from Colman *et al.* (1983) with permission from *Nature* [http://www.nature.com/].)

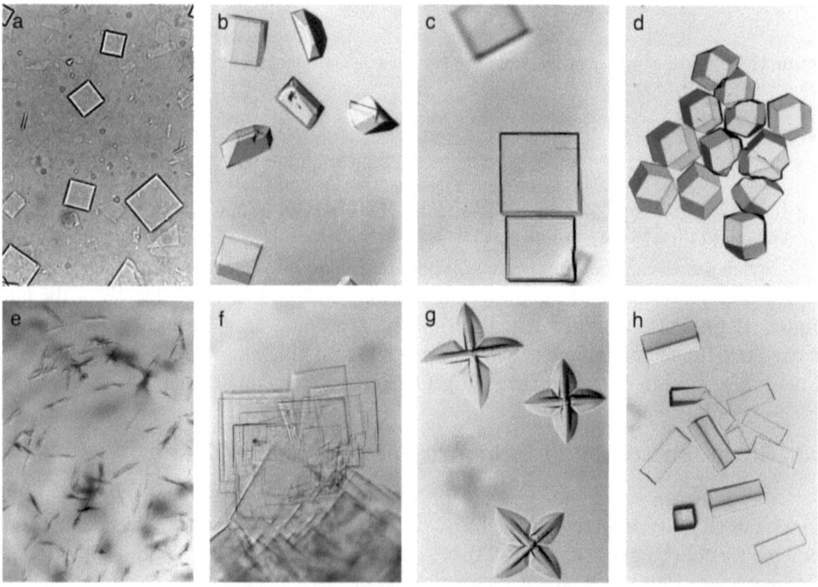

Figure 17.7. Crystals of influenza neuraminidase; (a) N2, (b) N6, (c) N8, (d) N9 used for X-ray structure determination, and (e) N1, (f) N3, and (g) N5 which were unsuitable for such studies, and (h) whale N9 in complex with 32/2 antibody (Fab). Crystal sizes range from 0.6 mm in the largest dimension (N9) to 0.15 mm (N6).

X-ray structure determination pivotally depends on obtaining a diffraction quality crystal that is a well-ordered array of protein molecules which will scatter X-rays in a coherent manner. The crystal is mounted on a motorized stage (a "goniometer") which allows it to be rotated in the X-ray beam in small angular increments. The diffraction patterns of the scattered X-rays are now collected on image plate or CCD (Charged-Coupled Devices) detectors, replacing the photographic film of old. Dedicated computer software is used for analyzing the diffraction images and extracting the intensities of the reflections and further experiments are required to obtain their phases (for further reading see Blow, 2002).

Experimental methods for crystallography have advanced dramatically in the last 15 years because of major technical developments. These include the advent of intense tunable synchrotron X-ray sources and fast read-out area detectors, a huge increase in computing power and the development of techniques for flash-cooling crystals to cryotemperatures (below 130 K) to substantially reduce radiation damage by the beam during data collection (Garman and Schneider, 1997).

2.4. Hemagglutinin Activity of Neuraminidase

Rosettes of isolated intact neuraminidase molecules of the N9 subtype also had hemagglutinin activity (Laver *et al.*, 1984). The hemagglutinin site was shown to be quite separate from the catalytic site in a 1.7 Å resolution X-ray structure which located a second sialic acid binding site situated about 21 Å from the catalytic site on N9 neuraminidase after a 4°C soak (as opposed to the usual 18°C soak). The residues in contact with the sialic acid come from three different loops in the structure. These residues are mostly conserved in avian strains of influenza, but not in those of human and swine. It is thus possible that the hemagglutinin site on the neuramindase has some as yet undiscovered biological function in birds (Varghese *et al.*, 1997).

3. Function of Influenza Virus Neuraminidase

In 1966, Seto and Rott showed that the function of neuraminidase was probably associated with the release of virus from host cells (Seto and Rott, 1966). It was then found that antibody directed specifically against flu neuraminidase, and which abolished the activity of the enzyme for large substrates, did not prevent the infection of susceptible cells, but blocked the release of newly formed virus particles (Webster and Laver, 1967).

The role of neuraminidase in the release of virus particles from infected cells was demonstrated most elegantly by Palese, Compans, and their colleagues in 1974 (Palese *et al.*, 1974a). Electron micrographs were made of surfaces of cells infected with temperature sensitive (ts) neuraminidase mutants of influenza virus at the permissive temperature and at the restrictive temperature (where the virus replicated but where the neuraminidase lacked enzyme activity). These showed virus particles budding normally from the cells and going off to infect other cells at the permissive temperature. However in cells infected with the ts mutants at the restrictive temperature, virus particles budded from the cell in the normal manner, but then remained attached to each other and to the surface of the infected cells, forming great clumps of virus particles. These were clearly not going anywhere, and the infection was effectively terminated (Figure 17.8).

It is believed, therefore, that the function of flu virus neuraminidase is to remove sialic acid receptors for the virus from the host cells, and also, perhaps more importantly, from the

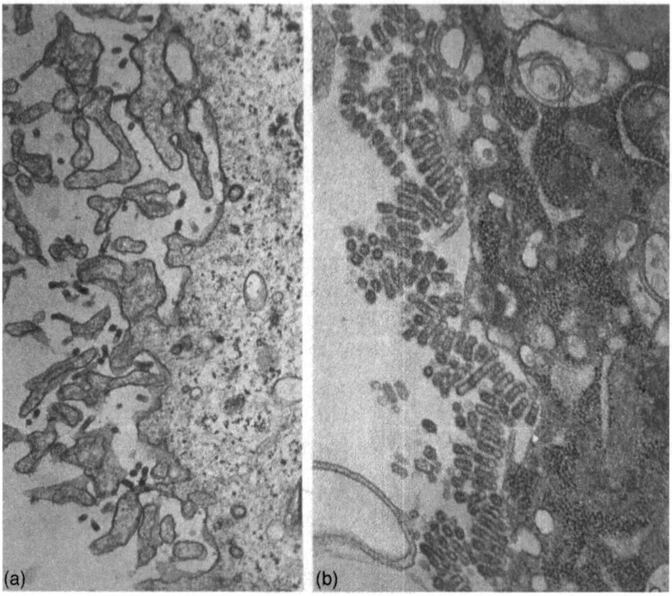

Figure 17.8. Electron micrographs of the surface regions of MDCK cells infected with temperature sensitive (ts) neuraminidase mutants of influenza virus after inoculation and incubation for 12.5 hr at the permissive temperature of 33°C ((a) left) and at the restrictive temperature of 39.5°C ((b) right). The aggregates of virus particles which accumulated at the restrictive temperature could be dispersed by incubation with bacterial neuraminidase. Staining experiments showed that the aggregated virus particles formed at the restrictive temperature were covered in sialic acid residues, while this was absent on those well-dispersed particles formed at the permissive temperature. Magnification approximately ×30,000. (Reprinted from Palese *et al.* [1974a] with permission from Elsevier.)

newly formed virus particles themselves. The two surface antigens on the influenza virus particle, the hemagglutinin and the neuraminidase, are themselves glycoproteins and possess carbohydrate side chains with terminal sialic acid receptors for other virus particles. The main function of the neuraminidase therefore might be to remove receptors for influenza virus from newly formed virus particles so allowing these to be released and spread the infection (Palese *et al.*, 1974a). Another function of flu virus neuraminidase might be to destroy sialic acid containing inhibitors for the virus in the mucous secretions of the respiratory tract, so enabling the virus to more easily infect cells, and there may be other functions as yet undiscovered.

Chickens vaccinated with pure neuraminidase "heads" were protected from death by lethal avian influenza viruses. But whether this protection was due to inhibition of neuraminidase activity or to enhanced clearance of the virus by the immune system was not established (Webster *et al.*, 1988).

3.1. Antigenic Properties of Influenza Virus Neuraminidase

Both of the surface antigens of influenza virus undergo extensive antigenic variation. This is of two types, antigenic drift and major antigenic shifts. Drift is the result of mutations in the genes coding the hemagglutinin and neuraminidase which lead to amino acid sequence changes in the antibody binding sites (epitopes) on these virus proteins.

The major shifts, on the other hand, involve complete replacement of the genes for one or both of the surface antigens as a result of reassortment between human and animal (or avian) influenza viruses, or by mutations in one of these latter viruses which results in their ability to infect humans (Garman and Laver, 2003).

Nine serologically distinct subtypes of Type A influenza have been discovered in nature. Of these, N1 and N2 have been found in viruses infecting humans. All of the nine subtypes have been found in viruses infecting wild water birds.

Neuraminidase of subtype N9 was isolated from a white-capped noddy tern on North West Island of Australia's Great Barrier Reef in 1975 (Downie *et al.*, 1977). Crystals of N9 neuraminidase (Figure 17.7) are of particularly high quality and this enzyme has been used to investigate the antigenic topology of flu neuraminidase and the way the antibody binding sites (epitopes) change during antigenic drift.

Until recently, the structure of epitopes on protein molecules was a matter of some controversy. Attempts to characterize the sites on proteins which bound antibodies involved a plethora of diverse methods (Laver *et al.*, 1990).

These included the use of protein fragments to absorb antisera, the production of antipeptide antibodies and their reaction with intact proteins, and proteolytic digestion of protein–antibody complexes in an attempt to discover protected peptide bonds. One claim was also made that the complete and precise determination of all the antigenic sites on lysozyme had been achieved (Atassi, 1980). It was stated that there were three precisely defined antigenic sites on the lysozyme molecule and that each comprised six to seven amino acids contained within sharp boundaries. However, when crystals of antibodies bound to lysozyme were analyzed by X-ray crystallography, none of the predicted sites was found to be involved in the binding (Davies *et al.*, 1989).

Furthermore X-ray crystallography showed that about sixteen amino acids on the surface of lysozyme were in contact with about the same number on the antibody, a far cry from the four to seven residues in the epitopes described by various authors (Laver *et al.*, 1990).

A number of complexes of antibodies bound to influenza virus neuraminidase have now been crystallized and the structures determined by X-ray crystallography. The structure of one of these complexes, N9 neuraminidase–NC 41 Fab, determined at 2.9 Å resolution, showed that the epitope on the neuraminidase is discontinuous, being composed of five separate peptide segments involving about 17 amino acid residues (Figure 17.9) which were in contact with a similar number of amino acid residues on the antibody molecule (Colman *et al.*, 1987). It has subsequently been shown that only about three or four of these residues in the epitope contribute to the energy of binding, the others simply having to show complementarity with residues on the antibody.

3.2. Antigenic Drift in Influenza Virus Neuraminidase

Antibodies to flu neuraminidase do not directly neutralize virus infectivity, but if the cells in which the virus is growing are bathed in antisera to the neuraminidase, most of the virus is prevented from exiting the cells and the infection is effectively terminated. However, if the cells are bathed in a monoclonal antibody to the neuraminidase, mutant virus particles with changes in the epitope recognized by the antibody will "escape" from the inhibiting effect of the antibody and continue to grow unhindered.

(a)

(b)

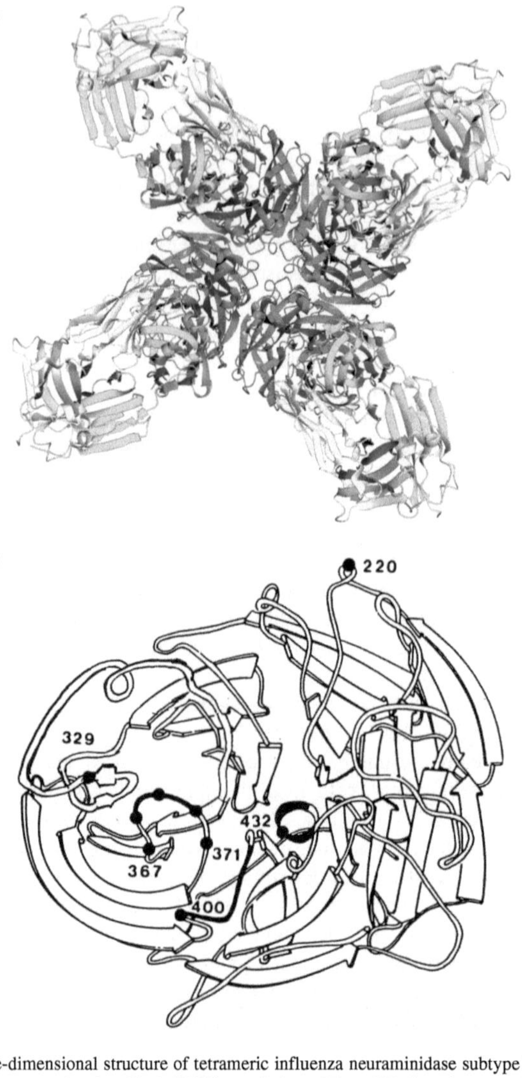

Figure 17.9. (a) Three-dimensional structure of tetrameric influenza neuraminidase subtype N9 complexed with an Fab fragment from the monoclonal antibody NC41 consisting of F_c and F_v from heavy and light chains which recognize an antigenic determinant (epitope) on the neuraminidase tetramer [PDB entry 1NCA] (Tulip *et al.*, 1992). (b) Schematic diagram showing a monomer of neuraminidase viewed down the 4-fold axis. The epitope recognized by NC41 antibody involves the three loops shown in heavy black and also part of the 329 (N2 numbering) loop (Laver *et al.*, 1987). The side chains of amino acids 368–370 point towards the viewer, while that of Arg 371 (an active-site residue) points away and into the catalytic site located above and to the right of C_α 371. Mutations at positions 367, 369, 370, 372, 400, and 432 abolish the binding of NC41 antibody to neuraminidase, whereas mutations at 368 and 329 reduce binding. A mutation at residue 220 (outside the NC41 epitope) has no effect on binding of NC41 to neuraminidase.

Many such neuraminidase escape mutants of N2 and N9 have been analyzed and in each case single amino acid sequence changes were found in the neuraminidase polypeptide (Air and Laver, 1986). These single changes were enough to completely abolish binding of the monoclonal antibody which was used to select the particular escape mutant analyzed.

Most of these sequence changes occurred on the top of the neuraminidase "head" on the rim surrounding the active-site crater, suggesting that neutralizing epitopes were situated in this region. Other epitopes almost certainly exist at the base of the tetramer, but escape mutants of these have never been obtained, presumably because antibodies binding in this region do not "neutralize" infectivity.

How do single amino acid sequence changes totally abolish antibody binding when only one out of about seventeen contact residues in the epitope is altered? This question was addressed structurally by Tulip *et al.* (1991) who determined the structure of five N9 antibody escape mutants. The mutations were all situated within 5–10 Å of the N9 catalytic site. Only local structural changes associated with the site of the amino acid substitution or residues on either side of it were found; no large scale rearrangements were observed. Although the precise basis for the abolition of antibody binding is still not clear, changes in charge and shape complementarity between the two interacting surfaces no doubt play a part.

4. Inhibition of Influenza Virus Neuraminidase

4.1. Design and Synthesis of Novel Inhibitors of Influenza Virus Neuraminidase

4.1.1. Relenza

Relenza was the first inhibitor to be synthesized which specifically inhibited the neuraminidase of both Type A and Type B influenza viruses and was effective in controlling influenza infections in people. Its design was based on the crystal structure of flu neuraminidase and a sialic acid scaffold. Sialic (neuraminic) acid (Figure 17.10a) is itself a mild inhibitor of flu neuraminidase, but the dehydrated derivative, deoxy dehydro *N*-acetyl neuraminic acid, DANA, Neu5Ac2en (Figure 17.10b) the transition state analog, is a much better inhibitor. This was convincingly demonstrated by Peter Palese and his colleagues in the 1970s (Palese *et al.*, 1974b). DANA inhibited influenza virus replication in tissue culture but failed to prevent disease in flu infected animals (Palese *et al.*, 1977).

In using the three-dimensional structure of flu neuraminidase for the rational design of antiviral drugs, manual inspection of the active site with the aid of computer graphics was complemented by probing the active-site interactive surfaces with various chemical substituents using the computer software program GRID (Goodford, 1996) to calculate energetically favorable substitutions on the sialic acid scaffold.

The following precise account of the design of Relenza by Mark von Itzstein and his colleagues is given by Dr. Wen Yang Wu.

"Structural studies where sialic acid was soaked into flu neuraminidase crystals showed that there was a negatively charged zone in the neuraminidase active site which aligned with the 4-position of the bound sialic acid (Colman *et al.*, 1983). This led to the suggestion that the introduction of a positively charged group, such as an amino group, to the 4-position of sialic acid should enhance its binding to the active site.

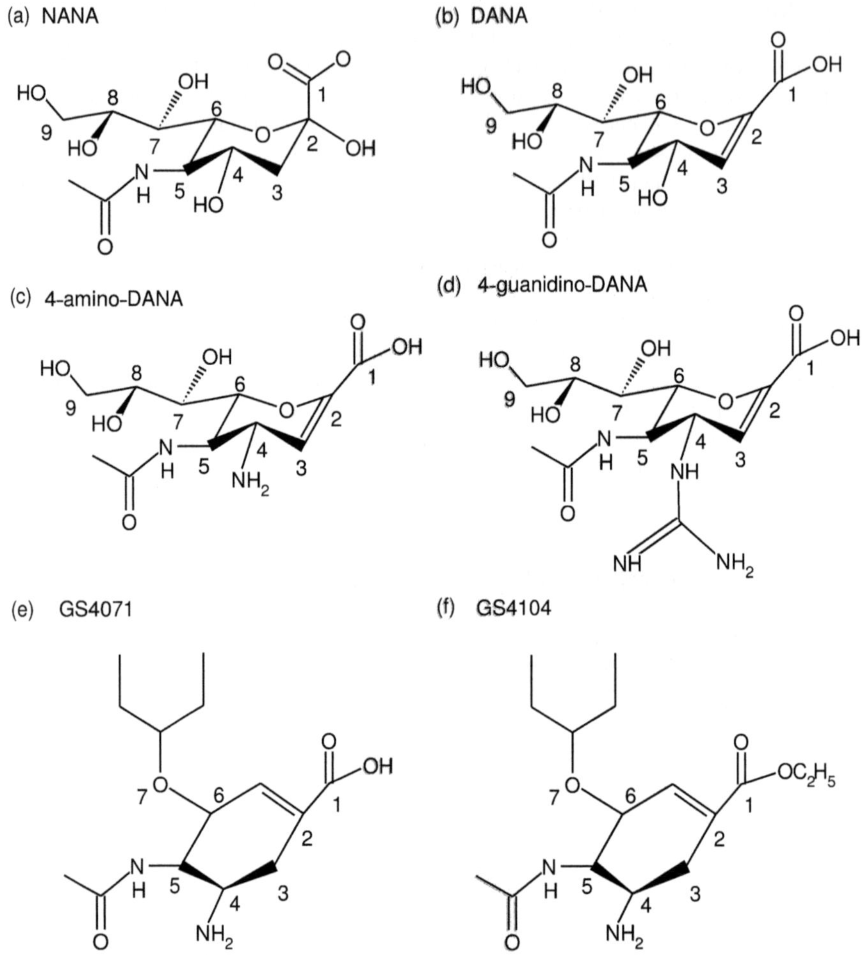

Figure 17.10. Chemical structure of (a) *N*-acetyl neuraminic acid, NANA, (b) 2-deoxy 2,3-dehydro-*N*-acetyl neuraminic acid, DANA, (c) 4-amino-DANA, (d) 4-guanidino-DANA, (Relenza, zanamivir), (e) (3*R*,4*R*,5*S*)-4-acetamido-5-amino-3-(1-ethylpropoxyl)-1-cyclohexane-1-carboxylic acid (GS4071), and (f) ethyl ester derivative of (GS4104, tamiflu).

4-Amino DANA was therefore synthesized and, as predicted, bound more tightly to the active site. There was a 100-fold increase in inhibitory activity for flu neuraminidase of 4-amino DANA compared to the unsubstituted DANA.

Furthermore, the 4-amino DANA was specific for influenza virus neuraminidase and did not inhibit mammalian neuraminidases. This strong inhibitory activity and high specificity suggested that the approach being used might lead to a safe and effective anti-influenza drug.

Further modifications at the 4-position were therefore explored. In this, the synthetic chemistry focused on the introduction of additional positive charges at the 4-position. A number of 4-substituted amino-DANA analogs were therefore prepared. All showed good inhibitory activity and specificity.

It was then proposed to synthesize an analogue with a guanidino group at the 4-position of DANA, because of its increased positive charge and bigger size, compared to the amino group. Although it appeared that the bulky 4-guanidino-DANA (Figure 17.10c) would not fit into the neuraminidase active site, after one water molecule was expelled from the active site, it would fit in perfectly.

After a few synthetic chemistry challenges were overcome, the 4-guanidino-DANA analogue was prepared and tested. It was found to be 1,000-fold better inhibitor of flu neuraminidase than DANA and did not inhibit mammalian neuraminidases (Von Itzstein *et al.*, 1993). Although subsequently many derivatives of 4-guanidino-DANA were prepared and tested, ultimately the 4-guanidino-DANA analogue (Figure 17.10d) was chosen for clinical trials. It is now marketed as "Relenza" by Glaxo-Smith Kline Ltd."

However, because of the guanidino group, Relenza is not orally bioavailable and is given as a powder which is puffed into the lungs.

A second generation Relenza is being developed. This is a dimer in which two molecules of 4-guanidino-DANA are linked via their 7-hydroxyl groups by an appropriate spacer such as a benzene ring or aliphatic chain. The dimer exhibits cooperativity in binding so that the inhibitory activity for flu neuraminidase is 100-fold greater than that of Relenza. Moreover, after administration, the dimer remains in the respiratory secretions for up to a week. This suggests that one dose of the dimer every 5 days should be effective, compared to the therapeutic regime for Relenza and Tamiflu of 2 doses/day for a period of 5 days (Tucker, 2002).

4.1.2. Tamiflu

In order to produce a neuraminidase inhibitor which was orally bioavailable and which flu sufferers could swallow as a pill, Choung Kim and his associates at Gilead Sciences in California synthesized a carbocyclic compound which fulfilled this requirement (Kim *et al.*, 1997). They noticed the presence of a large hydrophobic pocket in the active site region of flu neuraminidase that accommodated the glycerol side chain of the substrate, sialic acid, and exploited this pocket in the synthesis of carbocyclic sialic acid analogs with hydrophobic alkyl side chains. These carbocyclic compounds are not sugars and have no oxygen in the ring.

X-ray crystallography of flu neuraminidase with DANA bound in the catalytic site showed that the C7 position of the glycerol side chain had no interactions with any of the amino acids in the neuraminidase catalytic site. This suggested that the C7 hydroxyl could be eliminated from the glycerol side chain of the carbocyclic system without losing binding affinity to the neuraminidase.

The CHOH group at the C7 position of the glycerol side chain was therefore replaced by an oxygen atom. Then, in order to create a molecule with hydrophobic groups which would interact well with the amino acids Glu 276, Ala 246, Arg 224, and Ile 222 (N9 subtype numbering) in the large hydrophobic pocket occupied by the glycerol side chain of DANA, various lipophilic side chains were attached to the oxygen linker that had been introduced at the 7 position.

The carboxylate and acetamido groups corresponding to the same groups on DANA were retained on the new carboxylic compound and an amino group was introduced at position C4. An amino rather than a guanidino group was chosen at C4 since the latter, while giving a more tightly binding inhibitor, would create a molecule having the same disadvantage as Relenza, in that it would not be orally bioavailable.

The final compound chosen, GS4071, with a 3-pentyl side chain (Figure 17.10e) was a potent and specific inhibitor of Type A and Type B influenza virus neuraminidase with an IC50 of 1–2 nM. The X-ray crystallographic structure of GS4071 bound in the catalytic site of flu neuraminidase is shown in Figure 17.11.

Because of the zwitterionic nature of GS4071 imposed on the molecule by the carboxylate and amino groups, GS4071 was not orally bioavailable. This problem was overcome by converting the carboxylate to the ethyl ester. The resulting compound, GS4104 (Figure 17.10f) now marketed as "Tamiflu", can be swallowed as a pill. Following absorption of this prodrug from the gut, the ester is hydrolyzed in the liver and the resulting active neuraminidase inhibitor finds its way into the respiratory tract. It is not clear why GS4071 is able to cross membranes in the respiratory tract when it was unable to cross membranes in the gut.

Following clinical trials Relenza and Tamiflu are now being used worldwide for the treatment of influenza. They are safe and effective drugs, provided they are used correctly. They need to be given very soon after infection and they are effective against influenza only, and not against any other respiratory pathogens, viral or bacterial. Their effectiveness in preventing death in cases of severe influenza has not been established, but anecdotal evidence suggests that this may indeed be an important property of these drugs.

4.1.3. Other Inhibitors of Influenza Virus Neuraminidase

Two other potent and specific inhibitors of flu neuraminidase have been developed. One, invented at BioCryst, is BCX-1812 (1*S*,2*S*,3*R*–4*R*,1′*S*)-3-(1′-acetylamino-2′-ethyl) butyl-4-[(aminoimino)-methyl]amino-2-hydroxycyclopentane-1-carboxylic acid) (Babu *et al.*, 2000).

The second inhibitor, made by Abbott Labs, is A315675 (−)-(2*R*,4*S*,5*R*,1′*R*,2′*S*)-5-(1-acetylamino-2-methoxy-2-methyl-pentyl)-4-propenyl-pyrrolidine-2-carboxylic acid (Kati *et al.*, 2001). So far, neither of these drugs has been approved for human use.

The way these two compounds, as well as Relenza and Tamiflu, bind in the catalytic site of flu neuraminidase is shown in Figure 17.11.

4.2. Drug Resistance

One of the unanswered questions is; if Relenza and Tamiflu are used widely in the community to treat influenza, how easily will drug resistant mutants of influenza virus arise?

Experiments so far have suggested that the virus might have difficulty in escaping from the neuraminidase inhibitors. Influenza viruses resistant to the neuraminidase inhibitors have been selected *in vitro* by growing virus in the presence of sublimiting concentrations of the drugs. These experiments revealed the existence of two classes of resistant mutants (Roberts, 2001).

Some mutants had amino acid sequence changes in the hemagglutinin and none in the neuraminidase, while others had changes in the neuraminidase but not in the hemagglutinin. It is thought that the hemagglutinin mutants had a reduced capacity to bind to sialic acid receptors, and so had little need for these to be destroyed by neuraminidase for the virus to "escape." The fact that the hemagglutinin mutants have so far been found to be as susceptible to the neuraminidase inhibitors as the wild-type viruses in animal experiments, suggests that the neuraminidase may play some vital role other than receptor destruction in the infection process. Possibly the enzyme is required to facilitate the movement of virus particles through respiratory secretions, and thus if it is blocked, the virus may be trapped and immobilized.

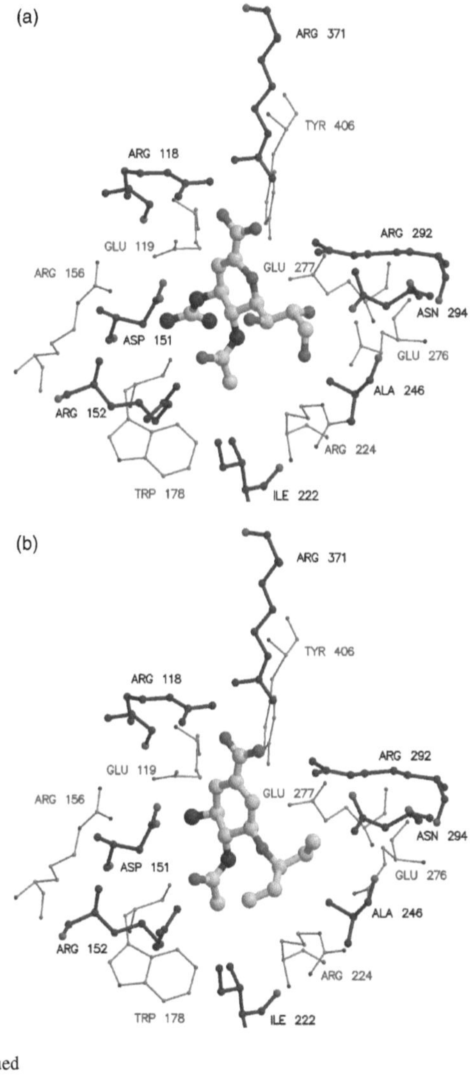

Figure 17.11. Continued

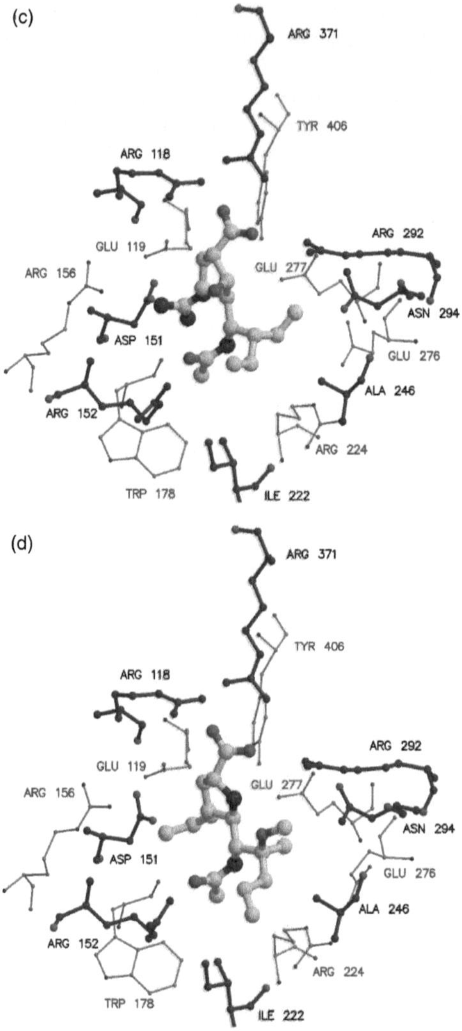

Figure 17.11. Crystallographic structures of influenza virus neuraminidase (N9 subtype) with four different rationally designed inhibitors bound in the active site of the enzyme. The inhibitors are shown as atom-colored ball and stick models. The catalytic site of the enzyme is shown with the closer carbon atoms shown darker than those shown further away. This catalytic site is conserved across all flu neuraminidases. (a) and (b) are antiflu drugs approved for use: (a) is Relenza (4-guanidino-Neu5Ac2en); (b) is de-esterified Tamiflu (4-acetamido-5-amino-3 (1-ethylpropoxyl)-1-cyclohexane-1-carboxylic acid); (c) and (d) are two further drugs which are being developed: (c) is BCX-1812 [BioCryst] (1*S*,2*S*,3*R*,-4*R*,1′*S*)-3-(1′-acetylamino-2′-ethyl)butyl-4-[(aminoimino)-methyl]amino-2-hydroxycyclopentane-1-carboxylic acid), and (d) is A315675 [Abbott] (−)-(2*R*,4*S*,5*R*,1′*R*,2′*S*)-5-(1-acetylamino-2-methoxy-2-methyl-pentyl)-4-propenyl-pyrrolidine-2-carboxylic acid. The figures were drawn with Molscript and rendered with Raster3d.

Sequence changes in the neuraminidase mutants selected by the neuraminidase inhibitors occurred in active-site residues of the enzyme. This makes sense, as it is those residues which are involved in binding the inhibitors. In particular, there were changes to Arg292 (to lysine), one of the three active-site arginines which interact with the natural substrate carboxylate group (which is also present in 4-guanidino-DANA and GS4071), and to Glu119 (to glycine) which lies in the pocket occupied by the 4-guanidino group of Relenza. A structural study of the Arg292Lys N9 mutant bound to various inhibitors (Varghese *et al.*, 1998) revealed that the structural and binding effects of the mutation were most marked for those inhibitors which were least like the natural ligand. For binding of GS4071 in neuraminidase, residue Glu276 rotates to form a salt link with Arg224, creating a hydrophobic pocket. In the Arg292Lys mutant, Glu276 appears to be anchored by an ionic link to Lys292 not present in the wild-type enzyme, and cannot rotate to form the necessary salt link with Arg224, thus reducing the binding of GS4071.

The active-site residues which have been found to mutate are also involved in catalysis, and the mutant neuraminidases were found to be "crippled" in some way, making the mutant virus less able to infect animals. These findings suggest that mutants resistant to Relenza and Tamiflu might arise infrequently in the human population.

5. Conclusions

The four drugs described above represent the first example of antiviral drugs rationally designed from knowledge of the X-ray crystal structure of the target protein of the virus.

These safe drugs, which are most effective if given within a day of symptoms appearing, have had a rocky ride being accepted into the market place. For example, currently in Britain, the normal time taken to see a General Practitioner is about two days. Because Relenza and Tamiflu are only available on prescription, by the time they are administered it is too late for efficacy. A further problem is that they are only effective against influenza viruses. This would be remedied if a cheap, rapid, and accurate diagnostic test for flu were available.

However, in the event of an influenza pandemic, it is generally accepted that antiviral drugs will provide the first line of defense against the new virus. Of these, the neuraminidase inhibitors are the drugs of choice (WHO, 2003).

Acknowledgments

We thank Drs. Kim and Wu for their personal insights into the inhibitor discovery process, and Stephen Lee, Martin Noble, Atlanta Cook, and James Murray for help with the figures.

References

Air, G.M. and Laver, W.G. (1986). The molecular basis of antigenic variation in influenza virus. *Adv. Virus Res.* **31**, 53–102.

Atassi, M.Z. (1980). Molecular immune recognition of proteins: The precise determination of protein antigenic sites has led to synthesis of antibody combining sites and other types of protein binding sites. In Laver, W.G. and Air, G.M. (eds.), *Structure and Variation in Influenza Virus*, Elsevier North Holland, Inc, pp. 241–271.

Babu, Y.S., Chand, P., Banta, S., Kotian, P., Dehghani, A., El-Katan, Y. *et al.* (2000). BCX-1812 (RWJ-27021). Discovery of a novel, highly potent, orally active and selective influenza virus neuraminidase inhibitor through structure-based drug design. *J. Med. Chem.* **43**, 3482–3486.

Baker, A.T., Varghese, J.N. Laver, W.G. Air, G.M. and Colman, P.M. (1987). Three-dimensional structure of neuraminidase of subtype N9 from an avian influenza virus. *Proteins* **2**, 111–117.

Blow, D. (2002). *Outline of Crystallography for Biologists.* OUP, Oxford, United Kingdom. 0198510519

Burmeister, W.P., Ruigrok, R.W.H., and Cusack, S. (1992). The 2.2 Å resolution crystal structure of influenza B neuraminidase and its complex with sialic acid. *EMBO J.* **11**, 49–56.

Burnet, F.M. (1948). quoted in: *Aust. J. Exp. Biol. Med. Sci.* **26**, 410.

Colman, P.M., Laver, W.G., Varghese, J.N., Baker, A.T., Tulloch, P.A., Air, G.M. *et al.* (1987). Three dimensional structure of a complex of antibody with influenza virus neuraminidase. *Nature* **326**, 358–363.

Colman, P.M., Varghese, J.N., and Laver, W.G. (1983). Structure of the catalytic and antigenic sites in influenza virus NA. *Nature* **303**, 41–44.

Davies, D.R., Sheriff, S., Padlan, E.A., Silverton, E.W., Cohen, G.H., and Smith-Gill, S.J. (1989). Three dimensional structures of two Fab complexes with lysozyme. In S. Smith-Gill and E. Sercarz (eds.), *The Immune Response to Structurally Defined Protein: The Lysozyme Model.* Adenine Press, New York, ISBN 0-940030-27-6, pp. 125–132.

Downie, J.C. Hinshaw, V., Laver, W.G. (1977). The ecology of influenza. Isolation of type "À" influenza viruses from Australian pelagic birds. *Aust. J. Exp. Biol. Med. Sci.* **55**, 635–643.

Els, M.C., Air, G.M., Murti, K.G., Webster, R.G., and Laver, W.G. (1985). An 18-amino acid deletion in an influenza neuraminidase. *Virology* **142**, 241–247.

Fujii, Y., Goto, H., Watanabe, T., Yoshida, T., and Kawaoka, Y. (2003). Selective incorporation of influenza virus RNA segments into virions. *PNAS* **100**, 2002–2007.

Garman, E., E. Rudino-Pinera, P. Tunnah, S.C. Crennell, R.G. Webster, and W.G. Laver (1995). Unpublished structure of N6 neuraminidase.

Garman, E.F. and Laver, W.G. (2004). Controlling influenza by inhibiting the virus's neuraminidase. *Curr. Drug Targets* **5**, 119–136.

Garman, E.F. and T.R. Schneider, (1997). Macromolecular cryocrystallography. *J. Appl. Cryst.* **30**, 211–237.

Goodford, P. (1996). Multivariate characterisation of molecules for QSAR analysis. *J. Chemometrics* **10**, 107–117.

Gottschalk, A. (1957). The specific enzyme of influenza virus and *Vibrio cholerae. Biochem. Biophys. Acta* **23**, 645–646.

Hirst, G.K. (1941). The agglutination of red cells by allantoic fluid of chick embryos infected with influenza virus. *Science* **94**, 22–23.

Kati, W.M., D. Montgomery, C. Maring, V.S. Stoll, V. Giranda, X., Chen *et al.* (2001). Novel α- and β-amino acid inhibitors of influenza virus neuraminidase. *Antimicrob. Agents Chemother.* **45**, 2563–2570.

Kim, C.U., Williams, M.A., Lui, H., Zhang, L., Swaminathan, S., Bischofberger, N. *et al.* (1997). Influenza neuraminidase inhibitors possessing a novel hydrophobic interaction in the enzyme active site: Design, synthesis and structural analysis of carbocyclic sialic acid analogues with potent anti-influenza activity. *J. Am. Chem. Soc.* **119**, 681–690.

Laver, W.G. (1964). Structural studies on the protein subunits from three strains of influenza virus. *J. Mol. Biol.* **9**, 109–124.

Laver, W.G. (1978). Crystallization and peptide maps of neuraminidase "heads" from H2N2 and H3N2 influenza virus strains. *Virology* **86**, 78–87.

Laver, W.G. and Valentine, R.C. (1969). Morphology of the isolated hemagglutinin and neuraminidase subunits of influenza virus. *Virology* **38**, 105–119.

Laver, W.G., Air, G.M., Webster, R.G., and Smith, G.S. (1990). Epitopes on protein antigens: Misconceptions and realities. *Cell* **61**, 553–556.

Laver, W.G., Colman, P.M., Webster, R.G., Hinshaw, V.S. and Air, G.M. (1984). Influenza virus neuraminidase with hemagglutinin activity. *Virology* **137**, 314–323.

Laver, W.G., Webster, R.G., and Colman, P.M. (1987). Crystals of antibodies complexed with influenza virus neuraminidase show isosteric binding of antibody to wild type and variant antigens. *Virology* **156**, 181–184.

Mayron, L.W., Robert, B., Winzler, R.J. and Rafelson, M.E. (1961). Studies on the neuraminidase of influenza virus 1. Separation and some properties of the enzyme from Asian and PR8 strains. *Arch. Biochem. Biophys.* **92**, 475–483.

Mitnaul, L.J., Castrucci, M.R., Murti, K.G., and Kawaoka, Y. (1996). The cytoplasmic tail of influenza A virus neuramindase (NA) affects NA incorporation into virions, virion morphology, and virulence in mice but is not essential for virus replication. *J. Virol.* **70**, 873–879.

Noble, M. (1995). Unpublished computer program, AESOP.

Noll, H., Aoyagi, T., and Orlando, J. (1962). The structural relationship of sialidase to the influenza virus surface. *Virology* **18**, 154–157.

Osterhaus, A.D.M.E., Rimmelzwaan, G.F., Martina, B.E.E., Bestebroer, T.M., and Fouchier, R.A.M. (2000). Influenza B virus in seals. *Science* **288** (5468), 1051–1053.

Palese, P., Ueda, M., Tobita, K., and Compans, R.W. (1974a). Characterization of temperature sensitive influenza virus mutants defective in neuraminidase. *Virology* **61**, 397–410.

Palese, P., Schulman, J.N., Bodo, G., and Meindl, P. (1974b). Inhibition of influenza and parainfluenza virus replication in tissue culture by 2-deoxy-2,3-dehydro-N-trifluoroacetylneuraminic acid (FANA). *Virology* **59**, 490–498.

Palese, P. and Schulman, J.N. (1977). Inhibitors of viral NA as potential antiviral drugs. In J.S. Oxford (ed.), *Chemoprophylaxis and Virus Infections of the Respiratory Tract*, Vol. I, pp. 189–205. CRC Press, Cleveland.

Roberts, N.A. (2001). Treatment of influenza with neuramindase inhibitors: Virologic implications. *Phil. Trans. Royal Soc. Lond.* **B356**, 1893–1895.

Seto, J.T. and Rott, R. (1966). Functional significance of sialidase during influenza virus multiplication. *Virology* **30**, 731–737.

Stone, J.D. (1949). Tryptic inactivation of the receptor-destroying enzyme of *V. Cholerae* and of the enzymic activity of influenza virus. *Aust. J. Exp. Biol Med. Sci.* **27**, 229–244.

Taylor, G., Garman, E., Webster, R., Saito, T., and Laver, G. (1993). Crystallisation and preliminary X-ray studies of influenza A virus neuraminidase of subtypes N5, N6, N8 and N9. *J. Mol. Biol.* **230**, 345–348 and unpublished results.

Tucker, S.P. FLUNET. (2002). A new approach for influenza management. 15th International conference on anti-viral research, Prague, March 20, 2002.

Tulip, W.R., Varghese, J.N., Baker, A.T., Van Donkelaar, A., Laver, W.G., Webster R.G. et al. (1991). Refined atomic structures of N9 subtype influenza virus neuraminidase and escape mutants. *J. Mol. Biol.* **221**, 487–497.

Tulip, W.R., Varghese, J.N., Laver, W.G., Webster, R.G., and Colman, P.M. (1992). Refined crystal structure of the influenza virus N9 neuraminidase–NC41 Fab complex. *J. Mol. Biol.* **227**, 122–148.

Varghese, J.N., Laver, W.G., and Colman, P.M. (1983). Structure of the influenza virus glycoprotein antigen neuraminidase at 2.9 Å resolution. *Nature* **303**, 35–40.

Varghese, J.N., Colman, P.M., Van Donkelaar, A., Blick, T.J., Sahasrabudhe, A., and McKimm-Breschkin, J.L. (1997). Structural evidence for a second sialic acid binding site in avian influenza virus neuraminidases. *PNAS* **94**, 11808–11812.

Varghese, J.N., Smith, P.W., Sollis, S.L., Blick, T.J., Sahasrabudhe, A., McKimm-Breschkin J.L. *et al.* (1998). Drug resistance against a shifting target: A structural basis for resistance to inhibitors in a variant of influenza virus neuraminidase. *Structure* **6**, 735–746.

von Itzstein, M., Wu, W.-Y., Kok, G.B. Pegg, M.S. Dyason, J.C. Jin *et al.* (1993). Rational design of potent sialidase-based inhibitors of influenza virus replication. *Nature* **363**, 418–423.

Webster, R.G. and Laver, W.G. (1967). Preparation and properties of antibody directed specifically against the neuraminidase of influenza virus. *J. Immunol.* **99**, 49–55.

Webster, R.G., Reay, P.A., Laver, W.G., and Colman, P.M. (1988) Protection against lethal influenza with neuramindase. *Virology* **164**, 230–237.

WHO. (2003). WHO guidelines on the use of vaccines and antivirals during influenza pandemics. www.who.int/emc/diseases/flu/annexe4.htm

Wrigley, N.G., Charlwood, P.A. Skehel, J.J., and Brand, C.M. (1973). The size and shape of influenza virus neuraminidase. *Virology* **51**, 525–529.

Structure, Function, and Inhibition of Influenza Virus Neuraminidase 267

18

Interaction of HIV-1 Nef with Human CD4 and Lck

Dieter Willbold

1. Introduction

The genome of the human immunodeficiency virus (HIV) codes not only for structural proteins of the virus particle, but also for several regulatory proteins (Figure 18.1). Two of them (Tat and Rev) are proteins that are absolutely essential for replication of HIV in cell culture. Others (Nef, negative factor; Vif, virus infectivity factor; Vpu, virus protein U; Vpr, virus protein R) are not essential for viral replication *in vitro* and are, thus, named "accessory proteins." They play, however, a decisive role for infectivity and pathogenesis of HIV. Not a single therapeutically applied agent is directed against one of these regulatory proteins though.

Besides serving as a basis for rational drug design, structure determination of these proteins will yield new insights into their molecular mechanisms and, thus, may open new approaches to antiviral therapies. Nuclear magnetic resonance (NMR) spectroscopy is well suited to study three-dimensional structures of these proteins, as can be seen in several other chapters of this book for example HIV Vpu, and in a variety of publications for Vpr (Roques *et al.*, 1997; Schüler *et al.*, 1999; Wecker and Roques, 1999; Engler *et al.*, 2001, 2002; Morellet *et al.*, 2003), Tat (Sticht *et al.*, 1993, 1994; Willbold *et al.*, 1993, 1994, 1996; Mujeeb *et al.*, 1994; Bayer *et al.*, 1995; Metzger *et al.*, 1996, 1997; Rösch *et al.*, 1996), and Nef (Grzesiek *et al.*, 1996a, 1997; Geyer *et al.*, 1999). NMR spectroscopy shows its advantages over other structural techniques especially for investigations on interactions between proteins, for example, cellular and viral proteins.

Of the above mentioned HIV regulatory proteins, at least two are membrane-associated. Vpu is an integral type-1 membrane protein. Nef can be anchored in the membrane via a myristoyl residue at its amino-terminal end. The amino-terminal part of Nef is cleavable by the HIV protease, thereby detaching the so called "core" domain of Nef from the membrane (Freund *et al.*, 1994a,b). Several models of Nef function dependent on its membrane association are being discussed (Arold and Baur, 2001).

Dieter Willbold • Institut für Physikalische Biologie, Heinrich-Heine-Universität, Düsseldorf, Germany und Forschungszentrum Jülich, IBI-2, 52425 Jülich, Germany.

Viral Membrane Proteins: Structure, Function, and Drug Design, edited by Wolfgang Fischer.
Kluwer Academic / Plenum Publishers, New York, 2005.

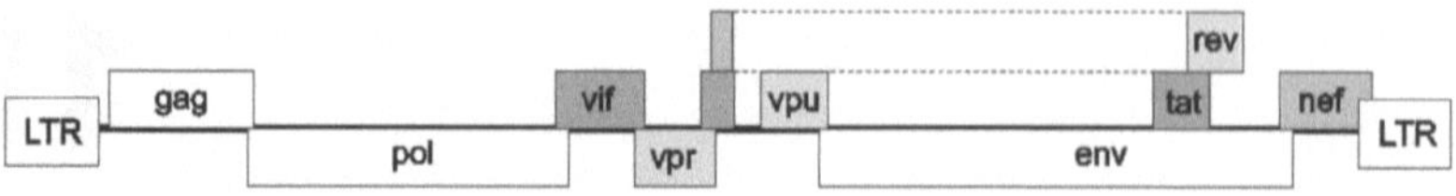

Figure 18.1. Scheme of the HIV-1 genome. Each rectangle represents a gene within one of the three possible reading frames. Regulatory genes are highlighted in gray.

The Nef protein of HIV type 1 (HIV-1) is important for the pathogenesis of HIV infection. This has been shown and confirmed in a number of reports, which included animal models and studies of humans infected with Nef-deleted HIV strains.

HIV-1 Nef is a protein containing roughly 200 amino acid residues. It is a membrane-associated protein that is produced at the earliest stage of viral gene expression (Cullen, 1994) and is a component of viral particles (Welker *et al.*, 1996). Nef has been reported to have diverse effects on cellular signal transduction pathways. It interacts with various cellular protein kinases and acts both as a kinase substrate and as a modulator of kinase activity (Greenway *et al.*, 1996; Baur *et al.*, 1997; Harris, 1999). In addition, Nef has been demonstrated to downregulate cell-surface receptors, cluster determinant 4 (CD4) and MHC I (Garcia and Miller, 1991; Anderson *et al.*, 1993; Benson *et al.*, 1993; Harris and Coates, 1993; Mariani and Skowronski, 1993; Marsh, 1999; Renkema and Saksela, 2000). Nef-mediated downmodulation of CD4 is well-understood now and appears to involve a whole set of factors. At least two distinct motifs in a long loop region of the protein were found to bind adaptins (AP 1/2/3) (Greenberg *et al.*, 1997; Bresnahan *et al.*, 1998; Craig *et al.*, 1998, 2000; Lock *et al.*, 1999). One of these motifs was additionally reported to interact with the regulatory unit of a vacuolar proton pump also involved in CD4 downregulation (Lu *et al.*, 1998). The β-subunit of COPI coatomers (β-COP) was shown to bind Nef subsequently to adaptins and seems to direct CD4 to a degradation pathway (Benichou *et al.*, 1994; Piguet *et al.*, 1999).

Also, at the plasma membrane, Nef interacts with signaling proteins from the T cell receptor environment, including not only CD4, but also Zeta, Lck, Vav, Pkc, Pak, and PI-3 kinase (Arold and Baur, 2001). These findings have implicated that Nef is part of and acts through a TCR-associated signaling complex (Fackler and Baur, 2002). A confirmation of this view came, particularly through two studies. Development of an AIDS-like disease in a HIV transgenic mouse model correlated with Nef-mediated activation of mouse T cells (Hanna *et al.*, 1998). A comparison of gene expression profiles of inducible T cell lines revealed that Nef- and anti-CD3 mediated T cell activation largely overlap (Simmons *et al.*, 2001). The molecular mechanism of how Nef activates T cells, however, is obscure.

The present chapter tries to summarize structural data of what is known about the interaction between HIV-1 Nef and two of its cellular target proteins, CD4 and Lck.

2. Interaction of Nef with Human CD4

2.1. The CD4 Receptor

The CD4 is a type I transmembrane glycoprotein with a molecular weight of 58 kDa and consists of an extracellular region of 370 amino acids, a short transmembrane region, and a cytoplasmic domain of 40 amino acids at the C-terminal end. The CD4 T lymphocyte coreceptor belongs to the IgG-superfamily and participates in T cell activation and signal

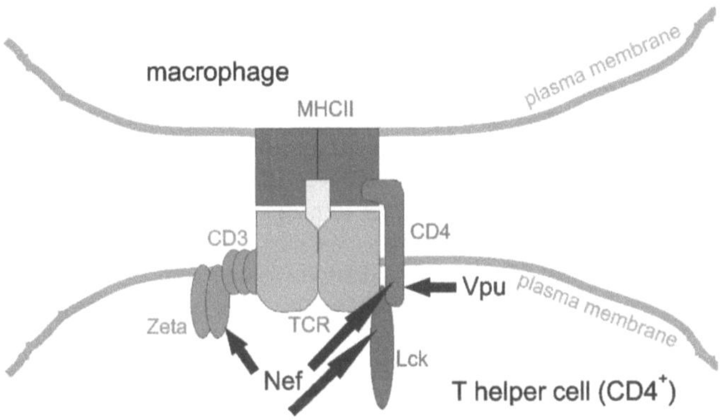

Figure 18.2. Sketch of MHC II antigen recognition by a TCR complex of a T helper cell. Sites of viral interference by Nef and Vpu are marked by arrows. Nef is known to directly bind Lck SH3 domain and CD4 cytoplasmic part. Zeta binding was reported for SIV and HIV-2 Nef. Vpu is known to directly bind CD4 cytoplasmic domain.

transduction. Surface CD4 is expressed on T lymphocytes that recognize antigens presented on class II major histocompatibility complex (MHC II) molecules (Maddon *et al.*, 1987) (Figure 18.2). This specificity of $CD4^+$ T cells for MHC II-expressing targets is probably based on direct interaction between CD4 and MHC II (Biddison *et al.*, 1984). CD4 associates with the T cell receptor (TCR) during T cell activation (Gallaher *et al.*, 1995). The mechanism by which CD4 participates in T cell activation is thought to involve transduction of intracellular signals. The interaction of the lymphocyte specific kinase (Lck), an Src-homologous tyrosine kinase, with the cytoplasmic part of CD4 is a crucial step of the T cell signaling pathway (Veillette *et al.*, 1988).

2.2. CD4 and HIV

In addition to these functions, CD4 serves as the major receptor for HIV infection (Dalgeish *et al.*, 1984; Klatzmann *et al.*, 1984a,b). The virus is internalized after binding of the viral gp120 to the extracellular domain of CD4. This leads to infection of the respective T helper cell, and the production and release of new virions.

Once HIV has successfully entered and, thereby, infected the cell, it is extremely advantageous for its replication to clean the cell surface of all remaining CD4 receptor molecules. CD4 molecules remaining on the cell surface would increase the risk of the cell to be repeatedly infected by HIV particles. Such superinfection and the risk of syncytia formation by fusion of several $CD4^+$ cells to a single virus particle usually leads to death of all cells involved. This, of course, results in the termination of all virus particles contained in these cells, too. Further, budding of newly synthesized HIV particles from an infected cell is facilitated if no CD4 receptor molecules are present on the cell surface.

In the case of HIV-1, at least two regulatory proteins have the function to downregulate CD4 molecules of infected cells. Nef protein is contained in the virus particles (Welker *et al.*, 1996) to induce internalization of surface CD4 receptors right after the infection event. Vpu is expressed at later stages of the infection cycle and one of its functions is to block supply of newly synthesized CD4 receptors from the endoplasmatic reticulum. A T helper cell that does not carry CD4 receptors on its surface is not able to fulfill its function. It is actually, by definition, no T helper cell ($CD4^+$) anymore.

CD4 interacts via its cytoplasmic domain with viral proteins, Nef and Vpu. Vpu induces degradation of CD4 molecules in the endoplasmatic reticulum. This process requires both proteins to be inserted into the same membrane compartment. The CD4 sequence relevant for this activity is located between amino acids 402–420 (Chen *et al.*, 1993).

In contrast, Nef acts at the cell surface to mediate the internalization and lysosomal degradation of CD4 (Aiken *et al.*, 1994; Anderson *et al.*, 1994; Sanfridson *et al.*, 1994). Nef dependent downregulation of CD4 is well-understood on a cellular level. It appears to involve a whole set of factors (Benichou *et al.*, 1994; Greenberg *et al.*, 1997; Bresnahan *et al.*, 1998; Craig *et al.*, 1998, 2000; Lu *et al.*, 1998; Lock *et al.*, 1999; Piguet *et al.*, 1999). From mutational analysis, it is known that residues 407–418 in the cytoplasmic tail of CD4 are necessary and sufficient for downregulation of CD4 by Nef (Garcia *et al.*, 1993; Aiken *et al.*, 1994; Anderson *et al.*, 1994; Salghetti *et al.*, 1995). Especially, the dileucine motif at sequence positions 413 and 414 is required for binding and downmodulation of CD4 by Nef.

2.3. Three-Dimensional Structures of CD4 Cytoplasmic Domain and HIV-1 Nef

Three-dimensional structures are known from CD4 cytoplasmic domain in trifluoroethanol-free aqueous solution (Willbold and Rösch, 1996) and in trifluoroethanol-containing solution (Wray *et al.*, 1998). Also, structures are known for Nef proteins with N-terminal and partially additional deletions (Grzesiek *et al.*, 1996, 1997; Lee *et al.*, 1996; Arold *et al.*, 1997) (Figure 18.3). This so-called Nef "core domain" consists of a type II poly-proline helix, three alpha-helices, a 3(10) helix, and a five-stranded antiparallel beta-sheet.

In trifluoroethanol-free aqueous solution, CD4(403–419) exhibits an α-helical secondary structure for residues 403–412, followed by a more extended conformation (Willbold and Rösch, 1996). This helix ends at position 412. Leucines 413 and 414 are known to be important as a "dileucine" motif necessary for internalization of transmembrane proteins, such as CD4, IgG Fc receptor, and CD3 γ and δ chains. The most remarkable point about the structure of CD4(403–419) is the fact that there is a defined secondary structure in this small peptide at all.

2.4. Nef Residues that are Important for CD4 Binding Map to the "Core Domain"

NMR investigations on the interaction between CD4 and Nef using a 13-residue peptide of CD4 (residues 407–419) and several Nef mutants ($Nef^{\Delta 2-39}$, $Nef^{\Delta 2-39,\Delta 159-173}$) elucidated residues W57, L58, E59, G95, G96, L97, R106, and L110 to be affected by CD4(407–419) binding (Grzesiek *et al.*, 1996). The dissociation constant (K_D) of this complex, however, was found to be only in the range of 0.5–1 mM.

Although N-terminal amino acid sequences among Nef proteins are not conserved, some residues therein are known to be essential for downregulation of CD4 expression (Aiken *et al.*, 1996; Hua *et al.*, 1997; Iafrate *et al.*, 1997). Moreover, a study employing the yeast-two-hybrid-system suggests that residues important for CD4 binding are scattered all over the Nef sequence (Rossi *et al.*, 1996).

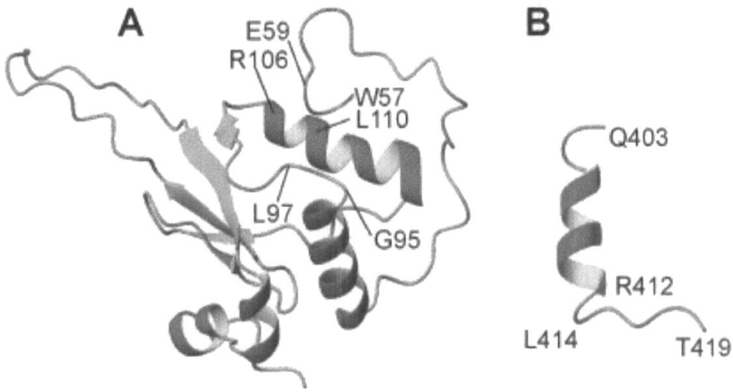

Figure 18.3. (A) Three-dimensional structures of Nef core domain and (B) CD4(403–419) contains in trifluoroethanol-free aqueous solution an α-helical secondary structure for residues 403–412, followed by a more extended conformation (Willbold and Rösch, 1996). The helix in CD4(403–419) ends at position 412. Leucines 413 and 414 are known to be important as a "dileucine" motif necessary for internalization of transmembrane proteins such as CD4, IgG Fc receptor, and CD3 γ and δ chains. Nef "core domain" consists of a type II poly-proline helix, three alpha-helices, a 3(10) helix, and a five-stranded antiparallel beta-sheet (Grzesiek *et al.*, 1997). The figures were prepared from the PDB entries 1WBR and 2NEF using MOLMOL (Koradi *et al.*, 1996). Residues 407–419 of CD4 were shown to be important for Nef binding by mutation experiments. Further, the presence of helical secondary structure in the amino-terminal part of CD4(407–419) seems to increase affinity to Nef. On the other side of the complex, residues W57, L58, E59, G95, G96, L97, R106, and L110 of Nef and some not yet identified parts of the amino-terminal part of Nef are important for binding to CD4 (Grzesiek *et al.*, 1996).

2.5. Amino-Terminal Residues of Nef are also Important for CD4 Binding

Because Nef variants employed in previous CD4 *in vitro* binding studies were lacking substantial amino-terminal parts, and the hereby deduced dissociation constant was unexpectedly high, additional investigations were carried out on direct *in vitro* binding between essentially complete binding partners. A chemically synthesized peptide comprising the 31 C-terminal residues (403–433) of human CD4 and a recombinantly expressed full-length Nef protein from HIV-1 strain SF2 yielded a significantly higher affinity of Nef to CD4 (Preusser *et al.*, 2001). Fluorescence titrations were used to determine the K_D to be 0.87 μM (Figure 18.4).

The observed K_D value for binding of full-length Nef to CD4(403–433) is about 1,000-fold lower than that observed for Nef mutants, $Nef^{\Delta 2\text{–}39}$ and $Nef^{\Delta 2\text{–}39,\Delta 159\text{–}173}$, and CD4(407–419) (Grzesiek *et al.*, 1996). Differences in those studies are the length of the CD4 peptide and the completeness of Nef protein. The possibility that the C-terminal tail of the CD4 cytoplasmic domain is involved in Nef binding can be neglected from mutation experiments (Garcia *et al.*, 1993; Aiken *et al.*, 1994; Anderson *et al.*, 1994; Salghetti *et al.*, 1995; Rossi *et al.*, 1996) and from control binding measurements with shorter peptides. Indeed, the dissociation constant of 1.4 μM obtained for full-length Nef and CD4(403–419) suggests only a minor role of residues 420–433 in CD4 for Nef binding (Figure 18.4).

The remaining difference between the CD4 peptides used in both *in vitro* binding studies were residues 403–406 that were either missing or not. The three-dimensional structure of CD4(403–419) exhibits an α-helix for residues 403–412 (Willbold and Rösch, 1996)

(Figure 18.3B). Thus, residues 403–406 form the N-terminal cap of an α-helix. This helix N-cap could not form in the CD4 peptide (407–419) used by Grzesiek and coworkers. Gratton and coworkers (Gratton *et al.*, 1996) concluded from their mutational studies, a correlation between the presence of this α-helix in CD4 and the susceptibility of CD4 to downregulation by Nef. All these data suggest that the existence of a preformed α-helix in CD4 supports binding to Nef. To determine the contribution of the four residues that form the helix N-cap to Nef binding, the dissociation constant of CD4 peptide (407–419) and full-length Nef was measured, again by fluorescence titrations (Preusser *et al.*, 2001). The obtained value of 3.3 μM (Figure 18.4) indeed indicates that the presence of residues 403–406 forming the helix N-cap increases CD4 affinity to Nef by a factor of, roughly, two.

The CD4(407–419) peptide yielding a K_D of 3.3 μM for Nef binding has exactly the same sequence as that used in earlier studies reporting a K_D of 1 mM (Grzesiek *et al.*, 1996). However, the amino acid sequence of the Nef protein used in the present study was completely in contrast to that used in earlier studies lacking 38 N-terminal residues (Grzesiek *et al.*, 1996). This strongly suggests that an intact N-terminal region of Nef is important for high affinity binding to CD4.

2.6. Leucines 413 and 414 of CD4 are Essential for Nef Binding

The role of leucines 413 and 414 in CD4 for Nef binding were assayed in a binding study with Nef and a CD4 peptide (403–419) having leucines 413 and 414 exchanged with alanines (Figure 18.4). This mutation is reported to render CD4 refractory to Nef-induced downregulation (Aiken *et al.*, 1994). No dissociation constant could be determined from the data points measured within the Nef concentration range between zero and more than 13 μM, suggesting that mutation of leucines 413 and 414 to alanines drastically reduces affinity of CD4 to Nef (Preusser *et al.*, 2001). This observation is in perfect accordance with published mutational data (Garcia *et al.*, 1993; Aiken *et al.*, 1994; Anderson *et al.*, 1994; Salghetti *et al.*, 1995).

2.7. High Affinity Between CD4(403–433) and Full-Length Nef can be Confirmed by NMR Spectroscopy

In order to confirm the observed high affinity binding of Nef and CD4 by an additional method, NMR spectroscopy was employed. Observation of chemical shift changes in a protein upon ligand binding is a sensitive method for measuring the strength of an interaction and for defining the protein's interaction surface (Otting *et al.*, 1990; Görlach *et al.*, 1992). Especially useful and sensitive are, for example, heteronuclear single quantum correlation (HSQC) spectra. To carry out such experiments, uniformly ^{15}N-isotope labeled protein is required. This can easily be obtained by expression of the protein in bacteria that grow in medium containing ^{15}N-ammonium chloride as sole nitrogen source.

A ^{1}H-^{15}N-HSQC experiment correlates the chemical shift of a ^{15}N-nitrogen nucleus of an NH_x group with the chemical shift of a directly attached proton. Each resonance signal in the HSQC spectrum, thus, represents a proton that is directly bound to a ^{15}N-nitrogen atom. The spectrum contains, therefore, the signals of the H^N protons and ^{15}N-nitrogens in the protein backbone (Figure 18.5A for HIV-1 Nef as an example). Since there is exactly one backbone H^N per amino acid (except for prolines), each HSQC signal represents one single amino acid. To be more exact, the HSQC can also contain signals from several side chains,

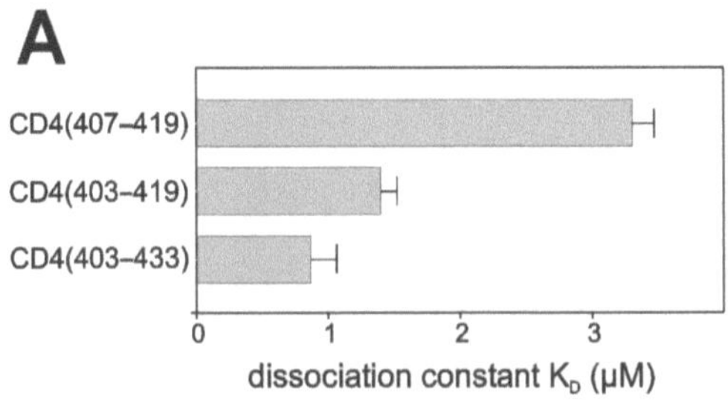

B

```
                 403 407            419          433
                  |   |              |            |
CD4(403-433)     QAERMSQIKRLLSEKKTCQCPHRFQKTCSPI
CD4(403-419)     QAERMSQIKRLLSEKKT
CD4(407-419)         MSQIKRLLSEKKT
CD4(LL-AA)       QAERMSQIKRAASEKKT
```

Figure 18.4. (A) Binding of full-length HIV-1 Nef to CD4 peptides of different lengths. Shown is the dissociation constant (K_D) of each peptide with HIV-1 Nef (Preusser *et al.*, 2001). Note that a higher K_D value means lower binding affinity. Fluorescence titrations were used to determine K_D of full-length Nef and several fluoresceine labeled CD4 peptides. Fluorescence was measured using excitation and emission wavelengths of 495 and 520 nm, respectively, with increasing amounts of Nef. As a control, the same titrations were performed with buffer devoid of Nef. Assuming a simple bimolecular reaction between Nef and CD4, analysis by nonlinear curve fitting yielded a K_D value of 0.87 ± 0.19 μM. Fluorescein isothiocyanate (FITC-I, SIGMA) as a control did not bind to Nef. An independent evaluation employing a Scatchard-plot analysis with linear regression analysis confirmed the K_D value to be 0.84 μM. (B) Overview of N-terminal fluoresceinylated CD4 peptides used for Nef binding studies. Amino acid sequences for the CD4 peptides named on the left are given using the one-letter-code. In addition, the residue numbers corresponding to the respective sequence positions in CD4 are shown in the top line.

for example, the amide groups of Asn and Gln, the amino group of Lys, the guanidinium group of Arg, and the aromatic H^N protons of Trp and His.

Because in a ^1H-^{15}N-HSQC two-dimensional spectrum, roughly each amino acid residue of the protein under investigation appears with one resonance, it is particularly useful to map ligand interaction sites on protein surfaces (Görlach *et al.*, 1992). Because the chemical shifts of the nuclei whose resonances appear in the HSQC are sensitive to their chemical environment, any binding of a ligand molecule in their vicinity induces changes in chemical shifts ("chemical shift perturbation") of the HSQC cross-resonances of the respective amide protons and nitrogens. Therefore, one can conclude that those amino acid residues that show changes in the chemical shifts of their resonances upon addition of a ligand are somehow affected by the ligand binding. One may further conclude that these residues build up the ligand binding site, although this is not exactly the same statement. Other experiments to map ligand binding sites are, for example, cross-saturation experiments (Takahashi *et al.*, 2000; Lane *et al.*, 2001).

Chemical shift perturbation experiments are based on the existence of two different conformations of the protein under investigation, namely the ligand-bound and the ligand-free conformations. A resonance of an amide affected by ligand binding, therefore, is composed of resonances coming from both conformations. The appearance of the resulting signal depends on the time scale of the interconversion between both conformations. The time scale

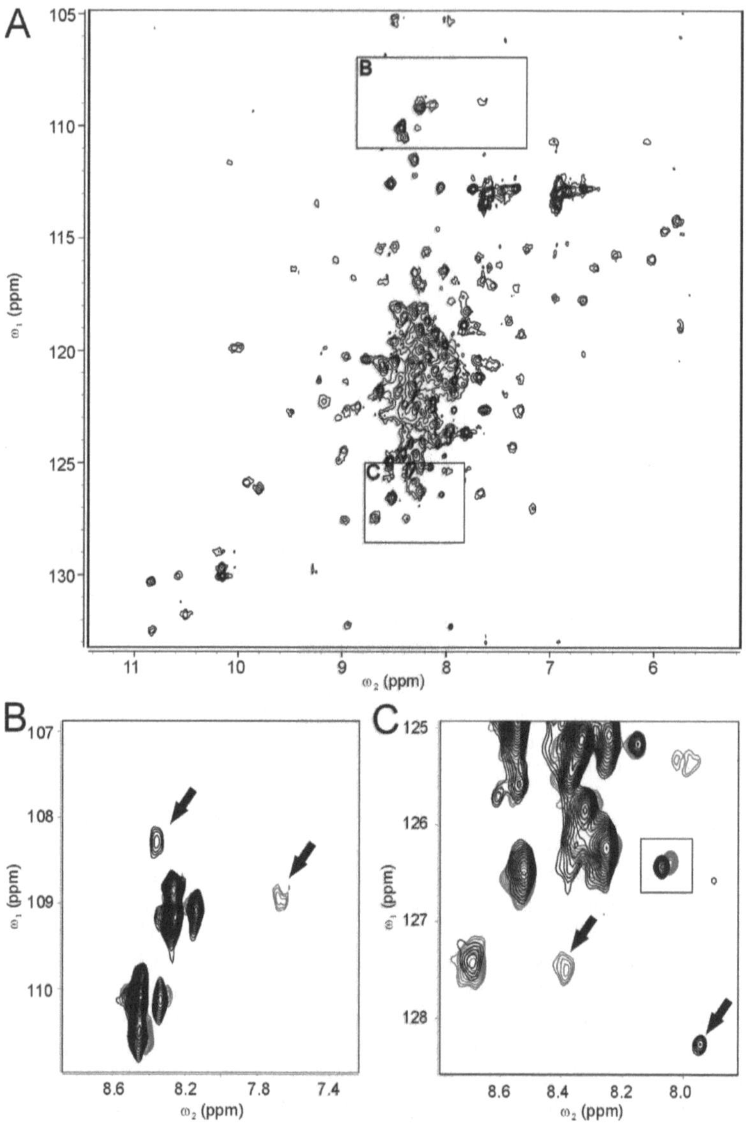

Figure 18.5. (A) Overview of a ^{1}H-^{15}N-HSQC spectrum of ^{15}N-isotope labeled HIV-1 full-length Nef protein. (B and C) Two selected regions of superimposed ^{1}H-^{15}N-HSQC spectra of HIV-1 Nef in absence (gray contour lines) and presence (black contour lines) of equimolar concentration of CD4(403–433) peptide. Note that for reasons of clarity contour levels are not identical in A, B, and C. During titration the gray colored peaks indicated by arrows did not shift, but their intensities decreased with ongoing titration (data not shown) and new peaks (black contour lines, indicated by arrows) appeared. The resonance highlighted by the rectangle (in C) shifted during titration. NMR-samples contained 180 μM uniformly ^{15}N-labeled full-length Nef protein in PBS buffer with 10% D_2O. All NMR spectra were recorded at 298 K on a Varian Unity INOVA spectrometer working at 750 MHz proton frequency.

is set up by the chemical shift (resonance frequency) difference of the resonances in the ligand bound and free state. This difference is given in hertz. If the rate of the interconversion between both conformations is slow compared to the chemical shift difference time scale, the resonance signal intensity of the ligand-free state will decrease during ongoing titration and the new resonance signal of the ligand-bound state will appear and increase. In case of such a "slow" exchange, the resonance will not shift during titration from the ligand-free resonance to the ligand-bound resonance. If the rate of the interconversion between both conformations is fast compared to the chemical shift difference time scale, then the respective resonance will shift during ongoing titration. If both time scales are similar, then an intermediate effect is observed leading to an increase of line width of the respective resonance that can ultimately lead to the disappearance of the resonance signal.

In order to confirm the observed high affinity binding of Nef and CD4 by an additional method, ^{1}H-^{15}N-HSQC spectra of ^{15}N-labeled full-length Nef protein with increasing amounts of CD4(403–433) peptide were recorded (Preusser *et al.*, 2001). For K_D of about 1 μM or even below, dissociation rates of less than 100 Hz can be expected even in the case of diffusion controlled association rate (10^7–10^8 HzM^{-1}). Thus, exchange between free and CD4-bound Nef should be slow on the NMR chemical shift time scale for at least some of the ^{1}H-^{15}N amide resonances. Indeed, intensities of some of the amide resonances in the ^{1}H-^{15}N-HSQC spectra decreased without shifting, while new resonances appeared with increasing intensities during ongoing titration with CD4 peptide (Figure 18.5). Assuming that the resonance pairs shown in Figure 18. 5 belong to the same amide cross-resonances of Nef in presence and absence of CD4 peptide, their proton chemical shift distances of 510 and 330 Hz, respectively, indicate that exchange between bound and unbound Nef is significantly slower than 300 Hz. A number of other resonances (one can be seen in Figure 18.5C) shifted during titration with CD4 peptide up to 30 Hz suggesting the dissociation of the complex to be fast compared to this time scale. Both observations confirm that the dissociation rate of the complex is about 100 Hz and, given the association rate to be diffusion controlled ($<10^8$ HzM^{-1}), the resulting dissociation constant is 1 μM or less, which is in perfect agreement with the results from the fluorescence titration. Most of the amide resonances in the ^{1}H-^{15}N-HSQC spectra did not show significant changes indicating that the overall three-dimensional structure of Nef does not dramatically change upon CD4 binding. Because resonances of the Nef variant (SF2) used in this study were not sequence specifically assigned, it was not possible at the present stage to directly identify Nef residues involved in CD4 binding.

2.8. The Presence of a Helix in Human CD4 Cytoplasmic Domain Promotes Binding to HIV-1 Nef Protein

The presence of residues 403–406 forming a helix N-cap in CD4(403–419) increases its affinity to Nef by a factor of, roughly, two (Figure 18.4). Helix N-cap structures are important for helix formation and stability. To investigate whether presence or absence of any helix N-cap forming residues N-terminal to the CD4(407–419) peptide is responsible for the observed difference in binding studies between Nef and CD4, binding experiments were carried out with peptides that do or do not have helix N-caps. Those experiments were able to show that binding between HIV Nef and CD4 correlates with the helix content of CD4 (Preusser *et al.*, 2002).

2.9. Summary of the CD4–Nef Interaction

Residues 407–419 of CD4 were shown to be important for Nef binding by mutation experiments. This was confirmed by direct *in vitro* binding experiments. Furthermore, the presence of helical secondary structure in the amino-terminal part of CD4(407–419) seems to increase affinity to Nef. On the other side of the complex, residues W57, L58, E59, G95, G96, L97, R106, and L110 of Nef and some not yet identified parts of the amino-terminal part of Nef are important for binding to CD4.

It is important (especially for the scope of this book) to mention that, so far, all studies were carried out with CD4 peptide and Nef protein not anchored on the same side of a membrane as it would be the case *in vivo*. Thus, in a living cell, binding affinity between Nef and CD4 can be expected to be even higher than observed in the above described *in vitro* studies due to a much more favorable entropic term of the binding energy.

3. Interaction of Nef with Human Lck

Nef has been reported to have diverse effects on cellular signal transduction pathways (Wolf *et al.*, 2001). It interacts with various cellular protein kinases and acts both as a kinase substrate and as a modulator of kinase activity (Greenway *et al.*, 1996; Baur *et al.*, 1997; Harris, 1999). A highly conserved proline repeat motif PxxP was found in all Nef variants (Saksela, 1997), which is known as the minimal consensus sequence defining the ligands of SH3 domains (Cicchetti *et al.*, 1992; Ren *et al.*, 1993). Binding of Nef to various members of the Src family of protein kinases, namely Hck and Lyn (Lee *et al.*, 1995; Saksela *et al.*, 1995), Lck (Collette *et al.*, 1996; Greenway *et al.*, 1996), Fyn (Greenway *et al.*, 1996), and Src (Arold *et al.*, 1998) has been demonstrated.

3.1. Lymphocyte Specific Kinase Lck

Protein tyrosine kinases are involved in signal transduction pathways that regulate cell growth, differentiation, activation, and transformation. Human lymphocyte specific kinase (Lck) is a 56-kDa protein involved in T cell- and IL2-receptor signaling. Lck is a typical member of the Src-type tyrosine kinase family and consists of four functional domains, namely unique, SH3, SH2, and kinase. Whereas amino acid sequences of the other domains are highly conserved among different kinases, those of the unique domains are not. Lck unique domain is thought to serve as a membrane anchor, but also plays a role in function and specificity of the other domains, for example, SH2 and SH3 (Carrera *et al.*, 1995). Because of its key role in T cell signaling and activation (Isakov and Biesinger, 2000), it is not surprising that pathogenic factors like HIV and *Herpesvirus saimiri* have evolved effector molecules that target Lck to ensure their own replication and persistence. In particular, HIV-1 Nef and *H. saimiri* Tip proteins directly bind to Lck SH3 domain. Tip binding to Lck SH3 domain was shown to be based on additional contacts to those typically observed for poly-proline peptides bound to SH3 domains (Schweimer *et al.*, 2002).

While Nef binding to most other SH3 domains assayed so far was shown to be based on K_D in the micromolar range (Arold *et al.*, 1998), Nef affinity to Hck SH3 was remarkably higher (Lee *et al.*, 1995; Arold *et al.*, 1998).

3.2. X-ray Structures of Nef–SH3 Complexes

Several X-ray crystallographic studies have revealed the complex structures of so-called core-Nef ($Nef^{\Delta 2-39,\Delta 159-173}$) with Fyn and Hck SH3 domains (Lee *et al.*, 1995; Arold *et al.*, 1997, 1998). Residues 71–77 (the poly-proline helix), Lys82, Ala83, Asp86, Leu87, Phe90, Trp113, Thr117, Qln118, and Tyr120 of Nef core domain were identified to be important for binding to Fyn SH3 domain. Because at least for HIV replication in T helper cells, Nef interaction with Lck certainly is more relevant than with Hck, and because Lck is associated with CD4, which in turn binds Nef, too, investigations to map the Lck–Nef-interaction site were of particular interest.

3.3. NMR Spectroscopy is a Suitable Tool to Map Nef–Lck Interaction Sites

A study to map the full-length Nef interaction site on Lck was recently carried out by NMR spectroscopy (Briese *et al.*, 2004). Because significant differences between core-Nef and full-length Nef proteins were already demonstrated for binding to CD4 cytoplasmic domain (Preusser *et al.*, 2001), full-length Nef was used for this study, too. Further, to investigate any contribution of the Lck unique domain to Lck–Nef interaction, a recombinant protein consisting of the 120 amino-terminal residues of Lck comprising the unique and SH3 domains of Lck (LckU3) was used. Although the unique part of LckU3 was shown to be absent of any stable structural elements (Briese and Willbold, 2003), it seemed possible to be involved in Nef binding, because Lck uniquely is responsible for Lck binding to human CD4 receptor cytoplasmic domain (Veillette *et al.*, 1988; Shaw *et al.*, 1990; Turner *et al.*, 1990; Huse *et al.*, 1998; Lin *et al.*, 1998), which in turn binds directly to Nef (Grzesiek *et al.*, 1996; Preusser *et al.*, 2001, 2002).

To elucidate the details of Nef binding to Lck(1–120), chemical shift perturbation mapping by NMR spectroscopy was employed, as already described in the section "High affinity between CD4(403–433) and full-length Nef can be confirmed by NMR spectroscopy." Thus, ^{1}H-^{15}N-HSQC spectra of ^{15}N-labeled Lck(1–120) with increasing amounts of full-length Nef protein were recorded (Figure 18.6). In contrast to the above-described NMR titration study with Nef and CD4 cytoplasmic domain, all resonances of LckU3 are known (Briese *et al.*, 2001). Therefore, it was possible to identify those LckU3 residues involved in Nef binding (Briese *et al.*, 2004).

From a reported dissociation constant, K_D, of about 10 μM (Arold *et al.*, 1998), dissociation rates of several hundred up to 1,000 Hz can be expected in the case of a diffusion controlled association rate. Thus, exchange between free and Nef-bound Lck were expected to be fast or intermediate on the NMR chemical shift time scale for ^{1}H-^{15}N amide cross-resonances of the affected residues. Indeed, a small number of resonances shifted during titration with Nef while some other resonances disappeared with increasing Nef concentration. Both observations confirm that the dissociation rate of the complex is within the expected region of several hundred hertz and, given the association rate to be diffusion controlled ($<10^{8}$ HzM^{-1}), the resulting dissociation constant is 10 μM or less, which is in perfect agreement with previously reported results (Arold *et al.*, 1998).

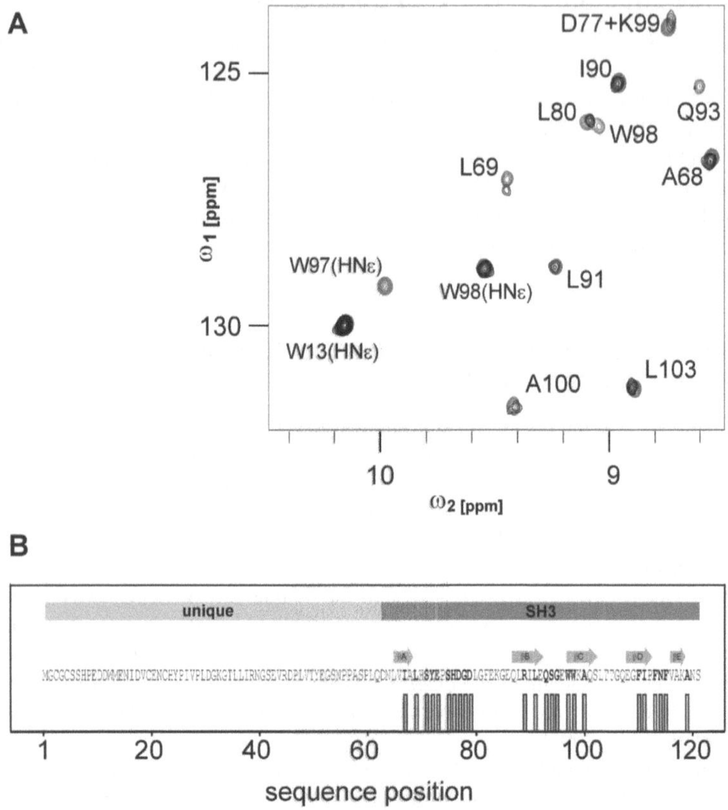

Figure 18.6. Mapping of LckU3 residues involved in HIV-1 Nef binding. Selected region (A) of superimposed ^{1}H-^{15}N-HSQC spectra of 15N-isotope labeled LckU3 in absence (gray contour lines) and presence (black contour lines) of full-length HIV-1 Nef protein (Briese *et al.*, 2003). During titration, some peaks did neither shift nor disappear (e.g., side-chain ε-imino groups of Trp13 and Trp98, backbone amide groups of Ala68, Ile90, and Leu103), some peaks disappeared completely or to a large extent (e.g. side chain ε-imino group of Trp97, backbone amide groups of Leu69, Asp77, Leu80, Leu91, Trp98, Lys99, and Ala100), and some peaks shifted (e.g., backbone amide groups of Leu69). NMR-samples contained 192 μM uniformly ^{15}N-labeled LckU3 in PBS buffer with 10% D_2O. At the titration end point, the samples contained a 1.5-fold molar excess of Nef protein. All NMR spectra were recorded at 298 K on a Varian Unity INOVA spectrometer working at 750 MHz proton frequency. Amino acid sequence and domain representation of LckU3 (B) with the unique and SH3 domains indicated as horizontal bars. Sequence position is given by numbers at the bottom. Locations of β-sheet secondary structure elements are indicated by horizontal arrows. Residues of LckU3 that disappeared or shifted upon addition of full-length Nef proteins are indicated with vertical bars at the appropriate sequence position and the respective letters in the amino acid sequence are highlighted in bold italic letters.

3.4. The Unique Domain of Lck is Not Involved in Nef Binding

Most of the amide resonances in the HSQC spectra did not show significant changes indicating that the overall three-dimensional structure of Lck(1–120) does not dramatically change upon Nef binding. More importantly, no residue of the unique domain is affected by

Nef binding (Briese *et al.*, 2003). This is interesting, because the unique domain is essential for Lck–CD4 interaction, and the question whether dissociation of Lck and CD4 is induced upon Nef binding is controversially discussed (Salghetti *et al.*, 1995; Gratton *et al.*, 1996). The NMR shift perturbation study clearly shows that residues of the unique domain are not required for or involved in Nef binding. Therefore, disregarding potential downstream events, binding of Lck to CD4 and Nef does not seem to be competitive. Previously, based on mutational analysis, non-overlapping binding sites were identified for Lck and Nef on CD4 (Gratton *et al.*, 1996).

3.5. Mapping of the Nef Interaction Site on Lck SH3

LckU3 residues that are affected upon Nef binding (Figure 18.7) are essentially those reported in similar investigations of SH3 domain interactions with poly-proline peptides (Larson and Davidson, 2000), and are in general accordance (Tyr72, Ser75 to Asp79, Glu96, Trp97, Asn114, and Phe115) with the analysis of the crystallographic complex structure of Fyn SH3 domain with Nef deletion variant $Nef^{\Delta2-39,\Delta159-173}$ (Arold *et al.*, 1997, 1998). Fyn and Lck SH3 domains show a sequence identity of 50% (Arold *et al.*, 1997). Differences between Fyn-SH3–$Nef^{\Delta2-39,\Delta159-173}$ (Arold *et al.*, 1997) and Lck-SH3–Nef (Briese *et al.*, 2003) are observed in that residues Ile67, Leu69, Glu73, Leu91, Gln93 to Gly95, Trp98 to Ala100, Phe110, Ile111, Phe113, and Ala117 seem to be affected upon Lck SH3 domain complex formation with full-length Nef but the respective homolog residues in Fyn SH3 were not identified to be important for interaction with $Nef^{\Delta2-39,\Delta159-173}$. Only His70 (homolog to Tyr91 of Fyn SH3), which was identified to play a role for Fyn-SH3–$Nef^{\Delta2-39,\Delta159-173}$ interaction, was not found to be involved in binding of full-length Nef to Lck-SH3 in the present study. Pro112 (homolog to Pro134 of Fyn SH3), which also was described to be involved in Fyn-SH3–$Nef^{\Delta2-39,\Delta159-173}$ interaction, was not detectable in the HSQC titration experiment we used to map the interaction surface, due to the lack of an observable amide proton in proline residues. These differences between both studies may be due to the use of full-length Nef in the present study compared to a Nef deletion variant ($Nef^{\Delta2-39,\Delta159-173}$) used in the X-ray analysis (Arold *et al.*, 1997). It may also reflect a higher sensitivity of the NMR

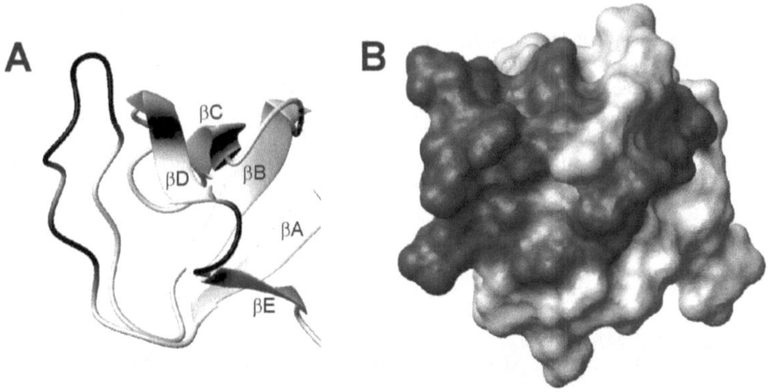

Figure 18.7. (A) Visualization of the Nef binding site on Lck SH3 domain as ribbon diagram and (B) surface view. Dark regions represent those residues that were affected by binding of full-length Nef to LckU3 (Figure 18.6B). The figures were prepared using MOLMOL (Koradi *et al.*, 1996).

titration approach because it also detects indirectly induced changes in Lck SH3 domain conformation upon Nef binding. In any case, however, both studies agree that the interaction between Nef and Src-type tyrosine kinase SH3 domains is basically of a canonical SH3–poly-proline type.

3.6. Summary of the Lck–Nef Interaction

Residues 71–77 (the poly-proline helix), Lys82, Ala83, Asp86, Leu87, Phe90, Trp113, Thr117, Qln118, and Tyr120 of Nef core domain were identified to be important for binding to Fyn SH3 domain (Arold *et al.*, 1997). Lck residues that are involved in Nef binding could be mapped to those typically found in poly-proline peptide binding. The unique domain of Lck was not affected upon Nef binding (Briese *et al.*, 2004). Whether this is different for myristylated Nef and myristylated and palmitylated Lck proteins anchored on the same side of a membrane, remains to be investigated. Thus, in a living cell binding affinity between Nef and Lck can be expected to be even larger than observed in our *in vitro* system due to a much more favorable entropic term of the binding energy. Also, Lck may exert a different binding behavior to Nef when CD4 is attached to its unique domain via a zinc ion.

Additional efforts will certainly yield better insight into interference of host signal transduction proteins by viral proteins. Exploring new drug target proteins in HIV infection aside from reverse transcriptase and protease is becoming increasingly important. Nef could be a very interesting target.

References

Aiken, C., Konner, J., Landau, N.R., Lenburg, M.E., and Trono, D. (1994). Nef induces CD4 endocytosis: requirement for a critical dileucine motif in the membrane-proximal CD4 cytoplasmic domain. *Cell* **76**: 853–864.

Aiken, C., Krause, L., Chen, Y.L., and Trono, D. (1996). Mutational analysis of HIV-1 Nef: identification of two mutants that are temperature-sensitive for CD4 downregulation. *Virology* **217**, 293–300.

Anderson, S., Shugars, D.C., Swanstrom, R., and Garcia, J.V., (1993). Nef from primary isolates of human immunodeficiency virus type 1 suppresses surface CD4 expression in human and mouse T cells. *J. Virol.* **67**, 4923–4931.

Anderson, S.J., Lenburg, M., Landau, N.R., and Garcia, J.V. (1994). The cytoplasmic domain of CD4 is sufficient for its down-regulation from the cell surface by human immunodeficiency virus type 1 Nef [published erratum appears in *J. Virol.* **68**(7): 4705. *J. Virol.* **68**, 3092–3101.

Arold, S., Franken, P., Strub, M.P., Hoh, F., Benichou, S., Benarous, R. *et al.* (1997). The crystal structure of HIV-1 Nef protein bound to the Fyn kinase SH3 domain suggests a role for this complex in altered T cell receptor signaling. *Structure* **5**, 1361–1372.

Arold, S., O'Brien, R., Franken, P., Strub, M.P., Hoh, F., Dumas, C. *et al.* (1998). RT loop flexibility enhances the specificity of Src family SH3 domains for HIV-1 Nef. *Biochemistry* **37**, 14683–14691.

Arold, S.T. and Baur, A.S. (2001). Dynamic Nef and Nef dynamics: how structure could explain the complex activities of this small HIV protein. *Trends Biochem. Sci.* **26**, 356–363.

Baur, A.S., Sass, G., Laffert, B., Willbold, D., Cheng Mayer, C., and Peterlin, B.M. (1997). The N-terminus of Nef from HIV-1/SIV associates with a protein complex containing Lck and a serine kinase. *Immunity* **6**; 283–291.

Bayer, P., Kraft, M., Ejchart, A., Westendorp, M., Frank, R., and Rösch, P. (1995). Structural studies of HIV-1 Tat protein. *J. Mol. Biol.* **247**, 529–535.

Benichou, S., Bomsel, M., Bodeus, M., Durand, H., Doute, M., Letourneur, F. *et al.* (1994). Physical interaction of the HIV-1 Nef protein with beta-COP, a component of non-clathrin-coated vesicles essential for membrane traffic. *J. Biol. Chem.* **269**, 30073–30076.

Benson, R.E., Sanfridson, A., Ottinger, J.S., Doyle, C., and Cullen, B.R. (1993). Downregulation of cell-surface CD4 expression by simian immunodeficiency virus Nef prevents viral super infection. *J. Exp. Med.* **177**, 1561–1566.

Biddison, W.E., Rao, P.E., Talle, M.A., Goldstein, G., and Shaw, S. (1984). Possible involvement of the T4 molecule in T cell recognition of class II HLA antigens. Evidence from studies of CTL-target cell binding. Distinct epitopes on the T8 molecule are differentially involved in cytotoxic T cell function. *J. Exp. Med.* **159**, 783–797.

Bresnahan, P.A., Yonemoto, W., Ferrell, S., Williams-Herman, D., Geleziunas, R., and Greene, W.C. (1998). A dileucine motif in HIV-1 Nef acts as an internalization signal for CD4 downregulation and binds the AP-1 clathrin adaptor. *Curr. Biol.* **8**, 1235–1238.

Briese, L. and Willbold, D. (2003). Structure determination of human Lck unique and SH3 domains by nuclear magnetic resonance spectroscopy. *BMC Struct. Biol.* **3**, 3.

Briese, L., Hoffmann, S., Friedrich, U., Biesinger, B., and Willbold, D. (2001). Sequence-specific 1H, 13C and 15N resonance assignments of lymphocyte specific kinase unique and SH3 domains. *J. Biomol. NNR* **19**, 193–194.

Briese, L., Preusser, A., and Willbold, D. (2004). Determination of HIV-1 Nef binding site on human Lck unique and SH3 domains. *Virology* submitted.

Carrera, A.C., Paradis, H., Borlado, L.R., Roberts, T.M., and Martinez, C. (1995). Lck unique domain influences Lck specificity and biological function. *J. Biol. Chem.* **270**, 13385–3391.

Chen, M.Y., Maldarelli, F., Karczewski, M.K., Willey, R.L., and Strebel, K. (1993). Human immunodeficiency virus type 1 Vpu protein induces degradation of CD4 *in vitro*: the cytoplasmic domain of CD4 contributes to Vpu sensitivity. *J. Virol.* **67**, 3877–3884.

Cicchetti, P., Mayer, B.J., Thiel, G., and Baltimore, D. (1992). Identification of a protein that binds to the SH3 region of Abl and is similar to Bcr and GAP-rho. *Science* **257**, 803–806.

Collette, Y., Dutartre, H., Benziane, A., Romas, M., Benarous, R., Harris, M. *et al.* (1996). Physical and functional interaction of Nef with Lck. HIV-1 Nef-induced T-cell signaling defects. *J. Biol. Chem.* **271**, 6333–6341.

Craig, H.M., Pandori, M.W., and Guatelli, J.C. (1998). Interaction of HIV-1 Nef with the cellular dileucine-based sorting pathway is required for CD4 down-regulation and optimal viral infectivity. *Proc. Natl. Acad. Sci. USA* **95**, 11229–11234.

Craig, H.M., Reddy, T.R., Riggs, N.L., Dao, P.P., and Guatelli, J.C. (2000). Interactions of HIV-1 nef with the mu subunits of adaptor protein complexes 1, 2, and 3: role of the dileucine-based sorting motif. *Virology* **271**, 9–17.

Cullen, B.R. (1994). The role of Nef in the replication cycle of the human and simian immunodeficiency viruses. *Virology* **205**, 1–6.

Dalgeish, A.G., Beverly, P.C.L., Clapham, P.R., Crawford, D.H., Greaves, M.F., and Weiss, R.A. (1984). The CD4(T4)antigen is an essential component of the receptor for the AIDS retrovirus. *Nature* **312**, 763–767.

Engler, A., Stangler, T., and Willbold, D. (2001). Solution structure of human immunodeficiency virus type 1 Vpr(13–33) peptide in micelles. *Eur. J. Biochem.* **268**, 389–395.

Engler, A., Stangler, T., and Willbold, D. (2002). Structure of human immunodeficiency virus type 1 Vpr(34–51) peptide in micelle containing aqueous solution. *Eur. J. Biochem.* **269**, 3264–3269.

Fackler, O.T. and Baur, A.S. (2002). Live and let die: Nef functions beyond HIV replication. *Immunity* **16**, 493–497.

Freund, J., R. Kellner, T. Houthaeve, and H.R. Kalbitzer. (1994a). Stability and proteolytic domains of Nef protein from human immunodeficiency virus (HIV) type 1. *Eur. J. Biochem.* **221**, 811–189.

Freund, J., Kellner, R., Konvalinka, J., Wolber, V., Krausslich, H.G., and Kalbitzer, H.R. (1994b). A possible regulation of negative factor (Nef) activity of human immunodeficiency virus type 1 by the viral protease. *Eur. J. Biochem.* **223**, 589–593.

Gallaher, W.R., Ball, J.M., Garry, R.F., Martin-Amedee, A.M., and Montelaro, R.C. (1995). A general model for the surface glycoproteins of HIV and other retroviruses. *AIDS Res. Hum. Retroviruses* **11**, 191–202.

Garcia, J.V. and Miller, A.D. (1991). Serine phosphorylation-independent downregulation of cell-surface CD4 by nef. *Nature* **350**, 508–511.

Garcia, J.V., Alfano, J., and Miller, A.D. (1993). The negative effect of human immunodeficiency virus type 1 Nef on cell surface CD4 expression is not species specific and requires the cytoplasmic domain of CD4. *J. Virol.* **67**, 1511–1516.

Geyer, M., Munte, C.E., Schorr, J., Kellner, R., and Kalbitzer, H.R. (1999). Structure of the anchor-domain of myristoylated and non-myristoylated HIV-1 Nef protein. *J. Mol. Biol.* **289**, 123–138.

Görlach, M., Wittekind, M., Beckman, R.A., Mueller, L., and Dreyfuss, G. (1992). Interaction of the RNA-binding domain of the hnRNP C proteins with RNA. *EMBO J.* **11**, 3289–3295.

Gratton, S., Yao, X.J., Venkatesan, S., Cohen, E.A., and Sekaly, R.P. (1996). Molecular analysis of the cytoplasmic domain of CD4: overlapping but noncompetitive requirement for lck association and down-regulation by Nef. *J. Immunol.* **157**, 3305–3311.

Greenberg, M.E., Bronson, S., Lock, M., Neumann, M., Pavlakis, G.N., and Skowronski, J. (1997). Co-localization of HIV-1 Nef with the AP-2 adaptor protein complex correlates with Nef-induced CD4 down-regulation. *EMBO J.* **16**, 6964–6976.

Greenway, A., Azad, A., Mills, J., and McPhee, D. (1996). Human immunodeficiency virus type 1 Nef binds directly to Lck and mitogen-activated protein kinase, inhibiting kinase activity. *J. Virol.* **70**, 6701–6708.

Grzesiek, S., Bax, A., Clore, G.M., Gronenborn, A.M., Hu, J.S., Kaufman, J. *et al.* (1996a). The solution structure of HIV-1 Nef reveals an unexpected fold and permits delineation of the binding surface for the SH3 domain of Hck tyrosine protein kinase. *Nat. Struct. Biol.* **3**, 340–435.

Grzesiek, S., Bax, A., Hu, J.S., Kaufman, J., Palmer, I., Stahl, S.J. *et al.* (1997). Refined solution structure and backbone dynamics of HIV-1 Nef. *Protein Sci.* **6**, 1248–1263.

Grzesiek, S., Stahl, S.J., Wingfield, P.T., and Bax, A. (1996b). The CD4 determinant for downregulation by HIV-1 Nef directly binds to Nef. Mapping of the Nef binding surface by NMR. *Biochemistry* **35**, 10256–10261.

Hanna, Z., Kay, D.G., Rebai, N., Guimond, A., Jothy, S., and Jolicoeur, P. (1998). Nef harbors a major determinant of pathogenicity for an AIDS-like disease induced by HIV-1 in transgenic mice. *Cell* **95**, 163–175.

Harris, M. (1999). HIV: a new role for Nef in the spread of HIV. *Curr. Biol.* **9**, R459–R461.

Harris, M. and Coates, K. (1993). Identification of cellular proteins that bind to the human immunodeficiency virus type 1 nef gene product in vitro: a role for myristylation. *J. Gen. Virol.* **74**, 1581–1589.

Hua, J., Blair, W., Truant, R., and Cullen, B.R. (1997). Identification of regions in HIV-1 Nef required for efficient downregulation of cell surface CD4. *Virology* **231**, 231–238.

Huse, M., Eck, M.J., and Harrison, S.C. (1998). A Zn^{2+} ion links the cytoplasmic tail of CD4 and the N-terminal region of Lck. *J. Biol. Chem.* **273**, 18729–18733.

Iafrate, A.J., Bronson, S. and Skowronski, J. (1997). Separable functions of Nef disrupt two aspects of T cell receptor machinery: CD4 expression and CD3 signaling. *EMBO J.* **16**, 673–684.

Isakov, N. and Biesinger, B. (2000). Lck protein tyrosine kinase is a key regulator of T-cell activation and a target for signal intervention by *Herpesvirus saimiri* and other viral gene products. *Eur. J. Biochem.* **267**, 3413–3421.

Klatzmann, D., Barre-Sinoussi, F., Nugeyre, M.T., Danquet, C., Vilmer, E., Griscelli, C. *et al.* (1984a). Selective tropism of lymphadenopathy associated virus (LAV) for helper-inducer T lymphocytes. *Science* **225**, 59–63.

Klatzmann, D., Champagne, E., Chamaret, S., Gruest, J., Guetard, D., Hercend, T. *et al.* (1984b). T-lymphocyte T4 molecule behaves as the receptor for human retrovirus LAV. *Nature* **312**, 767–768.

Koradi, R., Billeter, M., and Wuthrich, K. (1996). MOLMOL: a program for display and analysis of macromolecular structures. *J. Mol. Graph.* **14**, 51–5, 29–32.

Lane, A.N., Kelly, G., Ramos, A., and Frenkiel, T.A. (2001). Determining binding sites in protein-nucleic acid complexes by cross-saturation. *J. Biomol. NMR* **21**, 127–139.

Larson, S.M. and Davidson, A.R. (2000). The identification of conserved interactions within SH3 domain by alignment of sequences and structures. *Protein Sci.* **9**, 2170–2180.

Lee, C.H., Leung, B., Lemmon, M.A., Zheng, J., Cowburn, D., Kuriyan, J. *et al.* (1995). A single amino acid in the SH3 domain of Hck determines its high affinity and specificity in binding to HIV-1 Nef protein. *EMBO J.* **14**, 5006–5015.

Lee, C.H., Saksela, K., Mirza, U.A., Chait, B.T., and Kuriyan, J. (1996). Crystal structure of the conserved core of HIV-1 Nef complexed with a Src family SH3 domain. *Cell* **85**, 931–942.

Lin, R.S., Rodriguez, C., Veillette, A., and Lodish, H.F. (1998). Zinc is essential for binding of p56(lck) to CD4 and CD8alpha. *J. Biol. Chem.* **273**, 32878–32882.

Lock, M., Greenberg, M.E., Iafrate, A.J., Swigut, T., Muench, J., Kirchhoff, F., *et al.* (1999). Two elements target SIV Nef to the AP-2 clathrin adaptor complex, but only one is required for the induction of CD4 endocytosis. *EMBO J.* **18**, 2722–2733.

Lu, X., Yu, H., Liu, S.H., Brodsky, F.M., and Peterlin, B.M. (1998). Interactions between HIV1 Nef and vacuolar ATPase facilitate the internalization of CD4. *Immunity* **8**, 647–656.

Maddon, P.J., Molineaux, S.M., Maddon, D.E., Zimmerman, K.A., Godfrey, M., Alt, F.W., *et al.* (1987). Structure and expression of the human and mouse T4 genes. Proc. Natl. Acad. Sci. USA **84**, 9155–9159.

Mariani, R. and Skowronski, J. (1993). CD4 down-regulation by nef alleles isolated from human immunodeficiency virus type 1-infected individuals. *Proc. Natl. Acad. Sci. USA* **90**, 5549–5553.

Marsh, J. W. (1999). The numerous effector functions of Nef. *Arch. Biochem. Biophys.* **365**: 192–198.

Metzger, A.U., Bayer, P., Willbold, D., Hoffmann, S., Frank, R.W., Goody, R.S. *et al.* (1997). The interaction of HIV-1 Tat(32–72) with its target RNA: a fluorescence and nuclear magnetic resonance study. *Biochem Biophys. Res. Comm.* **241**, 31–36.

Metzger, A.U., Schindler, T., Willbold, D., Kraft, M., Steegborn, C., Volkmann, A. *et al.* (1996). Structural rearrangements on HIV-1 Tat (32–72) TAR complex formation. *FEBS Lett.* **384**, 255–259.

Morellet, N., Bouaziz, S., Petitjean, P., and Roques, B.P. (2003). NMR structure of the HIV-1 regulatory protein VPR. *J. Mol. Biol.* **327**, 215–227.

Mujeeb, A., Bishop, K., Peterlin, B.M., Turck, C., Parslow, T.G., and James, T.L. (1994). NMR structure of a biologically active peptide containing the RNA-binding domain of human immunodeficiency virus type 1 Tat. *Proc. Natl. Acad. Sci. USA* **91**, 8248–8252.

Otting, G., Qian, Y.Q., Billeter, M., Müller, M., Affolter, M., Gehring, W.J. *et al.* (1990). Protein-DNA contacts in the structure of a homeodomain-DNA complex determined by nuclear magnetic resonance spectroscopy in solution. *EMBO J.* **9**, 3085–3092.

Piguet, V., Gu, F., Foti, M., Demaurex, N., Gruenberg, J., Carpentier, J.L. *et al.* (1999). Nef-induced CD4 degradation: a diacidic-based motif in Nef functions as a lysosomal targeting signal through the binding of beta-COP in endosomes. *Cell* **97**, 63–73.

Preusser, A., Briese, L., Baur, A.S., and Willbold, D. (2001). Direct *in vitro* binding of full-length human immunodeficiency virus type 1 Nef protein to CD4 cytoplasmic domain. *J. Virol.* **75**, 3960–3964.

Preusser, A., Briese, L., and Willbold, D. (2002). Presence of a helix in human CD4 cytoplasmic domain promotes binding to HIV-1 Nef protein. *Biochem Biophys. Res. Comm.* **292**, 734–740.

Ren, R., Mayer, B.J., Cicchetti, P., and Baltimore, D. (1993). Identification of a ten-amino acid proline-rich SH3 binding site. *Science* **259**, 1157–1161.

Renkema, H.G. and Saksela, K. (2000). Interactions of HIV-1 NEF with cellular signal transducing proteins. *Front. Biosci.* **5**, D268–D283.

Roques, B.P., Morellet, N., de Rocquigny, H., Déméné, H., Schueler, W., and Jullian, N. (1997). Structure, biological functions and inhibition of the HIV-1 proteins Vpr and NCp7. *Biochimie* **79**, 673–680.

Rösch, P., Bayer, P., Ejchart, A., Frank, R., Gazit, A., Herrmann, F. *et al.* (1996). The structure of lentiviral tat proteins. In B. D. N. Rao and M. D. Kemple (eds), *NMR as Structural Tool for Macromolecules: Current Status and Future Directions.* Plenum Publishing Corporation, New York.

Rossi, F., Gallina, A., and Milanesi, G. (1996). Nef-CD4 physical interaction sensed with the yeast two-hybrid system. *Virology* **217**, 397–403.

Saksela, K. (1997). HIV-1 Nef and host cell protein kinases. *Front. Biosci.* **2**, 606–618.

Saksela, K., Cheng, G., and Baltimore, D. (1995). Proline-rich (PxxP) motifs in HIV-1 Nef bind to SH3 domains of a subset of Src kinases and are required for the enhanced growth of Nef+ viruses but not for down-regulation of CD4. *EMBO J.* **14**, 484–491.

Salghetti, S., Mariani, R., and Skowronski, J. (1995). Human immunodeficiency virus type 1 Nef and p56lck protein-tyrosine kinase interact with a common element in CD4 cytoplasmic tail. *Proc. Natl. Acad. Sci. USA* **92**, 349–353.

Sanfridson, A., Cullen, B.R., and Doyle, C. (1994). The simian immunodeficiency virus Nef protein promotes degradation of CD4 in human T cells. *J. Biol. Chem.* **269**, 3917–3920.

Schüler, W., Wecker, K., de Rocquigny, H., Baudat, Y., Sire, J., and Roques, B.P. (1999). NMR structure of the (52–96) C-terminal domain of the HIV-1 regulatory protein Vpr: molecular insights into its biological functions. *J. Mol. Biol.* **285**, 2105–2117.

Schweimer, K., Hoffmann, S., Bauer, F., Friedrich, U., Kardinal, C., Feller, S.M. *et al.* (2002). Structural investigation of the binding of a herpesviral protein to the SH3 domain of tyrosine kinase Lck. *Biochemistry* **41**, 5120–5130.

Shaw, A.S., Chalupny, J., Whitney, J.A., Hammond, C., Amrein, K.E., Kavathas, P. *et al.* (1990). Short related sequences in the cytoplasmic domains of CD4 and CD8 mediate binding to the amino-terminal domain of the p56lck tyrosine protein kinase. *Mol. Cell. Biol.* **10**, 1853–1862.

Simmons, A., Aluvihare, V., and McMichael, A. (2001). Nef triggers a transcriptional program in T cells imitating single-signal T cell activation and inducing HIV virulence mediators. *Immunity* **14**, 763–777.

Sticht, H., Willbold, D., Bayer, P., Ejchart, A., Herrmann, F., Rosin Arbesfeld, R. *et al.* (1993). Equine infectious anemia virus Tat is a predominantly helical protein. *Eur. J. Biochem.* **218**, 973–976.

Sticht, H., Willbold, D., Ejchart, A., Rosin Arbesfeld, R., Yaniv, A., Gazit, A. *et al.* (1994). Trifluoroethanol stabilizes a helix-turn-helix motif in equine infectious-anemia-virus trans-activator protein. *Eur. J. Biochem.* **225**, 855–861.

Takahashi, H., Nakanishi, T., Kami, K., Arata, Y., and Shimada, I. (2000). A novel NMR method for determining the interfaces of large protein–protein complexes. *Nat. Struct. Biol.* **7**, 220–223.

Turner, J.M., Brodsky, M.H., Irving, B.A., Levin, S.D., Perlmutter, R.M., and Littman, D.R. (1990). Interaction of the unique N-terminal region of tyrosine kinase p56lck with cytoplasmic domains of CD4 and CD8 is mediated by cysteine motifs. *Cell* **60**, 755–765.

Veillette, A., Bookman, M.A., Horak, E.M., and Bolen, J.B. (1988). The CD4 and CD8 T cell surface antigens are associated with the internal membrane tyrosine-protein kinase p56lck. *Cell* **55**, 301–308.

Wecker, K. and Roques, B.P. (1999). NMR structure of the (1–51) N-terminal domain of the HIV-1 regulatory protein Vpr. *Eur. J. Biochem.* **266**, 359–369.

Welker, R., Kottler, H., Kalbitzer, H.R., and Kräusslich, H.G. (1996). Human immunodeficiency virus type 1 Nef protein is incorporated into virus particles and specifically cleaved by the viral proteinase. *Virology* **219**, 228–236.

Willbold, D. and Rösch, P. (1996). Solution structure of the human CD4 (403–419) receptor peptide. *J. Biomed. Sci.* **3**, 435–441.

Willbold, D., Kruger, U., Frank, R., Rosin Arbesfeld, R., Gazit, A., Yaniv, A., *et al.* (1993). Sequence-specific resonance assignments of the 1H-NMR spectra of a synthetic, biologically active EIAV Tat protein. *Biochemistry* **32**, 8439–8445.

Willbold, D., Rosin Arbesfeld, R., Sticht, H., Frank, R., and Rösch, P. (1994). Structure of the equine infectious anemia virus Tat protein. *Science* **264**, 1584–1587.

Willbold, D., Volkmann, A., Metzger, A., Sticht, H., Rosin-Arbesfeld, R., Gazit, A. *et al.* (1996). Structural studies of the equine infectious anemia virus trans-activator protein. *Eur. J. Biochem.* **240**, 45–52.

Wolf, D., Witte, V., Laffert, B., Blume, K., Stromer, E., Trapp, S. *et al.* (2001). HIV-1 Nef associated PAK and PI3-kinases stimulate Akt-independent Bad-phosphorylation to induce anti-apoptotic signals. *Nat. Med.* **7**, 1217–1224.

Wray, V., Mertins, D., Kiess, M., Henklein, P., Trowitzsch-Kienast, W., and Schubert, U. (1998). Solution structure of the cytoplasmic domain of the human CD4 glycoprotein by CD and 1H NMR spectroscopy: implications for biological functions. *Biochemistry* **37**, 8527–8538.

Index

MIX
Papier aus verantwortungsvollen Quellen
Paper from responsible sources
FSC® C105338

If you have any concerns about our products,
you can contact us on
ProductSafety@springernature.com

In case Publisher is established outside the EU,
the EU authorized representative is:
Springer Nature Customer Service Center GmbH
Europaplatz 3, 69115 Heidelberg, Germany

Printed by Libri Plureos GmbH
in Hamburg, Germany